T0332596

# GLOBAL SUSTAINABILITY
## A Nobel Cause

Humankind faces an extremely complex set of challenges in the twenty-first century: climate stabilization, energy security, sustainable land use, and equitable development. If some nine billion people are to live a decent life on our crowded planet we will require, above all else, a steady and affordable supply of freshwater, food, fibres and fuel – the natural capital that forms the basis for a continuous generation of wealth. Yet we are unlikely to achieve global sustainability without a 'Great Transformation', making all facets of society more respectful of the existing planetary boundaries.

In an unprecedented attempt to address this transformation, the first Interdisciplinary Nobel Laureate Symposium on Global Sustainability in Potsdam brought together many of the world's pre-eminent thinkers. Nobel laureates in physics, chemistry, medicine, economics and peace, top-level political leaders, representatives of major NGOs, and renowned experts on sustainability met to discuss scientific and political strategies for reconciling human civilization with its physical and ecological support systems. The resulting book gives the reader access to highly stimulating discussions on some of the most important environmental, economic and ethical questions of our time. This publication features both advanced mainstream concepts and innovative transformational approaches presented by some of today's finest minds.

With its mixture of thought-provoking essays and scientific texts, this book will capture the attention and imagination of everyone interested in sustainability issues. It follows a radically interdisciplinary approach through a broad range of contributions, covering the latest findings of climate impact research, environmental economics, energy resource analysis, ecosystems science, and other crucial fields.

**Hans Joachim Schellnhuber** is Professor of Theoretical Physics at Potsdam University and Director of the Potsdam Institute for Climate Impact Research (PIK). He is also Chair of the German Advisory Council on Global Change and was appointed Chief Government Advisor on Climate and Related Issues during Germany's G8 and EU presidencies in 2007. From 2001–5 he was Research Director of the British Tyndall Centre for Climate Change Research. He is an elected member of, inter alia, the German National Academy (*Leopoldina*) and the US National Academy of Sciences. In 2004 he was awarded a CBE by Queen Elizabeth II, and in 2007 received the German Environment Prize.

**Mario Molina** studied physical chemistry and obtained his PhD at the University of California, Berkeley. In 1974, well before the first measurements of the Antarctic ozone hole, he co-authored a paper that described how chlorofluorocarbon (CFC) gases, widely used in industry at that time, destroy the atmospheric ozone layer. In 1995 Molina was honoured with the Nobel Prize in Chemistry for his work on ozone depletion. As Professor of Chemistry and of Earth, Atmospheric, and Planetary Sciences at the Massachusetts Institute of Technology, Molina continued his research on man-made changes in atmospheric chemistry. In 2004 he joined the faculty at the University of California in San Diego.

**Nicholas Stern** is IG Patel Professor of Economics and Government at the London School of Economics and Political Science, and Chairman of the Grantham Research Institute on Climate Change and the Environment. Stern was head of the UK Government Economic Service from 2003–7 and Chief Economist of the World Bank from 2000–3 and of the European Bank for Reconstruction and Development from 1994–99. He authored the Stern Review on the Economics of Climate Change, reporting to the UK Prime Minister and the Chancellor of the Exchequer from 2005–7. He was knighted for services to economics in 2004, and was appointed to the UK House of Lords as Lord Stern of Brentford in 2007.

**Veronika Huber** is Scientific Personal Assistant to Hans Joachim Schellnhuber, Director of the Potsdam Institute for Climate Impact Research. Previously, she worked on her doctoral thesis at the Institute for Freshwater Ecology and Inland Fisheries in Berlin and gained experience in climate policy at the UN Environmental Programme in Nairobi. She studied biology at École Normale Supérieure in Paris and at the universities of Konstanz and Potsdam.

**Susanne Kadner** has a research background in biology, chemistry and oceanography. She has worked for the Parliamentary Office of Science and Technology (POST) in London, the German Advisory Council on Global Change (WBGU), and as G8-consultant to Hans Joachim Schellnhuber in his position as German Chief Government Advisor on Climate and Related Issues. She now works in the Technical Support Unit of the Intergovernmental Panel on Climate Change's Working Group III.

# GLOBAL SUSTAINABILITY

## A Nobel Cause

*Edited by*

HANS JOACHIM SCHELLNHUBER

MARIO MOLINA

NICHOLAS STERN

VERONIKA HUBER

SUSANNE KADNER

CAMBRIDGE
UNIVERSITY PRESS

# CAMBRIDGE
## UNIVERSITY PRESS

University Printing House, Cambridge CB2 8BS, United Kingdom

One Liberty Plaza, 20th Floor, New York, NY 10006, USA

477 Williamstown Road, Port Melbourne, VIC 3207, Australia

314-321, 3rd Floor, Plot 3, Splendor Forum, Jasola District Centre, New Delhi - 110025, India

79 Anson Road, #06-04/06, Singapore 079906

Cambridge University Press is part of the University of Cambridge.

It furthers the University's mission by disseminating knowledge in the pursuit of education, learning and research at the highest international levels of excellence.

www.cambridge.org
Information on this title: www.cambridge.org/9780521769341

© Cambridge University Press 2010

This publication is in copyright. Subject to statutory exception and to the provisions of relevant collective licensing agreements, no reproduction of any part may take place without the written permission of Cambridge University Press.

First published 2010

*A catalogue record for this publication is available from the British Library*

ISBN 978-0-521-76934-1 Hardback

Cambridge University Press has no responsibility for the persistence or accuracy of URLs for external or third-party internet websites referred to in this publication, and does not guarantee that any content on such websites is, or will remain, accurate or appropriate.

Production management: Textpraxis Hamburg, Marion Schweizer
Copy-editing (except for foreword and chapters 5, 8, 11, 16, 21, 23 and 27): Stephen Roche
Layout and typesetting: Marina Siegemund

# Contents

## PART II

**Climate stabilization and sustainable development** .................. **65**

# Foreword

## Angela Merkel
## Federal Chancellor of Germany

Global climate change is one of the greatest challenges facing humanity in the twenty-first century. Our understanding of both the causes and consequences of global climate change has been profoundly influenced by the scientific work of the Intergovernmental Panel on Climate Change (IPCC), which was awarded the Nobel Prize for Peace in 2007. It is increasingly apparent that we must act now, as climate change is accelerating. Climate change threatens both our security and our economic development. Failure to take decisive action will have a dramatic impact.

Truly sustainable development requires global emissions to be cut at least by one half by the middle of this century. Only by reaching this goal can we keep global warming below the critical level of 2 °C, and avert the worst consequences of climate change.

The road towards achieving this aim needs to be mapped out today. This will involve transforming our energy production, transportation, manufacturing, and patterns of consumption to minimize future use of fossil fuels. We already have many of the necessary technologies and innovative ideas, and further advances are being pursued with strong commitment. However, for sustainable solutions to be broadly adopted, the right economic and political frameworks are needed.

The adoption of the Kyoto Protocol in 1997 was a first important step. But today, more than ten years later, we have to acknowledge that the advances we have made in climate protection are by no means sufficient. We must not lose another decade. The global community therefore needs to agree quickly upon a new post-Kyoto treaty that is both ambitious and effective.

It is clear where action is necessary. First, we need binding goals for reducing greenhouse gas emissions. Second, we need to increase our efforts to adapt to those consequences of climate change that are unavoidable. Third, we need to improve global cooperation in the areas of development and the application of sustainable technologies, such as renewable energies. And fourth, we need to create and expand financial mechanisms that encourage mitigation and adaptation strategies. Both public and private capital need to be mobilized. The expansion of the global carbon market is therefore of great importance to international climate policy.

The key to a successful and effective climate treaty is an approach that all countries acknowledge as being fair. We therefore need to take into consideration the 'polluter pays' principle, while also recognizing differing capacities to pay. In this respect industrialized countries need to lead the way. They are called upon to commit and adhere to ambitious goals for emission reductions. However, the global climate will only be stabilized if emerging economies contribute as well, namely by decoupling emissions from economic growth. In this way we can achieve convergence of global emissions per capita on a level that is commensurate with the goal of global climate protection. Such a process towards long-term convergence allows for sustainable development in all countries, while acknowledging the common but differentiated responsibility of each country.

However, the paradigm shift required globally will only be achieved if the brightest minds on our planet work together and advance it through new ideas. In this spirit the Nobel Laureate Symposium in Potsdam sent out an important message to scientists, politicians, and to all other citizens of the world. I greatly welcome this publication, which will enable a greater audience to follow the important discussions at the symposium.

# Preface

'In the tragic situation which confronts humanity, we feel that scientists should assemble in conference to appraise the perils that have arisen [...]'

These are the opening words of the Russell-Einstein Manifesto, which was issued in London on 9 July 1955. At the advent of the nuclear age, the eleven signatories of this historic document – all pre-eminent scientists and intellectuals, ten of them Nobel laureates – called for the scientific community to take responsibility and participate in the struggle for peaceful solutions. In their concluding statement, they urged the governments of the world to acknowledge the existential threats posed to humanity by the development of nuclear arms and to find peaceful means for the settlement of conflict. In response to the Manifesto, innumerable scientists from all over the world joined the public debate on the perils arising from weapons of mass destruction – independent of political persuasion. During times of the Cold War and until the present day, they have contributed decisively to developing strategies for disarmament and peace.

Today, at the beginning of the twenty-first century, humanity is similarly faced with an unprecedented threat. Dwindling energy sources, degrading terrestrial and marine ecosystems, accelerating climate change, and ongoing population growth could drive civilization to the brink of collapse. Migration and conflict on a massive scale could follow from our neglect and unpreparedness to change and to stop excessively exploiting the planet. The threat arising from approaching the limits of nature's capacities and resources must be confronted in a situation of currently one billion people in deep poverty. A strategy for managing climate change which runs counter to our attempts to fight poverty in the next few decades cannot build the coalition necessary to succeed. All of human ingenuity and imagination is needed to find solutions to this major crisis, and set the world on a path towards global sustainability, including a decent life for all on our crowded planet.

In this spirit, the Potsdam Nobel Laureate Symposium brought together some of the world's finest minds – Nobel laureates in physics, chemistry, medicine, economics, and peace, as well as leading scientific experts from various disciplines, top-level political actors, and important representatives of civil society. The meeting – under the banner of 'Global Sustainability: A Nobel Cause' – took place in autumn 2007, close to the locations that Albert Einstein and many other important scientists and intellectuals of his time frequently visited. For three intense days, the participants of the symposium conducted lively debates on climate stabilization and sustainable development, on energy security, on institutional and economic incentives, and on the responsibilities of the scientific community in these troubled times.

This publication aims at reflecting those discussions, placing them in the context of the year 2009 – the year of the Copenhagen Conference that could go down in history either as the moment humanity started putting words into action on climate change and sustainable development, or as another missed opportunity to change course. Acknowledging the importance of conveying scientific knowledge to the wider public, this compilation of essays invites a broad audience to take part in the stimulating and instructive debates of the symposium. In keeping with the course of the original discussions, it contains commentaries, in which various authors react to the statements of their colleagues with endorsement as well as with criticism.

Rather than providing detailed reading instructions here, we would like to let the texts speak for themselves. Each of the essays, even the commentaries, can stand alone and be read independently. The table of contents and the following 'walk through the hall of fame' will provide a minimum of guidance to help readers find their bearings within this intertwined debate.

The publication starts out with introductory remarks by the German head of government, Angela Merkel, who attended the symposium in 2007 – the year in which Germany held the EU and G 8 presidencies, and when she was dubbed the 'Climate Chancellor' for her commitment to climate change issues. Ian McEwan, the acclaimed British author and screenwriter, then takes the reader on a flight over the last vast empty spaces of the planet and down into the boot room of an Arctic research vessel to contemplate the fate of the Earth and the humble nature of man.

Well prepared by this cognitive journey, the reader will encounter some of the greatest authorities in science and society – the **Nobel laureates** who enrich the debate with their expertise and experience. Taking a comprehensive view of the problem, **Murray Gell-Mann** spells out the transitions needed if the world is to switch from present trends to greater sustainability, involving all facets of society: politics, economics, education, culture, and morality. Central to his thoughts is an interdisciplinary, holistic approach to science – what he calls 'a crude look at the whole' (CLAW). Like the other authors of the opening section, entitled *The Great Transformation,* he touches upon many of the challenges and responses related to sustainability that are discussed in greater depth in the subsequent sections.

The first topical section on *Climate Stabilization and Sustainable Development* builds upon the insight that the battle against climate change cannot be won without overcoming extreme poverty on this planet. As one of the first to bring the concept of a 'global deal' to the table of negotiations, Nicholas Stern discusses this cornerstone of international efforts for addressing the dual challenge posed by global warming and under-development. **Rajendra Pachauri,** Chairman of the Intergovernmental Panel on Climate Change (IPCC), which was awarded the Nobel Peace Prize in 2007, highlights crucial aspects of climate change mitigation and adaptation, with a special focus on the countries in the South. In a commentary to his essay, **Mario Molina,** who was among the first scientists to describe the chemical reactions depleting the stratospheric ozone layer, draws on his experience during the ozone-hole crisis and his recent involvement in efforts to combat climate change and contain air pollution in the developing world. **Wangari Maathai** shares her insights on climate change and development gained during decades of leadership in the women's tree planting initiative Green Belt Movement, for which she received the Nobel Peace Prize in 2004.

The following section on *Economic and Institutional Incentives* allows readers to take a closer look at proposed solutions to the global sustainability challenge. Putting a price on carbon emissions needs to be a salient feature of any effective solution, as most economists would agree. However, whether this price should arise through a cap-and-trade system for emission permits, through a tax on carbon, or through other innovative instruments is a controversial key policy issue. Some arguments of the debate are examined and discussed by **James Mirrlees**, who was awarded the Nobel Prize in economics for his analysis of incentives under imperfect information.

*Technological Innovations and Energy Security* is the topic of the closely related following section. **Walter Kohn** and **Alan Heeger,** who together produced the documentary film 'The Power of the Sun', write about renewable energy use and technology from different perspectives, but each with an intriguingly personal tone. In a general approach, Walter Kohn explores the possibility of powering the world entirely by wind and solar energy, while Alan Heeger highlights a specific renewable technology, expressing his enthusiasm about the prospects arising from efficient and low-cost plastic solar cells.

The overarching theme of the Potsdam Symposium – and also one of the main motivations for producing this publication – was to involve the scientific community in an educational effort that will enable individuals worldwide to contribute to finding sustainable solutions. This and related issues of democracy and participation are raised in the wrap-up section on a *Global Contract between Science and Society*. **John Sulston** powerfully makes the point that trust – a *sine qua non* if global sustainability is ever to be achieved in a world where incentives to free-ride persist – is impossible without open access to information and sharing of knowledge.

The book concludes with the *Potsdam Memorandum,* which was adopted by the participants of the Potsdam Symposium and which calls for a 'Great Transformation'.

This ubiquitous change in the human-environment interaction on Earth, introduced in the opening chapters of the book, is the unifying point of reference that binds this collection of essays together.

At this point, we would like to refer back to Bertrand Russell and Albert Einstein, who towards the end of their Manifesto observe:

'[...] what perhaps impedes understanding of the situation more than anything else is that the term *mankind* feels vague and abstract. People scarcely realize in imagination that the danger is to themselves and their children and their grandchildren, and not only to a dimly apprehended humanity.'

These sage words were valid back then and are perhaps even more valid today. We dearly hope that with this unique compilation of essays we can encourage readers to see themselves as part of this endangered humanity – this humankind which, for the first time in history, is acting as a truly global force, threatening the integrity of the Earth; but also possesses the power of comprehension and the ingenuity to save our precious planet.

*Hans Joachim Schellnhuber, Mario Molina, Nicholas Stern,*
*Veronika Huber, Susanne Kadner*

# Acknowledgements

It goes without saying that we are deeply grateful to all of the authors who contributed to this volume. We would also like to thank the reviewers – namely Nico Bauer, Thomas Bruckner, Steffen Brunner, Christian Flachsland, Hans-Martin Füssel, Dieter Gerten, Jan Christoph Goldschmidt, Hermann Held, Michael Jakob, Brigitte Knopf, Hermann Lotze-Campen, Wolfgang Lucht, Gunnar Luderer, Sylvie Ludig, Robert Marschinski, Fritz Reusswig, Jan Steckel, and Jan Strohschein – for their extremely valuable comments on preliminary versions of the essays. In addition, we acknowledge the help of numerous assistants and members of support staff who made communicating with the authors faster and easier.

In particular, we are grateful to Maiken Winter for supporting the editorial work with remarkable enthusiasm and for writing part of the glossary, and to Martin Wodinski for redrawing and refining many of the illustrations. Special thanks also go to Sabine Lütkemeier, Alison Schlums, Marcel Meistring, and Simone Groß from the Potsdam Institute for Climate Impact Research. All of their help was invaluable in bringing this publication about.

Last but not least, we would like to thank the event partner – the World Wide Fund for Nature (WWF) – and the sponsors – the German Federal Ministry of Education and Research, the Otto Group, Honda, and the Technologie Stiftung Brandenburg – for their generous support in hosting the 2007 Nobel Laureate Symposium in Potsdam.

# Prologue

## Save the boot room

---

## Ian McEwan

Ian McEwan was born in 1948 in Aldershot, England. His recent novels include *Atonement, Saturday,* and *On Chesil Beach.* His new novel, *Solar,* whose background is climate change, will be published in 2010. He is a Fellow of the Royal Society of Literature, and of the American Academy of Arts and Sciences.

The commonplace view of the Earth from an airplane at 12 000 metres – a vista that would have astounded Goethe or Darwin – can be instructive when we contemplate the fate of our Earth. We see faintly, or imagine we can, the spherical curve of the horizon and, by extrapolation, sense how far we would have to travel to circumnavigate, and how tiny we are in relation to this beloved home suspended in sterile space. When we cross the Canadian Northern Territories en route to the American west coast, or the Norwegian littoral, or the interior of Brazil, we are heartened to see that such vast empty spaces still exist – two hours might pass, and not a single road or track in view. But also large and growing larger is the great rim of grime – as though detached from an unwashed bathtub – that hangs in the air as we head across the Alps into northern Italy, or the Thames Basin, or Mexico City, Los Angeles, Beijing – the list is long and growing. These giant concrete wounds laced with steel, those catheters of ceaseless traffic filing towards the horizon – the natural world can only shrink before them. The sheer pressure of our numbers, the abundance of our inventions, the blind forces of our desires and needs, appear unstoppable and are generating a heat – the hot breath of our civilization – whose effects we are beginning to comprehend only too clearly. The misanthropic traveller, gazing down from his wondrous – and wondrously dirty – machine, is bound to ask whether the Earth might not be better off without us.

How can we ever begin to restrain ourselves? We appear, at this distance, like a successful lichen, a ravaging bloom of algae, a mould enveloping a soft fruit. We know enough now to understand in precise terms what we are doing to the Earth and its atmosphere. We have a fairly good idea what needs to be done, or what we need to stop doing. But can we agree among ourselves on how to proceed? We are, after all, a clever but quarrelsome species – in our public discourses we can sound like a rookery in full throat. In our cleverness we are just beginning to understand that the Earth – considered as a total system of organisms, environments, climates and solar radiation, biological and physical processes reciprocally shaping each other through hundreds of millions of years – is perhaps as complex as the human brain; as yet we understand only a little about that brain, and only just a little more about the home in which it evolved. Despite our ignorance, reports from a disparate range of scientific disciplines are overwhelming in their convergence, and are telling us with certainty that we are making a mess of the Earth, that we are fouling our nest and have to act quickly, decisively and against our immediate inclinations. For we tend to be superstitious, hierarchical and self-interested, just when the moment requires us to be rational, even-handed and altruistic. We are shaped by our history and biology to frame our plans within the short term, within the scale of a single lifetime; in democracies, governments and electorates that collude in tight cycles of promise and gratification. And in undemocratic regimes, and under tyrannies, ruling elites have no will and no reason to behave honourably or altruistically.

The present moment demands of us that we address the well-being of unborn individuals we will never meet and who, contrary to the usual terms of human interaction, will not be returning the favour. Perhaps we should take heart from the fact that there have been times in the past when people have done precisely that – looked to the well-being of future generations. Consider those who built Europe's great medieval cathedrals, or those who once thought to plant forests or lay out city parks.

To concentrate our minds, we have historical examples of civilizations that have collapsed through environmental degradation – the Sumerian, the Indus Valley, Easter Island. They feasted extravagantly on vital natural resources, and died. Those were test-tube cases, locally confined; when they failed, life continued elsewhere and new civilisations arose. Now, increasingly, we are one vast civilization, and we sense that it is the whole laboratory, the whole glorious human experiment, that is at risk. And what do we have on our side to avert that risk? Against all our deficits, we certainly possess a genetically inscribed talent for co-operation; we can take comfort from the memory of the Test Ban Treaty, drafted at a time of hostility and mutual suspicion between the Cold War super-powers. More recently, the discovery of ozone depletion in the upper atmosphere and world-wide agreement to ban CFC production should also serve as an example. Secondly, globalization, while it has unified economies, increased production and raised carbon dioxide levels, has also created global networks of expert opinion and citizens' demands that are placing pressure on governments to take action.

But above all, we have our rationality, which finds its highest expression and formalization in good science. The adjective is important. We need accurate representations of the state of the Earth. The environmental movement used to let itself down by making dire predictions, 'scientifically' based, which over the past two or three decades have proved inaccurate. Of itself, this does not invalidate dire predictions now, but it makes the case for scepticism – one of the engines of good science. We need not only reliable data, but their expression in the rigorous use of statistics. Well-meaning intellectual movements, from communism to post-structuralism, have a poor history of absorbing inconvenient data or challenges to fundamental precepts. We should not ignore or suppress good indicators on the environment – though they have become extremely rare now. It is tempting for the layman to embrace with enthusiasm the latest bleak scenario merely because it fits the darkness of our mood, the prevailing cultural pessimism. The imagination, as Wallace Stevens once said, is always at the end of an era. But we should be asking, or expecting others to ask, for the provenance of the data, the assumptions fed into the computer model, the response of the peer review community, and so on. The public – laymen like myself – are going to have to absorb the fundamental precepts of the scientific method. Pessimism is intellectually delicious, even thrilling, but the matter before us non-scientists is too serious for mere self-pleasuring. It would be self-defeating if the impetus that has built up in the world's democracies degenerated into a religion of gloomy faith (faith, ungrounded certainty, is no virtue). It was good science, not good intentions, that identified the ozone problem, and it led, fairly promptly, to good policy.

The wide view from the airplane suggests that whatever our environmental problems, they will have to be dealt with by international laws. No single nation is going to restrain its industries while its neighbours' are unfettered. Here too, an enlightened globalization might be of use. There has probably never before been a problem that was so wholly reliant for a solution on the apparently disparate fields of science and law. Of course, their common thread is, or should be, rationality. Good international law might need to use not our virtues, but our weaknesses (greed, self-interest) to leverage a cleaner environment; in this respect, the newly devised market in carbon trading is a good first move.

The climate change 'debate' was once hedged by uncertainties. Now the facts are stark. The record shrinking of the Arctic summer ice in 2007 is one cold fact that sets simpler questions before us: Are we at the beginning of an unprecedented era of international co-operation, or are we living in an Edwardian summer of reckless denial; is this the beginning, or the beginning of the end?

To find an answer to this question, I went with a group or artists and scientists in February 2005 to live on board a ship frozen into a fjord many miles north of Longyearbyen on the island of Spitsbergen, part of the Svalbard archipelago in the Arctic Ocean. We were a self-selected bunch, dedicated to understanding the effects of global warming on the remote poles, and on asking ourselves what we as artists might do. However, we reckoned without our nature – our all too human nature. I reflected on this journey in the following terms.

So, we have come to this ship in a frozen fjord to think about the ways we might communicate our concerns about climate change to a wider public; we will think about the heady demands of our respective art forms, we will consider the necessity of good science, and shall immerse ourselves in the stupendous responsibilities that flow from our stewardship of the planet, and the idealism and selflessness demanded of us as we subordinate our present needs to the welfare of unborn generations who will inherit the earth and thrive in it and love it – we hope – as we do. But first, we must remove our wet boots. Stepping out of minus 30 degrees, craving the warmth of the boat that is our home, we are obliged by our hosts to pause in a cramped and crowded space below the ship's wheel, and in near darkness, to try to bend over in our thick Arctic clothing to loosen our laces with numbed fingers. Then we must stand on a drenched cold floor in our socks and hang up our 'skidoo suits' – they resemble a toddler's splash suit – along with our helmets, and all the while keep track of our gloves, and the liners of our gloves, and our frosted goggles and frozen-mouthed balaclavas that gape at us from the floor in astonishment; we must do this against a flow of our fellows coming out of the boat, intent on putting all these items on, for it is our collective fate, to be going in and out all day. Naturally, we do all this with good cheer.

The whole world's population is to the south of us, and up here we are our species' representatives, making in the wilderness a temporary society, a social microcosm in the vastness of the Arctic. We are the beneficiaries and victims of our nature (social primates, evolved through time like wind-sculpted rock), merry and venal, co-operative and selfish; and as it happens, in this pure air and sunlit beauty, we find ourselves in a state of near-constant euphoria. When did we ever hear such shouts of laughter at breakfast? We are all so immensely tolerable. We potter about during the day with our little projects like contented infants in a day-care nursery.

And it is because we are gloriously imperfect, expelled from Eden, longing to return, that, on the second day, when you venture out into what I shall call the boot room, in your socks, in a hurry because your companions are waiting outside by the belching skidoos, ready to set off on yet another face-peeling punishment ride (oh God, seven more kilometres – when will it end?) across the cement-like floor of the fjord, you will find that someone has made off with your splash suit, or your helmet, or your boots, or your goggles, or all four. This person has his own stuff, but he has ruthlessly, or mistakenly, taken yours. In a moment's extravagance of self-pity, you might think all of history's narrative and all injustice is enacted here – this is how some people end up with three goats and nine hens while others have none. Why some live in palaces and others in cardboard boxes under bridges. The history books tell of little else – the filching of the neighbour's land, water, chattels or cattle, and, in reaction, war, revolutions, genocides.

Well, what are you going to do? Your impatient companions are stamping their feet on the ice. You might reflect that it is not evil that undoes the world, but small errors prompting tiny weaknesses – let's not call them dishonesty – gathering in rivulets, then cascades of consequences. In the golden age of yesterday, the boot room had finite resources, equally shared – these were the initial conditions, the paradise we are about to lose, the conditions before the Fall that we visitors are bound to re-enact. It could go something like this: the owner of size 43 boots left them last night in a remote corner he has already forgotten about. He comes out this morning, sees to hand another pair of 43s and puts them on. Half an hour later, their true owner comes out into the gloom of the boot room, cannot see his own boots, cannot see the 43s obscurely stowed, and empowered by a sense of victimhood, does exactly what you are doing now: reaching for the nearest 44s.

'Of Man's first disobedience', Milton blindly wrote, 'and the fruit of that forbidden tree …' – now you yourself are about try that 'mortal taste' that 'brought death into the world and all our woe, with loss of Eden …'. Ten minutes later, the owner of those size 44 boots appears. He's a good man, a decent man, but he must now take what is not his own. With the eighth Commandment broken, the social contract is ruptured too. No one is behaving particularly badly, and certainly everybody is being, in the immediate circumstances, entirely rational, but by the third day, the boot room is a wasteland of broken dreams. Who could be wearing five splash suits when they weigh twenty pounds each? Who needs more than one helmet? And where are the grown-ups to advise us that our boot room needs a system? Where was God, or even Matron? Hobbes would say we need a Common Power in which we might stand in awe. As things are, this is Chaos, just as Haydn conceived it, and tomorrow morning it will make us miserable. Meanwhile, as Arctic night gathers tightly around Tempelfjord, inside the toasty warmth of our Ark, elevated by the Vin de Pays, we discuss our plans to save a planet many, many times larger than our boot room.

We must not be too hard on ourselves. If we were banished to another galaxy tomorrow, we would soon be fatally homesick for our brothers and sisters and all their flaws: somewhat co-operative, somewhat selfish, loving and cruel, inventive and destructive – and very funny. But we will not rescue the Earth from our own depredations until we understand ourselves a little more, even if we accept that we can never really change our natures. All boot rooms need good systems so that flawed creatures can use them well. Good science will serve us well with diagnosis and prediction, but only good rules will save the boot room. Leave nothing to idealism or outrage, nor especially to good art – we know in our hearts that the very best art is entirely and splendidly useless.

On our last morning, when all the packing has been done and the last coldly reluctant skidoo has been started up, and as the pure northern air is rent by the howls and stink of our machines, our tirelessly tolerant hosts (as forgiving as God has not yet learned to be) come down the gang plank and deposit on the ice a huge plastic sack with all the lost gear retrieved from every corner of the ship. A few of us gather around this treasure, and poke about in it, not ashamed or even faintly embarrassed, but innocently amazed. Here is our stuff! Where has it been hiding all this time?

We barely know ourselves, and our collective nature is still a source of wonder to us – why else write fiction, why else read it? We haven't stopped surprising ourselves yet, and the fate of our largest boot room still hangs in the balance.

© Ian McEwan 2007

# Part I

The Great Transformation

# Chapter 1

# Transformations of the twenty-first century: transitions to greater sustainability

---

## Murray Gell-Mann

Murray Gell-Mann, born in 1929 in New York City, obtained his PhD in physics in 1951 at the Massachusetts Institute of Technology. He was awarded the Nobel Prize for Physics in 1969 for 'his contributions and discoveries concerning the classi-fication of elementary particles and their interactions'. Gell-Mann had found that subatomic particles such as neutrons and protons are composed of building blocks that he called 'quarks'. As Distinguished Fellow at the Santa Fe Institute and Pro-fessor Emeritus at the California Institute of Technology, Gell-Mann conducts theo-retical research in several fields of science.

A great deal of research and teaching in the sciences and the humanities, especially at universities, is confined to individual departments representing particular fields of knowledge. While specialization and sub-specialization are inevitable and necessary, they need to be supplemented by research and teaching that transcend sometimes narrow disciplinary boundaries. Many institutions, within or outside of universities, carry out such transdisciplinary activities (e.g., Hirsch *et al.,* 2008). The Santa Fe Institute, which I helped to found more than twenty years ago, and where I now work, is a place where it is the rule rather than the exception to have transdisciplinary problems studied by self-organized teams of people originally trained in many different specialties. These teams recognize and exploit similarities and connections between topics in very different fields. Similarly, the participants of the Potsdam Nobel Laureate Symposium may have started out as specialists in very different fields – in the physical sciences, the life sciences, the social and be-havioural sciences, or history – but they convened at the symposium to discuss a common concern for the future.

I have often spoken in public about the need for such research and about insti-tutional arrangements for making it happen. What I have to say here may sound similar, but it really concerns a very different topic, one not concerned with aca-demic disciplines in pure research and teaching, but rather with policy solutions that will affect the future of human societies. The Potsdam Symposium was largely concerned with policy-relevant studies, not only in relation to energy and global climate change but more generally in relation to the future of the human race and the biosphere of Planet Earth including all other species with which we share that biosphere.

In considering any very complex system, we tend to break it up into more man-ageable parts or aspects and to study these more or less separately. For example, when looking at the longer-term human future, we might divide the various issues into military and diplomatic issues (some might say security and foreign affairs); political issues, including domestic politics in each country or region; ideological issues; environmental issues, including those related to water, air, energy and bio-logical diversity; human health and wellness issues; family issues; demographic issues; economic issues, including the crucial socioeconomic challenge of reliev-ing extreme poverty; technology issues; scientific research issues; institutional or governance issues; issues of democracy and human rights; and so forth.

Other items might be added to such a list, but the general idea is clear. In each of these categories we have experts who have built their careers around the issues involved. Likewise, we have NGOs, promoting, for example, environmental protec-tion, human rights, arms control, or child health. In many cases, there are govern-ment departments or UN specialized agencies devoted to these issues.

As mentioned above, it is natural, when faced with a very complex system, to try to break it up into subsystems or aspects defined in advance, such as the categories just listed. The difficulty is that any attempt to understand a nonlinear system, especially a complex one, by assembling descriptions of various parts or aspects will work only if those parts or aspects interact weakly, so that the whole system is decomposable. But that is not true of the *world problématique*. In that sense there is truth in the old adage that the whole is more than the sum of its parts.

A look at the listed categories reveals the problem of isolated analysis. Can we really separate environmental issues from those involving population growth? Can we consider these issues in isolation from technological change or from economic policy? Can we think about the attempts to alleviate extreme poverty without considering the unwise, environmentally destructive projects that are sometimes carried out in the name of that worthy cause? Can we discuss issues of global governance without considering politics in the various countries and regions, or without looking at the competition and conflict between differing ideologies? If military and diplomatic policies fail and mankind is plunged into a hugely destructive war, can our other objectives be attained? Is economic growth not threatened by the widespread prevalence of fatal or debilitating diseases? Can we omit questions about democracy and human rights?

And what if democratic processes bring to power elements of society hostile to human rights and to tolerance or elements that favour environmental destruction or aggressive war? While separate consideration of the various aspects of the world situation is necessary and desirable, it very badly needs to be complemented by integrative thinking that not only combines studies of those aspects but also takes into account the strong interactions among them.

We must rid ourselves of the notion that careful study of a problem based on a narrow range of issues is the only kind of work to be taken seriously, while integrative thinking is to be relegated to cocktail party conversation. This prejudice exists in a great many places in our society, including academia and most bureaucracies. Some of my remarks on this subject were quoted near the beginning of Thomas Friedman's book *The Lexus and the Olive Tree* (1999). He came to a similar conclusion through his work for the New York Times. Before he became a columnist, he was assigned to cover first one set of issues, then a different set, and then yet another set. Each time he was reassigned he observed that the issue he was reporting was intimately connected with issues he had covered earlier.

What we need then, is not just detailed work on separate issues, but also the efforts of teams of brilliant thinkers, many of them specialists, devoted to considering the 'whole ball of wax'. It can, of course, be argued that this is too big a job for any single group of people, no matter how talented or erudite. This is true. Of course such an ambitious aim can be accomplished only crudely, and that is why I refer to it as taking a 'Crude Look at the Whole' (CLAW).

The chief of an organization – for example, a head of government or a CEO – has to act as if he or she is taking into account all aspects of policy, including the interactions among them. It is not easy, however, for the chief to take a 'CLAW' if everyone else in the organization is concerned only with a partial view. Even if some people are assigned to look at the big picture, it doesn't always work. A few years ago the CEO of a major corporation told me that he had a strategic planning staff to help him think about the future of the whole business, but that members of that staff suffered from three defects: they seemed largely disconnected from the rest of the company; no one could understand what they said; and everyone else in the company seemed to hate them. Unfortunately, this negative response to an attempt at integration seems to be the norm throughout our modern societies.

Despite such experiences, it is vitally important that we supplement our specialized studies of policy problems with serious attempts to unite them. For such an effort to succeed, some kind of simplification is naturally required. Certain things have to be treated in a cursory fashion and others in more detail. But that process (physical scientists would call it 'coarse graining') cannot be accomplished through pre-defined categories. It must follow from the nature of the world system itself. The required form of coarse graining must first be discovered.

Let us take, for example, the relationship between weather and climate: no clear results will emerge from examining the weather at each isolated location and each short interval of time, and if we ignore the strong interaction with other phenomena. But much can be learned from a study of weather suitably averaged over space and time and examined in tandem with certain information about ocean currents, the nature and quantities of atmospheric pollutants, fluctuations in solar radiation, and so forth. That is a simple example of a non-trivial form of coarse graining.

In trying to investigate future scenarios, however, it is necessary to go beyond averaging processes that produce relatively smooth trends. It is necessary to allow as well for less smooth effects, stemming from situations where chance plays a huge role or where major transitions may occur, like the interlinked transitions or transformations that will have to take place if anything like sustainability is to be achieved.

Sustainability is one of today's favourite catchwords. It is rarely defined in a careful or consistent way, so perhaps I can be forgiven for attaching to it my own set of meanings. Broadly conceived, sustainability refers to quality of human life and of the environment that is not gained at the expense of the future. But I use the term in a much more inclusive way than most people: sustainability is not restricted to environmental, demographic and economic matters, but refers also to political, military, diplomatic, social and institutional or governance issues. Ultimately, sustainability depends on ideological issues and lifestyle choices. As used here, the term sustainability refers as much to sustainable peace, sustainable global security arrangements, sustainable democracy and human rights, and sustainable communities and institutions as it does to sustainable population, economic activity, and ecological integrity. All of these are closely interlinked, and security in the narrow sense is a critical part of the mix. In the presence of highly destructive war, it is impossible to protect nature effectively or to keep certain human social ties from dissolving. Conversely, if resources are abused and human population grows rapidly, or if communities lose their cohesion, conflicts are more likely to occur. If great and conspicuous inequalities are present, people will be reluctant to restrain quantitative economic growth in favour of qualitative growth, as would be required to achieve a measure of economic and environmental sustainability. At the same time, great inequalities may provide the excuse for demagogues to exploit or revive ethnic or class hatreds, and to provoke deadly conflict.

In my book *The Quark and the Jaguar* (Gell-Mann, 1994) I suggest that we study possible paths towards sustainability (in this very general sense) during the course of this century, in the spirit of taking a 'CLAW'. The idea of such studies would be to seek out paths towards sustainability even if they may appear rather improbable. It is, of course, important that we not take these studies of possible future developments too seriously, but rather treat them as 'prostheses for the imagination'.

I employ a modified version of a scheme introduced by my friend James Gustave Speth, then President of the World Resources Institute, later head of the United Nations Development Program, and now Dean of the School of Forestry and Environmental Studies at Yale University (Speth, 2008). The scheme involves a set of interlinked transitions that must occur if the world is to switch from present trends to greater sustainability:

1. *A demographic transition* to a roughly stable human population worldwide and in each broad region. Without this, talk of sustainability seems pointless.
2. *A technological transition* to methods of supplying human needs and satisfying human desires with much lower environmental impact per person at a given level of conventional prosperity.
3. *An economic transition* to a situation where growth in quality gradually replaces growth in quantity, while extreme poverty, which cries out for quantitative growth, is alleviated. The economic transition must, of course, involve what economists call 'the internalization of externalities'. Prices will have to come much closer to reflecting true costs, including damage to the future.
4. *A social transition* to a society with less inequality, which, as pointed out earlier, should make the shift from quantitative to qualitative growth more acceptable. The social transition includes a successful struggle against large-scale corruption, which can vitiate attempts to regulate any human activity through law.

5. *An institutional transition* to more effective means of coping with conflict and with the management of the biosphere in the presence of human economic activity. We are now in an era of simultaneous globalization and fragmentation, in which the relevance of national governments is declining somewhat, even though the power to take action is still concentrated largely at that level. Most of our problems involving security – whether in the narrow or the broad sense – have global implications and require transnational institutions for their solution. We already have a wide variety of such institutions, formal and informal, and many of them are gradually gaining in effectiveness. However, they need to become far more effective. Meanwhile, national and local institutions need to become more responsive and, in many places, much less corrupt. Such changes require the development of a strong sense of community and responsibility at many levels, within a climate of political and economic freedom. Achieving the necessary balance between cooperation and competition and stabilizing commitments in the long run are difficult challenges at every level.

6. *An informational transition* in the acquisition and dissemination of knowledge and understanding. This will allow us to better cope at local, national, and transnational levels with technological advances, environmental and demographic issues, social and economic problems, questions of international security, and the strong interactions among all of them. Only if there is a higher degree of understanding, among ordinary people as well as elite groups, of the complex issues facing humanity is there any hope of achieving sustainability. But so far most of the debate on the new digital society focuses on the dissemination and storage of information, much of which is extremely useful but some of which is false or badly organized. We need to support and better reward the difficult work of converting that raw information into knowledge and understanding. This point illustrates particularly well the pervasive need for a 'Crude Look at the Whole'.

7. *An ideological transition* to a world view that combines local, sectarian, national, and regional loyalties with a 'planetary consciousness', a sense of solidarity with all human beings and, to some extent, all other living beings. Only by acknowledging the interdependence of all people and, indeed, of all life can we hope to broaden our individual outlooks so that they reach out in time and space to embrace vital long-term issues and worldwide problems in addition to immediate concerns close to home. This transition may seem even more utopian than some of the others. However if we are to reduce and eliminate conflict based on destructive particularism it is essential that groups of people that have traditionally been in conflict with one another acknowledge their common humanity. Such a progressive extension of the concept of 'us' has, after all, been

a theme in human history from time immemorial. One dramatic manifestation of this is the greatly diminished likelihood of armed conflict in Western Europe. That achievement has been aided by the long process of creating and developing European institutions, from the Coal and Steel Community to the European Union.

When studying these transformations, which are closely connected to the issues debated at the Symposium, it is especially important to devote considerable effort to integrative work. What actually happens in the world will depend not only on studies but on the behaviour of a huge multitude of human actors all over the globe, and in the end it is the nature of those actors that matters most. What will happen to human nature in the long-term future? Will it be changed artificially through the application of science and technology? Will it rather just continue to be gradually modified by culture and, if so, how? Will devotion to one's own country, ethnic group, religion or class really be supplemented by a planetary consciousness that defines 'us' as part of the entire human race and, to some extent, the other organisms with which we share the biosphere? Or will human cussedness lead us into disasters made ever worse by our advancing technology?

It is possible that disciplined yet imaginative speculation about the longer-term future can be of some help in seizing opportunities and in avoiding some of the worst catastrophes. But in thinking about the future let us take seriously the idea of a 'Crude Look at the Whole'.

## References

Friedman, T. L. (1999). *The Lexus and the Olive Tree.* New York.
Gell-Mann, M. (1994). *The Quark and the Jaguar.* New York.
Hirsch Hadorn, G., Hoffmann-Riem, H., Biber-Klemm, S. *et al.,* eds. (2008). *Handbook of Transdisciplinary Research.* Dordrecht.
Speth, J. G. (2008). *The Bridge at the Edge of the World.* New Haven.

# Chapter 2

# Integrated sustainability and the underlying threat of urbanization

---

## Geoffrey B. West

Geoffrey B. West, born in England in 1940, received his BA from Cambridge University in mathematics and physics in 1961, and his doctorate from Stanford University in 1966. He later joined the Stanford faculty in 1970. His primary interests have been in elementary particles, their interactions and cosmological implications. He was the founder of the high-energy physics group at Los Alamos National Laboratory. In 2003 West joined the Santa Fe Institute as a distinguished professor and was named its president in 2005. His interest in universal scaling laws led him to develop quantitative models of organisms based on universal principles. Recently he extended these ideas to studying quantitatively the structure and dynamics of cities and corporations, including the relationships between efficiency, growth, innovation and sustainability. He has received numerous awards and was included in Time Magazine's 2006 list of the 100 most influential people in the world.

*Note:* This chapter is a commentary on chapter 1.

In this essay I want to emphasize two major themes that have not received the attention I believe they deserve if we are to seriously tackle the question of long-term global sustainability in its broadest sense. In so doing I take for granted that it is important and fundamental to formulate the questions, problems and solutions relating to sustainability within a scientific paradigm. Such a paradigm can provide a credible platform for the socio-political leadership inevitably needed to effect change. An underlying motif will be a rallying call to recognize the importance of breaking down the boundaries between traditional academic disciplines. This includes the equally pressing question, not addressed here, of the inter-relationship between science, culture and politics – the 'three cultures problem'.[1] In this respect some of what I have to say builds on remarks made by my colleague Murray Gell-Mann, who emphasized transdisciplinarity, and encouraged the development of coarse-grained descriptions of complex systems. The two major themes I will address here are:

1.  The need for a broad, integrated scientific framework that encompasses a quantitative, predictive, mechanistic theory for understanding the relationship between human engineered systems, both social and physical, and the 'natural' environment. Somewhat whimsically, I shall refer to this conceptual framework as the *grand unified theory of sustainability.*
2.  As a corollary to this, the recognition that cities and the ever-expanding urbanization of the planet have played a seminal underlying role in bringing us to this critical point in the planet's history. Intimately related to this are questions of the dynamics of innovation, cycles of boom and bust, the seemingly inevitable increase in the pace of life, and the spectre of a planet of slums, pollution, disease, and conflict.

In recent years increasing worldwide attention has been paid to a multitude of threatening phenomena, such as global climate change and the incipient crises in food, energy and water availability. The recognition of an impending crisis has led to burgeoning national and international concern about questions of global sustainability, and has stimulated a proliferation of programmes focused on many of these issues. These have been promoted not only by leading governmental and international organizations but also by corporate and other non-governmental institutions.

However, most of these programmes and almost all existing approaches to the challenge of global sustainability have focused on relatively specific issues, such

---

[1] This is a take on the title of the lectures given by C. P. Snow in 1959 highlighting the divide between the sciences and the humanities (Snow, 1993). Interestingly, Snow himself embraced all three cultures; he was a scientist, a well-known author, and very much involved in politics and high-level governmental decisions in an advisory capacity.

as the environmental consequences of future energy sources, the economic consequences of climate change, and the social impact of future energy and environmental choices. While such focused studies are of obvious importance and, indeed, are where most of our research efforts should be directed, they are not sufficient. No overarching, integrated conceptual framework has yet been developed that can provide a long-term big picture uniting the many highly inter-related themes underlying sustainability. Existing approaches have, to a large degree, failed to come to grips with the essence of the long-term sustainability challenge; namely, *the pervasive interconnectedness and interdependency of energy, resources, environmental, ecological, economic, social, and political systems.* Early attempts along these lines include the well-known Club of Rome report in 1972 (Meadows *et al.*, 1972) and, more recently, the Stern report to the UK government in 2007 (Stern, 2007). However, there has not yet been any attempt to develop an explicit, overarching, systematic, conceptual scientific framework. Without such an integrated 'bigger picture', we risk repeating the classic mistake of developing short-term, highly-focused 'solutions' tailored to a narrow sector of the totality, ignoring the tight relationships between issues, and their dependence on a myriad of other problems. This approach inevitably leads to long-term and potentially disastrous consequences. A well-known, small-scale example currently under discussion is the advocacy of biofuels. This has generally ignored the potential consequences of vastly increased biomass production for water demand, biodiversity, or the increased cost and reduced availability of food for human consumption, and its effects on markets (see, for example, Inderwildi and King (2009); Creutzig and Kammen, this volume). For a comprehensive approach to sustainability, we need to overcome such 'stove-piping' and provide an integrated holistic framework that addresses the pervasive interdependencies and interconnectedness of different systems.

## A grand unified theory of sustainability?

It is becoming increasingly clear that one of the most profound challenges facing science and society today is the need for an *integrated* conceptual framework for understanding *sustainability* in its broadest sense. Such a framework is essential if we are to gain a comprehensive understanding of the multitude of strongly interacting factors that fall under the umbrella of sustainability, and which are typically treated as effectively independent. These include the following: (a) energy, food and resource production and consumption; (b) ecology, the environment, and climate change; (c) human population, health, and well-being; (d) the global economy, including the nature of risk and the dynamics of financial markets; and (e) the social, cultural, and political institutions and organizational structures upon which the preceding depend. Such a comprehensive understanding of the interacting and interdependent systems is critical if humankind is to make informed choices between the many competing 'solutions' to the energy, environment, economic, and social problems that constitute the sustainability challenge.

*A priori,* it is not at all clear whether such a lofty goal is indeed achievable. The extent of the problem is daunting and, until recently, very little attention had been given to thinking in these terms. Only now are tools and techniques being developed to address such questions, and it is far from apparent that anything like a serious quantifiable *unified theory of sustainability* is at all possible. Nevertheless, I would like to take a provocative position and suggest that such an exercise is worthwhile even if it fails, since it may, in any case, stimulate potentially new systemic ways of thinking, or even lead to an alternative, complementary paradigm. Perhaps most importantly, it may, at the very least, stimulate new questions, new areas of investigation or innovative ways of thinking that would otherwise not be quite so apparent when viewed only from a more restricted, highly focused perspective.

The concept of complex adaptive systems provides a potential framework for developing such an integrated, systemic, conceptual approach[2]. The ideas inherent in 'complexity science' have been developing slowly over the past 20 years and are now generally viewed as an exciting new paradigm for addressing the kinds of problems posed by the challenge of sustainability. Furthermore, the culture of complexity science has stimulated the emergence of a serious transdisciplinary approach, which is clearly required to address many of the key issues. The exploration of complex systems stemmed to a large degree from the realization that many mysteries of nature involve nonlinear behaviour. In these systems multiple feedback mechanisms play a major role, and the whole is substantially greater than, and often significantly different from, the sum of its parts. Many systems are composed of myriad relatively simple individual components. Yet, once aggregated they take on collective characteristics that are not manifested in, nor could be easily predicted based on, the components themselves. Such 'emergent phenomena' are typical of all social systems and characterize the kinds of interactions and problems associated with economies, markets, urban communities, the environment, the weather, the health system, and other complex systems. The study of complex systems has taught us to be wary of naively deconstructing the system into independently acting component parts, and that a small perturbation in one part of the system may have major unforeseen consequences elsewhere. This familiar phenomenon was spectacularly manifested last year by the meltdown of financial markets across the globe, apparently stimulated by misconceived dynamics in the relatively localized US mortgage industry, with potentially devastating social and commercial consequences worldwide.

---

[2] A good modern overview is provided by Mitchell (2008). See also Waldrop (1993).

The developing science of complexity embraces an integrated systemic approach that brings together a broad spectrum of powerful techniques and concepts. These include agent-based modelling, cellular automata, network theory, multi-scale thinking (both temporal and spatial), field theory, statistical physics, scaling theory, and the renormalization group. Furthermore, it addresses transdisciplinary questions and concepts that are central to any discussion of sustainability. These include adaptability, evolvability, robustness, resilience, regulation, and conflict. In addition, an important lesson learned in investigating many complex phenomena is that, while it is not typically possible to predict detailed aspects of the system, it is sometimes possible to derive a 'coarse-grained' description that allows for quantitative predictions of the generic, salient features of the system. The development of such a quantitatively predictive, coarse-grained theoretical framework encompassing the challenges of risk, financial markets, climate, the environment, health, pollution, urbanization, etc. would be a major accomplishment. It would allow not only an assessment of long-term questions of sustainability but would also provide the basis for cost-benefit analyses of alternative scenarios involving all of these highly-coupled phenomena.

As funding agencies and universities worldwide are beginning to recognize, complexity science coupled with a transdisciplinary approach will play an increasingly important role in the academic landscape of the twenty-first century. To quote Stephen Hawking[3]: 'Q: Some say that while the twentieth century was the century of physics, we are now entering the century of biology. What do you think of this? A: I think the next century will be the century of complexity.' What has yet to be appreciated, however, is that bringing such a perspective to the challenge of global sustainability and the long-term survival of our planet will be critical because it inherently recognizes the kinds of interconnectedness and interdependencies so frequently ignored in current discourse.

As an example, we need to develop a natural framework for understanding the fundamental and critical problem of how human social dynamics (manifested by the dominance of urban living – the source of and solution to most of our problems) drives the changing environment (usually in a negative way), as well as the reverse interaction of how the changing environment influences engineered and evolving human systems. A major challenge of the twenty-first century is the fundamental question as to whether human-engineered social systems, from economies to cities, which have only evolved over the past 5000 years, can coexist with the 'natural' biological world, which evolved over billions of years. We will only have a sustainable planet that can support over 10 billion people living in 'harmony' with

---

[3] Stephen Hawking quoted in an interview on January 23, 2000 in the San Jose Mercury News; see http://www.mercurycenter.com/resources/search.

the biosphere from which we evolved if we understand the principles and underlying dynamics of this social-environmental coupling.

The increasing worldwide attention paid to issues of sustainability is both gratifying and frightening; gratifying because it is one of the most critical issues facing humankind, and frightening because we risk the possibility of squandering huge financial investments and enormous social capital if we continue to pursue limited and single-system approaches to sustainability without developing a unifying framework. Now is the time to recognize that a broad, multi-disciplinary, multi-institutional initiative, guided by a broader, more integrated and unified perspective, is likely to play an important role in yielding sound scientific conclusions. A strong case can be made that now is the time to initiate a dedicated Manhattan-style project or Apollo-style programme for global sustainability in the integrated, systemic sense described here.

### The central role of cities and urbanization: can we avoid a planet of slums?

In this last section I would like to present an example of how the broad perspective implicit in 'complexity science' can be usefully applied to a key issue in global sustainability, namely urbanization and the role of cities. This will be a highly condensed overview of a large body of work, some of which is well established but parts of which are perhaps a little more speculative. Though justice cannot be done in just a few paragraphs, this illustrates a way of thinking inspired by ideas falling under the rubric of complexity.

The future of humanity and the long-term sustainability of the planet are inextricably linked to the fate of our cities. It is estimated that in 2005 an historic threshold was crossed with more than half of the world's population now living in urban centres (UN, 2005). This is in marked contrast to the situation that pertained for almost the entire time-span of human existence over the last several thousand years, when almost all human beings resided in non-urban environments. For example, even as recently as the birth of the United States at the end of the eighteenth century, only a small percentage of Americans were urban dwellers, whereas today more than 80% live in cities. By 2050, this will very likely be true for the entire planet. The extraordinary growth of cities is often associated with the rapid rise of standards of living, prosperity and quality of life. Indeed, the more urbanized countries are, on average, richer. Moreover, the world's two most populous countries – China and India – are undergoing unprecedented experiments in rapid urbanization, with unforeseeable consequences for their future resource consumption, their impact on the natural environment, and social stability.

Cities have traditionally been, and continue to be, the sources of creativity, innovation and wealth production. They are hubs of social activity, the magnets that attract creative individuals, vacuum cleaners that suck up innovation, and stimulants for ideas, growth and wealth production. Analyses of data confirm this (Bettencourt *et al.*, 2007); regardless of which indicator one looks at, the larger the city the more innovative 'social capital' is produced. For example, if a city doubles in size, then, on average, wages, wealth, the number of patents and number of educational and research institutions all increase by approximately 15% on a per-capita basis. We refer to this systematic phenomenon as 'superlinear scaling'; the larger the city, the more the average individual resident owns, produces and consumes, whether it be goods, resources or ideas. As urban creatures we all participate in the multiple networks of intense human interaction manifested in the metropolitan buzz of productivity, speed, and ingenuity. This is the good news about cities and why they have been so attractive and seductive.

Now to the bad news: similar analyses of data representing the negative side of urban life manifest an analogous 'superlinear' behaviour. By approximately the same degree as for the positive indicators, negative indicators of human social behaviour also systematically increase with city size: doubling the size of a city not only increases wages, wealth and innovation by approximately 15% but also increases the amount of crime, pollution and disease to the same degree (on a per-capita basis). Apparently, the good, the bad and the ugly seem to come hand in glove as an integrated, almost predictable, package. A person may move to a bigger city drawn by more innovation, a greater sense of 'action' and better wages, but they can also expect to confront an equivalent increase in garbage, theft, stomach flu, and AIDS.

Until the middle of the last century, this dual nature of cities as the origin of wealth and ideas and, at the same time, the source of pollution and disease was not perceived as a serious threat because cities were still sub-dominant in terms of population. As cities began to dominate, their entropy production inevitably led to degradation of the environment, non-linear consequences for the climate, severe stresses on resources and energy, and the beginnings of the multiple problems we face under the banner of sustainability as we enter the twenty-first century. *Cities have emerged as the source of the biggest challenges the planet has faced since humans became social, yet cities are also the source of the solution since they are the reservoir of creativity and ideas.*

This remarkable and seemingly inextricable link between the benefits and costs of community structure very likely has its origins in the 'universal' dynamic of the network structure and group clustering of human interactions; when humans began serious interpersonal interactions about 10 000 years ago, forming sizeable communities, discovering economies of scale and the fruits of wealth creation, they brought a fundamentally new dynamic to the planet, a dynamic that went beyond biology. The resource and energy networks that have evolved in the 'natural world' to sustain biological organisms and ecosystems are primarily dominated by economies of scale ('sublinear scaling') – roughly speaking, the larger the organism, the less energy is required per second to support each one of its cells (West *et al.*, 1997; West and Brown, 2005). The dynamics of such networks constrain the pace of biological life to decrease systematically with increasing size. For example, large mammals live longer, take longer to mature, have slower heart rates, and cells that work less hard than those of small mammals, all to the same degree; (doubling the mass of a mammal increases time-scales, such as its lifespan and time to maturity, by about 20 % on average and, concomitantly, decreases all rates, such as its heart-rate, by the same amount). Small creatures live life in the fast lane while large ones move ponderously, though more efficiently, through life (think of a mouse versus an elephant!). The social networks that underlie the 'superlinear scaling' of wealth creation, innovation, crime and pollution behave in exactly the opposite fashion to these biological networks; the larger the organization, the faster the pace of life (Bettencourt *et al.*, 2007). In large cities, disease spreads more quickly, business is transacted more rapidly and people walk faster, all approximately to the same degree and in approximately the same systematic, predictable fashion (as a rule by approximately 15 %).

In biology a further consequence of economies of scale and of sublinear scaling is that organisms like mammals eventually stop growing, reaching some approximately fixed size at maturity (West *et al.*, 2001). Over time-scales that are very long compared to human social time-scales, biological systems are relatively stable and sustainable, with major changes taking place over many thousands to many millions of years. On the other hand, in social organizations where growth is driven by the superlinear scaling associated with wealth creation and social innovation, growth is unbounded, never reaching an 'asymptotic' stable state, and proceeding at a rate that is faster than exponential (Bettencourt *et al.*, 2007). To sustain such growth in light of resource limitation requires continuous cycles of paradigm-shifting innovations such as those associated in human history with the discovery of iron, steam, coal, computation, and, most recently, digital information technology. Indeed, the litany of such discoveries is testament to the extraordinary ingenuity of the human social mind in overcoming the looming threat of running out of the perceived essential resource. However, there is a serious catch: theory dictates that, to sustain continuous growth – one of the primary assumptions upon which modern societies have evolved – *such discoveries must occur at an increasingly accelerated pace; the time between successive innovations must inevitably get shorter and shorter. So, if we insist on continuous growth driven by wealth creation, not only does the pace of life inevitably quicken, but we must innovate at a faster and faster rate!*

Until recently the period of time between major innovations far exceeded the productive life span of a human being. Beginning towards the end of the twentieth century this was no longer true; a typical human now lives significantly longer than the period between major innovations. The period between the most recent major shift from the 'Computer Age' to the 'Information and Digital Age' was only about 20 years, which is to be compared with the order of thousands of years between the Stone, Bronze and Iron Ages. Furthermore, the time differential to the next significant innovation is destined to be even shorter. This is surely not sustainable, and, if nothing changes, we are heading for a major crash and a potential collapse of the entire socio-economic fabric. The challenges are clear: Can we return to an analogue of the 'biological' phase whence we evolved and be satisfied with 'sublinear scaling' and its attendant natural limiting, or no-growth, asymptotically stable configuration? Is this even possible? Can we have the kind of vibrant, innovative, creative society driven by ideas and wealth creation as manifested by the best of our world's cities and social organizations, or are we destined to a planet of urban slums and the ultimate spectre of devastation raised by Cormac McCarthy's novel The Road (McCarthy, 2007)?

Given the special, unique role of cities as the originators of many of our present problems and their continuing role as the super-exponential driver towards potential disaster, understanding their dynamics, growth and evolution in a scientifically predictable, quantitative framework is crucial to achieving long-term sustainability on the planet. Perhaps of even greater importance for the immediate future is to develop such a theory within the context of a 'grand unified theory of sustainability' by bringing together the multiple studies, simulations, databases, models, theories and speculations concerning global warming, the environment, financial markets, risk, economies, health care, social conflict and the myriad other characteristics of man as a social being interacting with his environment.

## References

Bettencourt, L. M. A., Lobo, J., Helbing, D., Kühnert, C. and West, G. B. (2007). Growth, innovation, scaling, and the pace of life in cities. *Proceedings of the National Academy of Sciences of the United States of America,* **104**(17), 7301–6.

Inderwildi, O. R. and King, D. A. (2009). Quo vadis biofuels. *Energy & Environmental Science,* **2**(4), 343–6.

McCarthy, C. (2007). *The Road.* New York.

Meadows, D. H., Meadows, D. L., Randers, J. and Behrens, W. W. (1972). *The Limits to Growth: A Report for the Club of Rome's Project on the Predicament of Mankind.* New York.

Mitchell, M. (2008). *Complexity: a Guided Tour.* New York.

Snow, C. P. (1993). *The Two Cultures.* Cambridge.

Stern, N. (2007). *The Economics of Climate Change: The Stern Review.* Cambridge.

UN – United Nations (2005). *World Demographic Trends: Report of the Secretary-General.* United Nations Report E/CN.9/2005/8.

Waldrop, M. (1993). *Complexity: The Emerging Science at the Edge of Order and Chaos.* New York.

West G. B. and Brown, J. H. (2005). The origin of allometric scaling laws in biology from genomes to ecosystems: towards a quantitative unifying theory of biological structure and organization. *Journal of Experimental Biology,* **208**(9), 1575–92.

West, G. B., Brown, J. H. and Enquist, B. J. (1997). A general model for the origin of allometric scaling laws in biology. *Science,* **276,** 122–6.

West, G. B., Brown, J. H. and Enquist, B. J. (2001). A general model for ontogenetic growth. *Nature,* **413,** 628–31.

# Chapter 3

# Earth system analysis and taking a crude look at the whole

## Wolfgang Lucht

© Klaus Welp, Helsinki

Wolfgang Lucht is Chair of the research domain on 'Climate Impacts and Vulnerability' at the Potsdam Institute for Climate Impact Research in Germany. He also holds the Alexander von Humboldt Chair in Sustainability Science at the Department of Geography at Humboldt University in Berlin. A 'physicist turned geo-ecologist', Lucht's research addresses how climate and land use change affect global landscapes, the role of humans in shaping the Earth's future, and pathways toward improved sustainability of human societies.

*Note:* This chapter is a commentary on chapter 1.

## Global change in the twenty-first century

The collective outcome of global humanity in action is, in our time, worldwide environmental degradation of a magnitude not seen before. Climate change and land-use-driven planetary deforestation are the two tips of a dangerous iceberg that signals a deep crisis in the relationship of humans to their material environment (see Fig. 1). These changes herald a transformation of Planet Earth that is on par with a number of major fluctuations, interruptions and transitions in the Earth's history. The root cause is the explosive growth of human material turnover and population (see Kohn, this volume) in the last several decades.

The question that largely remains open at this point is whether the Earth's transition to a new state of operation will be largely suffered by humankind (and with it a great many other species that share the planet), as a consequence of humanity's myopic focus on short-term advantages. Or whether, instead, humanity will be able to collectively influence the ongoing transition, at least to some extent, or even divert it in ways that would allow human societies and the greater environment to continue through the transition phase with considerable, but still manageable losses (see Fig. 2). In other words, the question is whether human societies will be able to develop the collective cognitive power to re-order their affairs in a manner that reflects an understanding of the interconnected workings of the planetary system, and whether they can come to a common understanding of major desired and undesired developments and the associated required revisions in the functioning of today's societies.

Should this challenge one day be successfully met, it would impressively testify to an ability of human cultures to produce, explain and justify collective responsibilities that reach beyond the present; a mental and cultural ability that, one could argue, is in many ways at the root of the differentiation of humans from other higher life forms. If the challenge is not met, however, the ongoing evolutionary experiment of rational intelligence may have reached its planetary limits.

The global anthropogenic transformation that has been set in train will have fundamental consequences not only for the state of the atmosphere, oceans and land surfaces, but equally for human societies (Costanza *et al.,* 2007). There is no particular reason to believe that social structures are more resilient to change under systemic forcing than the environment. They will be equally, if not more, affected. Tipping points that may cause state changes in characteristic parameters or spatial patterns are known to exist in the Earth's climate system and in the biosphere interacting with it (Lenton *et al.,* 2008). Similarly, tipping points can be expected to exist in the even more complex networked systems of societies. Currently, however, little is known about them.

A number of recent crises within the cultural, social and economic systems of

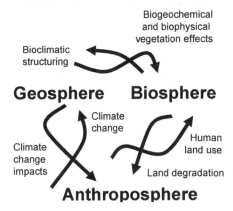

**Fig. 1.** Systemic interactions in the Earth system of the twenty-first century. (*Source:* W. Lucht)

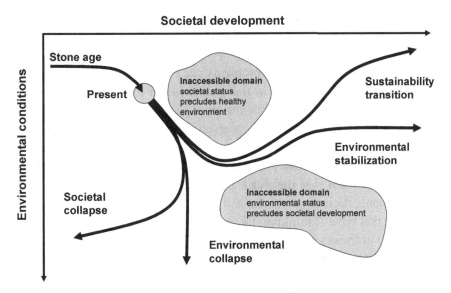

**Fig. 2.** Pathways in the future co-evolution of societies and the environment. (*Source:* W. Lucht, developed using ideas from Schellnhuber, 1999)

the world's societies have revealed intrinsic features of the self-organisation of these societies. In contrast, the influence of societies on the environment is still largely perceived as being external, and similarly the feedbacks of the ensuing changes on societies remain largely outside their self-reflection processes. The causal perception loop between societal and environmental dynamics is not closed in many important topical areas. However, thanks to recent efforts, the case of climate change is increasingly becoming a notable exception. The tipping points and disruptions that lurk in this loop are at the centre of the problems facing humanity in the twenty-first century.

Three prototypical solution pathways seem to be available and are supported by three strands of discourse: the technocentric, the value-oriented and the rationalistic-scientific strands.

## Technological pathways

There is a widespread belief, or rather a hope, that technological progress will outpace growth in such a way as to make possible a breakthrough to clean, green, environmentally friendly technology without interrupting economic and material growth. This is the paradigm favoured in many current discussions about global change and the prospects of sustainable development. The ultimately technological causes of the great environmental problems currently at hand will, according to this school of thought, also lead in the medium term to means of surmounting them, if only technological developments are wisely steered in the right direction. According to this view, the currently observed biodiversity losses, climate change and environmental pollution are merely a dirty bottleneck through which human civilisation and with it the planet has to pass before a sustainable high-tech future unfolds. This argument is widespread: the most important debates on recycling, dematerialization, efficiency increases and semi-closed material loops all make use of it. Without this type of thinking, the world would already be in a much worse state. But in all of these scenarios, primary energy use is set to triple by the end of the century. How credible is it that the projected increase in available energy will lead to a decrease in the volume of materials used?

The fundamental problem is that there is little historical evidence that technological progress in material use and waste per produced unit has, on a large scale, been effective in reducing the overall material throughput of societies. Despite very substantial advances across the board in efficiency and in the material and energetic intensities of industrial processes, economic growth has up to now mostly outstripped these gains. Both the net harvest of materials from the environment as well as net waste flows into the environment have increased with time, often dramatically, when viewed across large regions and many sectors. Achieving a transition to a lower level of socioeconomic metabolic turnover of materials would in fact be a first in human history; from hunter-gatherers to agricultural cultures to industrialized societies there has been a steady increase in the material throughput required to maintain, grow and reproduce human societies (Haberl and Fischer-Kowalski, 2007). History does not support the expectation that future technologies will, in their sum, be considerably more efficient and crucially less damaging to the environment in their production and implementation while economic growth continues. Such an expectation entails the crucial assumption that the future will somehow be qualitatively different from the past in this respect. This is almost completely speculative.

So while next-generation technology certainly will be an indispensable, immensely important factor in achieving a more sustainable future, unless the problem of growth is tackled it is very possible that a purely technological solution will, despite all progress, fall short. Since so much of the current world is based on growth, with the rich nations struggling to cope even with reduced growth, the rapidly industrializing nations greatly concerned about the robustness of their growth, and poor nations very justifiably aspiring to grow out of their poverty, this is a worrisome prospect.

## The world of values

A second approach to the transition problem is embodied in the wide-ranging discourses on values, justice, and generic rights of the natural world. In this approach the solution is not sought primarily in technology but in the cultural power of humans: to frame their lives through cultural identity constructions and societal orders, built upon political and ethical systems, religious understandings and spiritual relationships to the world. At the core of this approach are central questions concerning who we are as humans, who we should and can be, and what our place in this world is.

From these questions follow directions for societies. While many such systems have placed humans in a controlling, possessing position in the world, providing the ethical, religious, spiritual, tribal or national underpinning of environmental appropriations, many of the same and a number of alternative cultural systems, not only in indigenous cultures, emphasize respect for life in general, for the world and its inherited orders, for other humans and for the self as the best path towards a rich existence. In this view, the limits to growth are given where it impinges on the inherent rights of others, whether in this or a future generation, whether geographically close or afar, and whether in the human domain or in the wider domains of life. They are given where growth compromises the particular quality of the existence of the other. This world view appeals for a revision or even revolution of lifestyles, values and priorities driven by alternate cultural self-constructions. Justice is a core element of this debate.

There are two problems with such a value-oriented approach to achieving the reordering of human material relations to the world required for a sustainability transition. For one, history shows that for the mainstream of human cultures, appeals to become more responsible and to champion the good have too often lacked the power to overcome the material orders of societies, which people have often been reluctant to compromise. Humans seem disposed to put material wealth before mental well-being, though often the two are connected; a materially poor life is a happy one only with great difficulty, and often only in artificial monastic settings. Also, the strongly structured social orders characteristic of humans and most primates produce a close relationship between power and material control, making material production an element of deeply engrained social relations, and thus a difficult factor to overcome.

A second problem with this approach is that transforming value systems and thereby, to some extent, engineering a transition of cultural identities to a state that is compatible with sustainability is likely to conflict with the most fundamental of modern human values, that of individual freedom. Proactive cultural construction has too often been a tool of dictatorships and tyranny, with devastating consequences, for people not to be wary of consciously engineered value systems. Cultural construction is an ongoing human experiment that does not seem to be bound by a peaceful human inclination, once more for reasons probably rooted in the problematic but deeply constituting legacy of humans' primate past.

However, if a controlled transition in the interlinked social-environmental world system is to be achieved, transitional progress has to be made not just in the environmental domain, where the impacts have to be lessened, but also in the social domain, where the problems have their origin. The power of cultural re-invention should not be underestimated in this context. It is precisely what allowed humans to flourish in all corners of the world. When in the brains of early *homo sapiens* environmental and technological knowledge began to mix with their old and profound social intelligence, the foundations were laid for the experiment of nature unfolding in modern humans. Culture is elementary to our condition. Therefore, to ignore the powers of the cultural dimension in seeking solutions would amount to negating the core factor that has made modern humans what we are.

Perhaps for this reason, particularly in the American discourse, the solution for sustainability seems often to be sought in a combination of green technological breakthroughs and value changes (Raskin *et al.*, 2002). Unfortunately, as shown, technology probably will not be sufficient and value changes in a free world not a priority over material accumulation. This leaves this vision, despite the central position of culture for the human species, uncomfortably adrift of the workings of the real world.

## Pathways through rationality

The third proposed avenue for engineering the collective sustainability transition rapidly is to rely once again on the hope that the rational side of the human intellect will in the end overcome the intricate webs of human societal and technological identity constructions. Admittedly, it is a hope that may be as questionable as that concerning values. The progress of rational thought since the Enlightenment has undoubtedly produced great improvements in the human condition, proving its power to transform, but it has also degraded the world by removing richness in cultural meaning and by tending to produce universalistic, dominating economic and technological structures. This has been called the totalitarian aspect of the dialectics of the Enlightenment (Horkheimer and Adorno, 1947). The hope is that, in the end, the intellect rooted in the human mind will understand the lock-in, transcend it, and, driven by the will to be something particular in the world, open up new avenues. We can question whether revolutions and cataclysmic crises are necessary to stimulate such breakthroughs. But certainly there is a deep conviction, particularly in European thought, that rational solutions to problems can be found, and implemented, even if the ultimate objectives of such rational action remain rooted in culturally formed self-understanding.

It is here that science enters the debate. Certainly humankind requires an analytical, diagnostic and prognostic science of the Earth system before the problems it faces can be adequately viewed and understood. Climate change is not a problem that can be described purely as socially constructed in the way some other aspects of human reality can; if emissions continue, climate change will occur irrespective of the prevalent social discourse. Planetary realities are impinging on the symbolic and discursive systems of humans in challenging new ways. It is only very recently that humans have even begun to see and appreciate the Earth as a physical, chemical and biological system. Concerning scientific insights into the world, it is worth remembering that a mere 200 years ago the meaning of prehistoric finds such as dinosaur bones or hand axes was unknown. Nobody knew how old the Earth was, that there were ice ages, where the sun obtains its energy from, or, how chemistry works. There was no knowledge of genes or epigenetics, no theory of evolution, and little to be called historical science.

It is only on the basis of this newly created scientific image of the Earth, rather than the earlier cosmological, religious, cultural images of the Earth, that a warning can now be sounded, perhaps just in time, about the consequences of human action on the planet; only computer models built with the knowledge of Earth system science are now able to project climate change and land-use scenarios and the resulting impacts on the world's ecosystems in a way that will affect political action. It is the system of rational analysis that has contributed this crucial element to human reflection. Despite the large remaining uncertainties and gaps in knowledge, what has become known is significant enough to have triggered the current global debates on climate change, land use and sustainability.

The question now is whether the expectation that rational insights have the power to influence and ultimately transcend cultural and economic practices is warranted. That is, whether the powers of the collective human brain will allow a narrow escape from the predicament by steering the tools of culture, economy and social relations in directions that sustainably support a future free of unmanageable tipping point transitions in either the environment or in societies. Is this a realistic prospect? Realities around the world are more strongly shaped by cultural and economic forces than by rational analysis. The deeper challenge, therefore, is how to integrate the findings of the sciences into the sometimes fast-changing, sometimes sluggish societal self-constructions that dominate human processes. If this is not successful, rational analysis will remain a marginal activity in the government of human affairs, its power of insight and foresight wasted.

## Looking at the Earth as a system

Murray Gell-Mann argues that a way forward might best become apparent if we take a 'crude look at the whole' as our starting point (Gell-Mann, this volume). This formulation encapsulates his analysis that relevant Earth system processes are firmly interconnected and that the 'whole' includes identifiable macroscopic properties, including transitional behaviour. His proposal is based in science, the rationalistic vein of analysis, but goes far beyond it by building on the realization that, in the end, it is the human mind that has to come to conclusions and has to find ways to bridge the gaps between the realities of social structures, cultures and sciences, and bring it all together in a mentally adequate manner. It is for this reason that the disciplinary segregation inherited from the history of science is not suited to the problem of climate change. A more comprehensive approach to applying the intellect to the problems of the world – a crude look at the whole – is needed.

Alexander von Humboldt championed a similar approach, depicting the complexities of world landscapes that he encountered by describing their natural history, geology, ecology and human colonization in narratives composed of well-selected details, arranged to provide insight into the larger whole (von Humboldt, 1807). They were meticulously accurate and highly selective in their depiction, and formed a whole of consciously aesthetic quality, as a means of facilitating the incorporation of scientific knowledge into the human mind.

James Lovelock (2003) has argued in a closely related vein against reductionism in Earth system science. He writes that reductionist disciplinary approaches, despite their indisputable successes, are fundamentally unsuited to explaining the major systemic interconnections that form the whole of the planet; and hence make it more difficult, if not impossible, to understand the change underway in that whole. By drawing analogies between the planetary and the human body Lovelock observes that just as the human phenomenon cannot be understood from the mere sum of its biochemical states, so the Earth as a whole cannot be understood from a merely reductionist summation of its physical and chemical states. He then describes the Earth as a self-regulating system in which humans are in danger of marginalizing themselves through their own actions.

Hans Joachim Schellnhuber (1999) has described the emergence of the modern scientific enterprise as a series of revolutions that have signalled the advent of systemic reflection in the life of the planet; the original Copernican revolution, looking out into the heavenly world, has been followed by a recent 'second Copernican revolution', looking inward into the workings of the planet. In both cases, optical instruments led the way, producing essential images that helped establish a coherent new science. The insights gained were not initially of immediate relevance to daily lives, but subsequently shifted perceptions of human identity in a most profound manner while also opening up new methodological avenues. Building on new knowledge, and using the tools of scientific Earth system analysis, humankind is now in the process of forming a disembodied, networked collective Global Subject that is attempting to order its affairs in the world while struggling with the intimidating complexity of the task.

## Earth system analysis

So how, then, can a crude look at the planetary whole be achieved? Based on Schellnhuber's analysis, three elements support an adequately reflective consciousness. First, a highly developed, comprehensive science of Earth system analysis is required, using medium-complexity computer simulation as an important synthetic tool for projecting the joint dynamics of geosphere, biosphere and anthroposphere into the past and into the future. Second, a comprehensive, global-scale Earth observation system is needed to provide the essential empirical links between the past or present states of the planet (including the many local realities in its regions), and theoretically constructed macroscopic images of these states. And third, a globally networked, multi-hubbed system of communication, negotiation and goal-setting is required to enable distributed, multifaceted communication, understanding and then management of a considerable number of processes relevant to the basic functioning of the Earth system. Together, these elements will constitute the distributed, collective, networked global consciousness that may steer planetary processes out of dangerous territory by influencing the powerful dynamics of the anthroposphere.

Current medium-complexity Earth system modelling already provides to some extent crude looks at the whole. Profound insights have been generated in the past 30 years about the functioning of the Earth system and its many interlinked biogeochemical cycles, geophysical balances and system feedbacks. Nonetheless, these models still treat the main cause of today's disruption of global biogeochemical and energetic balances – human action – as largely external. Neither the deeper social drivers nor the impacts of the change on these drivers are yet part of most modelling systems, partly because the processes of the anthroposphere cannot yet be systematically computerized. Again, one of the deeper reasons for this deficiency lies in the disciplinary structure of the sciences, out of which Earth system science has grown.

A similar gap is evident in current global Earth observation, which is required to provide humankind with sensory feedback on the Earth's history and current state. Current observations focus strongly on non-human systems. With the notable exception of global economic and related national statistics, the all-important human dimension is subject merely to weak, largely unsystematic or under-evaluated observation. A more comprehensive observation of the whole, particularly of the exchange processes between human societies and their environment, is urgently required if a crude look at the whole is to be achieved. One of the greatest challenges in sustainability research is to develop methods to identify the details on the basis of which a crude look at the Earth system and its interactions with humans can be achieved. The challenge is to bring local realities into the framework of global interconnections. That process involves more than creating a loose mosaic by reductionist summation of separate parts.

In terms of communication and decision-making structures working with crude looks at the whole, the global transitions to be managed in this century are of a magnitude that will require coordinated international, though not necessarily unified, approaches. Bodies such as the Security Council of the United Nations may recognize that it is in their remit to pro-actively anticipate the geopolitical dangers resulting from mismanagement of the looming food, climate, energy, industrialization, population, and resource crises. In order to avoid, limit, channel, or manage these dangers the world will need coordinated optimization of resource use, adherence to agreed bottom-line standards of international justice, joint financing of overarching countermeasures, stimulation of education and innovation, and the sustainable regulation of many resources that, under the prevalent economic and political paradigms, are not coherently managed. If the impending change is to be managed rather than suffered, human societies will need to adopt self-engineered paths to sustainability. These will have to lead to substantial reductions in worldwide human material extraction, emission and waste flows.

### New cosmologies

In summary, there is still widespread failure to appreciate that the methods of the twentieth century are not fully adequate to address the transformative crisis at hand. The assumption persists that somehow societies will be able to more or less continue on their current paths, with some adaptation to environmental changes, but little or only gradual alteration in basic functions. This may turn out to be one of the greatest misconceptions of our time. The adaptive powers of societies are certainly strong, but most likely are too slow to keep pace with environmental changes, and even at a slow pace they will likely transform societies. The challenges of the twenty-first century will be fundamentally different in quantity and quality from those of the twentieth century because the fundamentals of the problems to be tackled are very different. The question now is whether an investment can be made – intellectually, financially, and culturally – in finding pathways that will allow a future based on sustainable use rather than profligate consumption of resources.

It is only for this reason of urgency that we can probably not avoid adopting the very uncomfortable word 'engineering': the world can no longer avert dangerous change unless human societies actively engineer, or manage, a rapid way out of their predicament. This engineering or management will engage multiple sectors: technological engineering (as in new energy and production technologies), societal engineering (as is currently happening in the form of a politically agreed transition in the world's energy systems and in the creation of international institutional structures), environmental engineering (as in the world system of human-controlled nature reserves), and perhaps even – as a very last resort that is better avoided – limited, targeted geoengineering. Engineering, however, is by definition built on a rationalistic basis, and is subject to the fundamental cultural risks associated with that. To be truly effective it must pay particular attention to matters of design: it must by design be deeply embedded in a social and value-based analysis looking at consequences and pitfalls.

In this manner, the three strands of technological, value-oriented and rationalistic-scientific approaches are interwoven in this process of Earth system management and are not mutually exclusive. All of them must be applied in order for humankind to turn a potential dead-end into a bottleneck that in turn may lead to a sustainable opening. In fact, closer analysis shows that these three approaches operate on different levels. The rationalistic-scientific approach deals with understanding the control problem at hand. The technological and value-oriented strands have to do with means for exercising the necessary, albeit surely very partial, control in the technological and social domains. In addition, the value-based approach is concerned with the question of what the operating principles and directions should be, beyond merely avoiding the worst.

The global transformations now under way are the latest expression of a transformation that probably began with the advent of symbolic information processing in the brains of humans, or more precisely in the brains of the latest species of humans, *homo sapiens,* some 100 000 years ago. Ever since that transformation the human domain has been structured according to ideas of culture, religion, language, tribe, nation, place, personal identities and histories, leading ultimately to the still somewhat mysterious processes of agriculture and industrialization that are now causing such dangerous systemic side effects. In that sense, many of the dynamics of the anthroposphere are a cultural phenomenon. The ultimate root causes of the global transformation will be found in these intrinsic, still poorly understood processes of human culture. Ways forward will therefore, ultimately, have to also be anchored in social and cultural dimensions.

Environmental feedbacks from human action have always been an integral part of societal dynamics, but for the first time they now have to be considered on a global scale. For the first time in human history, a systemic understanding of the Earth system needs to enter cultural processes. Therefore, the crude narratives of the whole to emerge from this process will inevitably need to depart from socio-cultural narratives if they are to be effective since they must ultimately aim at affecting societal structures. Taking a crude look at the whole in this sense puts an immense responsibility on the human mind to develop well-founded narratives that are in full resonance with the latest scientific findings (Lucht and Pachauri, 2004).

The process of constructing such views is not without dangers. Human history is replete with societal visions of nature and the natural that have clouded the practice of human interaction with the environment. It takes a culturally embedded and reflective, yet highly capable science to prevent such misguided approaches. This will be all the more difficult as comprehensive, fundamental theories of ecological systems are not yet available (if they even exist). Attempts at a theoretical explanation of the historical evolution of human societies and their interactions with the environment are fragmentary at best and the underlying assumptions deeply controversial among historians, economists, sociologists and anthropologists. We lack guiding theories of society-environment interactions, let alone of society-environment co-evolution. Yet such insights are required to form a framework upon which new interpretations of the human as well as the planetary condition can be formulated.

I therefore propose that the key factor in taking a crude look at the whole is a belief, maybe even merely a hope, that the human mind – in this case the collective mind of networked humanity – will be able to construct mental images of the whole that are more than mere figments of cultural or scientific projection. Rather, they must be equally founded in rational analysis and cultural production. These mental images have to take the form, I propose, of new cosmologies, cosmologies that blend cultural narratives of the position of humans in the world with the findings of Earth system analysis, encompassing both the natural world and the human condition in its cultural expression. These new cosmologies will do one thing: they will once more describe the place of humans in the Earth system.

The required sustainability transition will certainly not happen by accident. It will not merely emerge – by chance or by necessity. Achieving it will require a collectively conscious societal effort based on a reflective Earth system science that takes into account the full extent of the human experience, expressed in new cosmologies. Such a transformation will likely open up interesting new pathways for human societies on our planet. It will require that humankind applies its unprecedented scientific knowledge determinedly to the problems facing the world, and shows a great openness to renewed discourses on values, priorities, justice and self-images, with consequences that will structure societies, through the power of narratives. Such societal self-engineering is not a small intervention, however the consequence of inaction will be equally transforming. It is, in the end, a question of identity because the damage done will be largely irreversible.

The ultimate question, I suggest, is: What will we do with our freedom? This question can only be answered by the human mind. This is what is meant by taking, through a science of Earth system analysis that is comprehensively embedded in sustainably re-empowered cultural practices, a crude but well-defined and essential look at the whole.

## References

Costanza, R., Graumlich, L., Steffen, W. *et al.* (2007). Sustainability or collapse: what can we learn from integrating the history of humans and the rest of nature? *Ambio,* **36,** 522–7.

Haberl, H. and Fischer-Kowalski, M., eds. (2007). *Socioecological Transitions and Global Change: Trajectories of Social Metabolism and Land Use.* Cheltenham.

Horkheimer, M. and Adorno, T. W. (1947). *Dialectic of Enlightenment: Philosophical Fragments.* Palo Alto.

Humboldt, A. von (1807). *Ansichten der Natur.* Tübingen.

Lenton, T. M., Held, H., Kriegler, E. *et al.* (2008). Tipping elements in the Earth's climate system. *Proceedings of the National Academy of Sciences of the United States of America,* **105,** 1786–93.

Lovelock, J. E. (2003). Gaia and emergence. *Climatic Change,* **57,** 1–3.

Lucht, W. and Pachauri, R. K. (2004). The mental component of the Earth system. In: H. J. Schellnhuber, P. J. Crutzen, W. C. Clark, M. Claussen and H. Held, eds., *Earth System Analysis for Sustainability.* Cambridge, 341–65.

Raskin, P., Banuri, T., Gallopin, G. *et al.* (2002). *Great Transitions. The Promise and Lure of Times Ahead.* PoleStar series report No. 10. Boston.

Schellnhuber, H. J. (1999). Earth System Analysis and the Second Copernican Revolution. *Nature* **402,** Suppl., C19–23.

# Chapter 4

## Making progress within and beyond borders

---

### Johan Rockström, Katrin Vohland,
### Wolfgang Lucht, Hermann Lotze-Campen,
### Ernst Ulrich von Weizsäcker, and Tariq Banuri

Johan Rockström is Associate Professor of Natural Resources Management at Stockholm University, Sweden, and Executive Director of the Stockholm Environment Institute and the Stockholm Resilience Centre. His research focuses on global environmental change and development issues, global water resources research, resilience and sustainability science, and integrated soil and water management in tropical agro-ecosystems. Rockström serves as a scientific advisor in several international organizations, and is involved in several research projects across scales, from resilience and global change processes to water management systems for improved food security among small-holder farmers in semi-arid savannah environments of sub-Saharan Africa.

*Note:* Photos and biographies of co-authors can be found in the appendix.

## Transformations of the twenty-first century

Climate change is now almost universally recognized as one of the gravest threats to life and well-being on this planet. Unfortunately, any potential response to this threat is complicated, if not hobbled, by four other factors.

First, there is another unfinished global policy agenda – the eradication of poverty and global inequality – whose only widely accepted solution – economic growth – conflicts directly with climate stabilization. Second, climate change has emerged as part of a complex mosaic of challenges, some of which are closely related to it. A short list of these challenges includes trans-national epidemics (such as HIV/AIDS, SARS, and Avian Flu), environmental degradation and biodiversity loss, accelerating water stress, increased frequency and/or intensity of various catastrophic events (floods, droughts, hurricanes, cyclones, tsunamis, and earthquakes), and threats to global security (especially from terrorism). Moreover, economic globalization has revived the spectre of runaway financial epidemics as manifested in the recent global financial crisis and subsequent economic recession. The current global economic crisis is not only the deepest since the 1930s, it occurs simultaneously with global climate change and ecological crises, all of which are closely interwoven; with unsustainable, excessive consumption and production patterns humanity has applied the logic of sub-prime lending not only to the housing sector but also to the global ecosystem. At the same time, economic globalization has eroded the capacity of states to cope with financial or other epidemics, or more broadly to protect social welfare and environmental resources by regulating financial and corporate capital.

Third, the human impacts of climate change are determined by the social and ecological resilience of human societies and the natural capital that supports them. The dramatic 'hockey-stick' pattern of temperature and greenhouse gas accumulation from anthropogenic emissions applies to virtually all critical ecosystem services of the Earth, as observed in land degradation, loss of biodiversity, deforestation, overfishing, and air pollution (see Fig. 1). Over half of the cumulative anthropogenic greenhouse gas emissions have been absorbed by terrestrial ecosystems (in forests and soils) and the oceans (Canadell *et al.*, 2007).

We can expect unforeseen positive feedbacks from climate change, when the warming interacts with the broad spectrum of hockey-stick patterns. It remains unclear though, what human-induced surprises could be triggered, even though several of the risks have been identified (e.g., abrupt change in the African and Indian monsoons, accelerated melting of glaciers, abrupt savannization of rainforests; Lenton *et al.*, 2008), and have even been observed (the abrupt collapse of the Arctic summer ice in 2007). A key element of this unknown is the global degradation of ecosystem functions (e.g., carbon sequestration) and services (e.g., food and fish

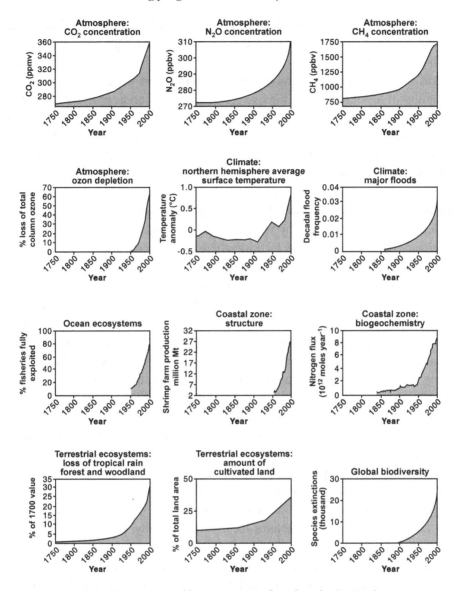

**Fig. 1.** Hockey-stick pattern of key ecosystem functions in the Earth system under pressure from human drivers. (*Source:* Steffen *et al.,* 2003, p. 133)

production). The UN Millennium Ecosystem Assessment, presented in 2005 (Millennium Ecosystem Assessment, 2005), was the first global health check on the state of the planet's ecosystems. It concluded that we have degraded 60 % of key ecosystem services, which are not only fundamentally important for human well-being, but particularly critical for poor communities, and a key feature of our capacity to adapt to climate change.

Fourth, climate change upsets the very foundation of modern society. The growth momentum of the industrial age since the middle of the eighteenth century has been built upon the harnessing of energy from fossil fuels, and the bulk of modern physical infrastructure and corporate profitability is premised on the continued availability of fossil fuels. An effective resolution to the crisis will imply a radical transformation of both the technological and corporate basis of industrial activity. This, in turn, will only occur if pushed by a fundamental social transformation.

But the problem goes deeper than fossil fuels. Climate change is the thin end of the wedge of an irresolvable conflict between finite resources and unending growth. Continued and unending economic growth has become the very definition of progress and the basis for social solidarity in industrial society. Ultimately, this conflict will be resolved only by weaning post-industrial society from its continued reliance on growth, and thus by critically reassessing growth itself and some of the core values that underpin it: competition, entrepreneurship and consumption.

Climate change is ultimately the visible face of an absolutely unprecedented challenge to the international community. This challenge forces us to simultaneously ask (a) how to sustain the process of economic development in poor countries (both fast- and slow-growing ones), (b) how to move existing infrastructure and economic institutions away from their almost exclusive reliance on fossil fuels, (c) how to continue to enhance social welfare while weaning modern society from its dependence on unending growth and resource use, (d) how to strengthen the conventional locus of policy making – the nation state – while creating effective institutions for local and global governance, and (e) how to do all this while simultaneously addressing other areas requiring immediate attention – health, environment, financial instability, and political conflict.

This will require novel instruments and institutions of global governance, a dramatic change of direction of technological progress towards resource productivity, and strong incentive structures locally and globally, encouraging all actors to abandon unsustainable technologies and habits and to work towards a sustainable future. Most of all, it will require enlightened and responsible global leadership that serves to unite people from all nations in a common cause rather than creating divisions, friction, and distrust.

### The climate challenge: crisis and opportunity

The climate community has long articulated the 2 °C limit (namely an average temperature increase of no more than 2 °C over preindustrial levels) as the safe threshold beyond which irreversible, costly and even catastrophic change becomes likely. The Fourth Assessment Report of the Intergovernmental Panel for Climate Change (IPCC) interprets this target as implying a stabilization of carbon concentration at

450 parts per million (ppm) $CO_2$ equivalent, which in turn means drastic reductions in global carbon emissions. The scientific assessment in the IPCC cautioned that even stabilization at 450 ppm $CO_2$ equivalent constitutes no less than a 30% risk of exceeding 2 °C, and more recent science suggests a need to keep the carbon dioxide concentration below 350 ppm, which would correspond to approximately 400 ppm $CO_2$ equivalent, to avoid accelerated and dangerous climate change (Hansen *et al.,* 2008). Today, in 2009, we have already reached 385 ppm $CO_2$ and almost 450 ppm $CO_2$ equivalent.

As mentioned already, the carbon stabilization goal has emerged at a time when the pre-existing common agenda of humankind, namely poverty eradication and reduction of global inequality, is still unfinished. The well-known 'champagne glass figure' (see Fig. 2) from the cover of the 1992 Human Development Report depicts this issue vividly. The poorest 20% of the global population earned only 1.4% of the global net income, while the richest 20% received 82.7%, a ratio of 1:60. This inequality appears to be widening rather than narrowing. In 2004, the corresponding ratio was estimated at 1:90.

The only sure way to reduce this inequality, and thereby also to address associated social ills – poverty, unequal access to basic human needs (nutrition, health, education, and right to due process and participation), and protection from predatory behaviour – is economic growth in poor countries. A few countries, especially in Eastern Asia, have taken off into what appears to be a robust growth pathway, but they still face enormous challenges that call for global cooperation: how to protect the momentum from getting derailed by external pressures, how to make it compatible with resource limits, and how to extend it to areas where poverty persists. Other countries and regions are showing slow or intermittent growth, and there too global cooperation is of paramount importance to increase the momentum of growth by addressing familiar obstacles of governance, institutions, and human resources. Economic growth, however, is an imperative not only in developing countries. It provides the foundation for the successful operation of a modern economy. While it is now becoming clear that our love affair with economic growth must come to an end, the means of achieving this transition are far from clear.

All this, in other words, represents an unfinished global responsibility. The only hope of obtaining the requisite political support in rich countries is to gradually de-couple welfare from growth so as to accommodate social needs within the resource portfolio of a finite biosphere. Likewise, the only hope of marshalling the energies of four-fifths of the world behind newer challenges that are assuming ever-greater importance is that the sustainable development agenda in poor countries continues to be viewed as a common global agenda until such time as the most glaring inequalities have been eliminated.

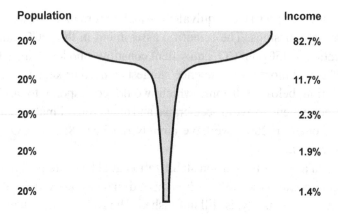

| Population | | Income |
|---|---|---|
| 20% | | 82.7% |
| 20% | | 11.7% |
| 20% | | 2.3% |
| 20% | | 1.9% |
| 20% | | 1.4% |

**Fig. 2.** The 'champagne glass' of global inequity. There is large and growing in-equity in the distribution of wealth, with 20% of the world's richest inhabitants receiving more than 80% of the world's income, while the poorest 20% receive approximately 1%. (*Source:* adapted from UN HDR, 1992)

In the absence of concerted global action, climate change, foremost among the newer challenges, will remain on a collision course with development and growth. With the existing technological portfolio, continuing growth in rich as well as poor countries would lead to a threefold increase in carbon dioxide emissions by the end of the century, with consequences that can only be described as catastrophic. On the other hand, without additional measures, many of which require visionary action, any serious response to the climate challenge will disable the growth process, under-mine societal welfare in rich as well as poor countries, and deal a severe blow to prospects of global solidarity.

## Crises as springboards for collective action

The Chinese pictogram for the word 'crisis' is a combination of two characters: threat and opportunity. The current financial and economic crisis represents not only a threat but also an opportunity. Long-term solutions will require fundamental change to the way financial markets and global financial institutions are regulated. The Bretton Woods institutions, set up to rebuild a war-torn world after the Second World War, are not configured to deal with the global social, economic, and eco-logical crises humanity faces today. The financial crisis has triggered a healthy insight that these and other institutions will require reform. The huge sums in-vested in various 'stimulus' packages, which amount to thousands of billions of US dollars, could be directed towards investment in low-carbon technologies and practices. The large investments now being generated to 'save' predominantly rich economies from collapse expose by comparison the ridiculously paltry amounts allocated to 'development' in poor countries in the world (with global development aid in the order of USD 80 billion compared to stimulus spending of more than USD 1000 billion in the United States alone). These factors constitute an opportunity. There is a risk, however, that the large stimulus funding will be re-invested in the old 'business-as-usual' economic system that was the original cause of the crisis, thereby stimulating more unsustainable consumption and growth (see Töpfer, this volume).

In the past, great crises have often served to unite people by creating a common cause of action. Through enlightened leadership and an engaged populace, crises have often regenerated societal trust and collective action on the basis of new visions, new institutions, and new laws and agreements. The example of the economic crises of the inter-war period, leading to the emergence of the welfare state, is often given. At the international level, acute problems have similarly served as springboards for testing and improving means of international coordination, balancing interests, sharing burdens, learning about and managing impacts, and expanding scientific understanding.

Such exploitation of opportunity has been evident in recent crises. Global epidemics have stimulated unprecedented international cooperation between countries and institutions that otherwise are not closely linked. Global dissemination of information on violent genocides has provoked the international community into developing new institutions to ensure dignity and human rights for everybody. The increased frequency as well as awareness of natural disasters (including earthquakes, floods, droughts, and storms) has led to charitable actions and solidarity as well as the beginnings of investments in institutional coping capacity. All of these developments have been pushed, supported, and monitored by global social movements for the environment, human rights, women's rights, and the rights of indigenous communities.

However, crises can also lead to more regressive responses. The rapid growth in international migration in recent decades, driven by economic, political, security, or environmental factors, has fostered a fortress response. Similarly, while the Cold War created a stimulus for the peace movement and arenas of international cooperation, the so-called 'War on Terror' has triggered a more paranoid response by governments as well as civil societies. Finally, globalization has weakened traditional institutions that protect the vulnerable, including the organs of the welfare state, and has undermined social solidarity, although the response of countries affected by the Asian financial crisis helped balance some of these trends.

### The great transition or a fortress world?

This, then, is the challenge for the leaders of the twenty-first century: how to pilot the world towards unity of action and common purpose on a sustainable pathway that builds resilience and steers away from undesirable tipping points, rather than to the erection of divisions, barriers, and fortresses? The pursuit of sustainability is deeply embedded in the agenda of global solidarity. Actions within borders impact and are impacted by those beyond borders, and all foreign policy has become, in essence, global domestic policy. Actions within one sphere affect and are affected by actions as well as omissions in others, and the major questions regarding the basis of human welfare have been reopened.

What follows is a brief list of issues thrown up by this challenge. While there are powerful forces that seek to divide and fragment, there are also equally powerful visions of a world that enable us to overcome differences, and unite all people in a common future. These visions include at minimum the following elements.

### *Democracy and participation*

One of the most powerful forces both in bringing people together and enabling a search for collective solutions is the institutionalization of democracy and participation at all levels. At local and national levels, it means the participation of the entire population, including women, children, the poor, and elderly people.

At a global level, it means strengthening the United Nations system, making it more effective, transparent, and responsible. It also means ensuring that markets work fairly in the service of global prosperity, welfare, and sustainability, and that market institutions support rather than subvert democracy. Finally, in the twentieth century we learned of the power of an engaged civil society to harness entrepreneurial energies, provide common visions, challenge conventional wisdom, and monitor and render transparent the workings of governments.

### *The development agenda*

After a long period of unfulfilled promise, there is evidence that the development momentum has picked up sufficiently to address the concerns of large numbers of poor people, especially in Asian countries. It is a matter of tremendous importance that this momentum be sustained and expanded.

Economic growth is a necessary but not sufficient condition for eradicating the worst aspects of poverty. The world community sought to address this in a targeted approach through the Millennium Development Goals. This initiative, which aims to reduce by half the number of people living in extreme poverty by 2015, supports funding programs and raises awareness of global poverty. However, to achieve global development targets, a change in rich countries' policy is urgently needed. Investments in innovative options are required to meet the needs of the poor, for whom traditional approaches are not appropriate. These options include community development and micro-credit schemes. There is also a need to shift towards more integrated approaches, which lead to sustainability in both resource management and service delivery systems.

### *The energy system revolution*

The climate challenge is associated closely with the energy system. The Industrial Revolution was based ultimately on the harnessing of increasing volumes of fossil fuels. The challenge now is to engender a transformation to a radically new structure that is not dependent on fossil fuels. However, the first energy revolution has yet to reach the vast majority of the world's population (see Nakicenovic, this volume). While the energy systems of industrial countries have reached a stable level, those in developing countries still have to grow considerably.

Most of the instruments being considered at a global level to address climate change are indirect in nature. They include national emission targets, trading schemes, and support for the emergence of an emissions market. All these have found much greater acceptance in industrialized countries than in poor countries, mainly because they are at best irrelevant and at worst inimical to the development agenda.

An early idea for incorporating development concerns into the emissions trading framework was that of equitable emission rights. It remained on the sidelines of the climate debate until the recent courageous statement by German Chancellor Angela Merkel that national emission entitlements should gradually converge towards equal per-capita levels (a proposal presented in August 2007 on the occasion of her visit to Japan). The idea of equal rights to the global commons represents the spirit within which a consensus solution could be found. A global climate regime for greenhouse gas emissions that builds on the principles of the UN Framework Convention on Climate Change, stating that burden-sharing must be based on capacity and responsibility, has been developed by the Stockholm Environment Institute with partners (Baer *et al.*, 2007, p. 95). This so-called Greenhouse Development Rights (GDR) framework couples climate science with the right to development among the world's poor. It clearly shows that if humanity is serious about solving the climate crisis in an equitable way that still allows room for development among the poor majority on the planet, emission reductions in many industrialized countries (essentially OECD countries) will have to already exceed 100 % by 2020. This is achievable if industrialized countries, in addition to reducing emissions domestically, commit to investing in emission cuts in developing countries.

By itself, however, the assignment of rights to development will not produce a miraculous transformation of existing energy systems and infrastructures. Immediate infrastructure investment in alternative energy systems is needed to set such a transformation in motion; it will also require the development of institutions that can help poor people to defend and benefit from their new rights. For purposes of immediate action, it might be necessary to shift from the language of 'rights and targets' to the language of 'investment and action' aimed at engendering a new energy revolution.

### *A change in values: long-term thinking and sustainable lifestyles*

Beyond government regulation and institutional settings, individual values will shape future developments. Teaching our children new ways to view the world may even have the strongest impact in the long run. A transition to more sustainable values and life styles will take place gradually. The example of the demographic transition is highly relevant. It represents a fundamental revision of the entire bases of traditional society: the notion of family and kin relationships, the basis for economic organization, the relation between men and women, parents and children, and between citizens and the state. This transition has occurred within the space of one generation in many developing countries.

### Placing climate policy in context

A number of elements of a potential response are being debated in the policy community. These include political/institutional interventions, and ecological, economic, technological, and discursive instruments.

### *Linking three disconnected UN processes*

Climate change, as clearly pointed out in the IPCC's Fourth Assessment Report, is already today impacting on the lives of poor communities. The most vulnerable are hardest hit, and are expected to bear the greatest burden of a climate crisis they have not caused (see Pachauri, this volume). Already the 2015 UN Millennium Development Goal targets of halving hunger, poverty and health threats are at risk due to climate change. At the same time, nowhere are ecosystem services so fundamental to human well-being as in the fight against poverty, and these ecosystems are negatively affected by climate change.

Despite these close relationships between climate change, ecosystems and development, there is a disconnect between the three UN processes supporting the governance and management of these domains: the UN Framework Convention on Climate Change (UNFCCC); the UN Convention for Biological Diversity (UN CBD), the UN Millennium Ecosystem Assessment (UN MA) and the follow-up process to establish an equivalent to the IPCC on biological diversity and ecosystem services (the Intergovernmental Platform on Biological Diversity and Ecosystem Services,

IPBES); and the UN Millennium Development Goals (UN MDGs) of halving hunger and poverty by 2015 and ensuring sustainable development among the world's poor (see Fig. 3). There is an urgent need, as well as an opportunity, for a comprehensive policy-coherent effort to connect these processes within the framework of the UN system.

### *Political/institutional instruments*

The challenge of global sustainability requires investment in institutions of democratic governance at all levels; local, national, and global. At the global level, the overriding imperative is to invest in the UN system. At the national level, a key goal is build political constituencies in all nations for effective and fair global engagement, expanding the reach of participatory and democratic institutions, and channelling support for strengthening development in poorer countries. At the local level, there is a need to establish participatory institutions of self governance. In rural areas, there has been considerable experience with community organization programmes led by visionary leaders from civil society and government. These programmes must be expanded in order to address the livelihood needs of the majority of poor and undernourished people from rural areas. An increasing share of the world's population lives in mega-cities that are difficult to manage. There is a need for concerted investment in the governance institutions of urban areas, and also to improve the basis of rural-urban exchange.

Other important areas where institutional investments are needed include education at all levels, economic justice and income distribution, law enforcement, property rights, damage compensation, (international) burden sharing, and political transparency and participation.

### *Technological instruments*

Technology is a broad term that includes not only the machines used in the production of goods and services, but also infrastructure and know-how for the organization of society.

Much of the discussion on climate change has focused on the deployment of renewable energy technologies on a large scale. However, the instruments that are being used to stimulate such deployment are mostly indirect in character. The ambivalence of global policy-makers sends conflicting signals to the private sector and the research community. The time has come for the global public sector to show its hand by committing itself to a large-scale infrastructure investment program, along the lines, for example, of the Apollo Programme, to help realize the potential of the technological portfolio. Such an investment would provide a clear

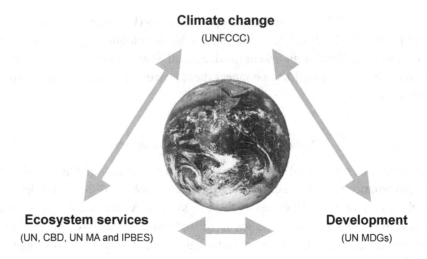

**Climate change**
(UNFCCC)

**Ecosystem services**
(UN, CBD, UN MA and IPBES)

**Development**
(UN MDGs)

**Fig. 3.** Three currently disconnected global UN policy and development processes that require urgent linking. The possibilities of stabilizing climate change and adaptation to unavoidable climate change (mandated to the UNFCCC) will require active stewardship of biological diversity and ecosystem services (mandated to the follow-up process of the UN Millennium Ecosystem Assessment (UN MA), the UN Convention for Biological Diversity (UN CBD), and the international initiative to establish the Platform on Biodiversity and Ecosystem Services (IPBES) – an equivalent to the IPCC on ecosystems). Ecosystem services are directly impacted by climate change. Climate change undermines the ability to reach the UN Millennium Development Goals (UN MDGs). Investment in development to support the majority on the planet living in poverty will determine the final outcome of anthropogenic climate change. Ecosystems form the fundamental basis for social and economic development, and therefore also the basis for achieving the MDGs. (*Source:* J. Rockström)

and unambiguous signal to the private sector and spur both the development and deployment of technological options.

However, the idea of technology goes far beyond renewable energy infrastructure. It includes concepts of ecological efficiency, social organization, and social control of technology.

The investment in energy efficiency will not take place without adequate public support. The nature of urbanization and urban infrastructure development reflects the current inappropriate incentive schemes, and alternative pathways will need clear and unambiguous support from governments. Moreover, the idea of social control of technology assumes even greater urgency in a situation that demands extensive and sustained intervention. It is absolutely critical that technological choices be subjected to sustained and persistent criticism from civil society, parliaments, mass media, and academia. The chances and risks provided by new technologies have to be assessed in a broad and continuous social discourse.

A final issue concerns technological extension. A good example is the Green Revolution, which saw the transfer of the knowledge of an emerging technological system from a few hundred scientists into the hands of several million farmers (most of them illiterate) within a span of a decade. This revolution was engendered by support for an expertly crafted and interlocked system, which included education, research, policy, extension, input supplies, credit, and a marketing infrastructure. Compared to this highly professional system, the new technological transition is being handled in an ad-hoc and unprofessional manner.

## Ecological instruments

Tragic as it is, under massive pressure from investors and market fundamentalists, many states worldwide have more or less given up on regulating resource use, water and energy markets, and even pollution. Some of the biggest problems, if not scandals, are biopiracy, patents on genes and other private appropriations of bio-diversity. The ecological agenda is linked inextricably with the agenda of reviving the developmental state, which can forge political consensus for sustainability, implement environmental regulations, and protect biodiversity against piracy. For example, a case could be made for placing a significant proportion of the world's land area (say 15%) under protection. As the conversion of land to agricultural uses is the most important factor in biodiversity loss, economic and political means have to be improved to make agriculture more ecologically sustainable.

Ecological instruments are based increasingly on solid and reliable research. However, there is enormous variation in research capacity between countries and regions. Indeed, the areas that are richest in biodiversity as well as in traditional knowledge of husbandry are often the ones with the least support from the organized research community. There is a need to build organized research capacity at national and local levels, and provide support for continuous investigation of impacts in priority areas: the maintenance of freshwater resources and soil functions, conservation of biodiversity, the management of environmental conflicts, and the protection of indigenous knowledge.

## Economic instruments

There is considerable controversy surrounding the strength and limitations of economic instruments. On the one hand, it is clear that measures that go against economic common sense are difficult to sustain over long periods. As such, it is widely accepted that policy measures should incorporate 'ecological and social truth' into economic activities by internalizing unwanted environmental, health and distributional impacts.

However, economic instruments suffer from some major shortcomings with regard to the agenda of sustainable development.

First, economic instruments are often found to be in conflict with the goal of equity. This is clearly visible in the controversy over climate change. Most economic instruments (including the volatility of oil prices and the unequivocal long-term trend towards higher oil prices) are highly regressive in nature, and subversive of the development and poverty agendas. In this case, it is wiser to rely on more direct policy approaches for engendering the transition in a fair and effective manner. Second, the issue of equity pertains especially to access to energy, industrial resources, financial markets, global public goods, and social infrastructure. A number of initiatives (e.g., micro-credit organizations) have tried to overcome the barriers created by the unfettered functioning of markets. These need to be supported. Third, volatile markets and a focus on short-term profitability must be rejected in favour of longer-term perspectives and higher predictability. Fourth, as already mentioned, in the absence of strong legal and political safeguards against the expropriation of the rights of poor and vulnerable groups, the exclusive reliance on market instruments will prove to be harmful.

## Discursive instruments

Communication is essential for meeting the challenges of the twenty-first century. This requires access to information exchange channels, together with expanded and improved observation systems in the social and environmental spheres. The Internet and mobile telephone networks have already started to improve this access in areas that were until recently excluded. Remote parts of poor African countries have become a part of 'online humanity'. If the gain of information and empathy is not to remain virtual, a global discourse on ethical and power issues is of vital importance. This can help to share values with respect to nature, justice, and the human position.

This is of particular relevance to the need for value change. The building of a global political constituency for a transition to a sustainable pathway requires that we move beyond the current situation in which people seem to be concerned only with very narrowly defined parochial interests. Current evidence suggests that the willingness to cooperate internationally in rich societies strongly depends on two things: direct involvement and impact, and available methods and technologies to react.

## Conclusions: a strategic vision

Today's challenges provide the chance to develop global mechanisms for sustainable development. They can act as springboards towards higher resource productivity and efficiency, environmentally friendly technologies, and sustainable habits and lifestyles.

The above discussion brings up a number of issues that require thorough consideration. However, the discussion has focused mostly on the individual components of the policy framework, not on the framework itself. It may be useful to provide a brief reflection on the strategic vision that can hold these diverse components together.

We must recognize that the response of the global leadership to the current crisis has been extremely slow. Even now, there is considerable scepticism both about the commitment and capacity of the global political system. The necessary response must bring together a global constituency for change. This will not happen through piecemeal or desultory interventions.

What is needed is a bold and strategic vision that can address the goals discussed here – economic development, biodiversity conservation, and climate stabilization – directly and in an integrated manner, instead of indirectly and disjointedly. For this, it may be necessary to shift from the language of targets and trading to the language of investment. A concrete example of a direct and integrated approach to climate and development would be a globally funded public investment programme in four areas: deployment of renewable energy technologies, institutions for promoting energy efficiency, governance of biodiversity and ecosystem services, and institutions and structures for enhancing adaptation capacity.

However, such a programme will test the limits of current governance arrangements. Existing means of international exchange and cooperation will have to be improved, and new global governance structures developed. Since large social and political transformations are inevitable, the world needs blueprints for action to sustain its struggle for universal goals – the eradication of poverty and inequity, reversing environmental degradation, protecting human security, and ensuring interregional and intergenerational justice. If these transformations are managed with skill, empathy and foresight in a globalizing multi-polar world, they can drive a broad agenda of sustainability and development within borders and beyond.

## References

Baer, P., Athanasiou, T. and Kartha, S. (2007). *The Right to Development in a Climate Constrained World: The Greenhouse Development Rights Framework.* Berlin.

Canadell, J. G., Le Quéré, C., Raupach, M. R. *et al.* (2007). Contributions to accelerating atmospheric $CO_2$ growth from economic activity, carbon intensity, and efficiency of natural sinks. *Proceedings of the National Academy of Sciences of the United States of America,* **104**(47), 18866–70.

Hansen, J., Sato, M., Kharecha, P. *et al.* (2008). Target atmospheric $CO_2$: where should humanity aim? *The Open Atmospheric Science Journal,* **2**, 217–31.

Lenton, T. M., Held, H., Kriegler, E. *et al.* (2008). Tipping elements in the Earth's climate system. *Proceedings of the National Academy of Sciences of the United States of America,* **105**(6), 1786–93.

Steffen, W., Sanderson, A., Tyson, P. D. *et al.* (2003). *Global Change and the Earth System: A Planet Under Pressure.* New York.

UN HDR (1992). *United Nations Human Development Report 1992: Global Dimensions of Human Development.* Oxford. Available at http://hdr.undp.org/en/reports/global/hdr1992.

# Chapter 5

## Towards a sustainable future

------------

### James P. Leape and Sarah Humphrey

James P. Leape has worked in conservation for more than three decades. Appointed Director General of WWF International in 2005, he is the Chief Executive of WWF International and leader of the global WWF Network, which is active in more than 100 countries. Leape previously directed the conservation and science initiatives of the David and Lucile Packard Foundation, one of the largest US philanthropies. Prior to that, he served as Executive Vice President of WWF-US.

Sarah Humphrey received a degree in environmental sciences from King's College, London. She began her career at the International Union for Conservation of Nature before moving to East Africa to work with the Western Indian Ocean Marine Science Association. She completed her PhD at Newcastle University, collaborating with the European Commission on the development of their coastal management policy. Humphrey joined WWF's Africa and Madagascar Programme in 2000, and since 2007 has worked on WWF's Living Planet Report series.

The essays in this book draw attention to the urgency of the sustainability challenge, which in the past decade has been brought to the forefront by the growing understanding of the nature and impacts of anthropogenic changes in atmospheric chemistry. They highlight some of the ways in which we can harness technology, human ingenuity and innovation to address critical sustainability issues particularly in managing the energy demands of the twenty-first century.

In this essay we present a broad perspective on sustainability to look at the wider set of direct and indirect pressures that humans are placing on the planet's renewable natural resources and biodiversity, drawing on data presented in the *2008 Living Planet Report* (WWF *et al.*, 2008). The Living Planet Report presents a stark picture of how humanity is living beyond our means as our consumption of natural resources exceeds their regenerative capacity and of the resulting decline in the Earth's biodiversity. If current trends are allowed to continue, by the mid 2030s we would need two planets to meet the demands we place on the planet's natural capital.

So what will the world look like in 2050? We believe that if we continue our current consumption patterns and development pathways, humankind may be facing ecological collapse on an unprecedented scale due to degradation of natural capital and loss in ecosystem services. Jared Diamond has explained how 'ecocide' – the loss in vital ecosystem services – has led to the collapse of past civilisations that were unable to adapt to environmental changes, whether man-made or natural (Diamond, 2006). In the modern world, examples of how we are eroding the planet's natural capital through overuse and misuse of natural resources are all around us. With widespread starvation, reduced life expectancy, environmental insecurity, and loss of social capital the consequences of ecological collapse at a global scale would eclipse our current concerns about rising food prices, water shortages and increased environmental risk.

Humanity has the capability to reverse the current trajectory of ecological decline, however, and to shape a future where humans live in harmony with nature. Such a 'Great Transformation' (see Potsdam Memorandum, this volume) will require bold action at a global scale to reduce our footprint and maintain or increase the resilience of natural systems. In this essay we will point out some of the major steps the global community should take in order to avoid a global environmental collapse.

## Challenges to sustainability

The *2008 Living Planet Report,* produced by the World Wide Fund for Nature (WWF) with its partners the Global Footprint Network and the Zoological Society of London (WWF *et al.,* 2008), provides a vivid picture of the path we are on. It offers three insights that define the challenge of sustainability. The first and most fundamental

is the sheer volume of humanity's consumption – we are devouring the world's natural capital to the point where we are endangering our future prosperity. The second
insight is interdependence – almost every country now depends upon the resources
of others; better management of the planet's natural resources has thus become a
shared responsibility. Finally, the *Report* charts the challenge of decoupling development and footprint – the relationships among human well-being, income, population, and sustainability.

### The ecological credit crunch

The *2008 Living Planet Report* offers two measures of sustainability. The *Ecological Footprint* measures our demand on the biosphere in terms of the area of biologically productive land and sea required to provide the resources we use and to
absorb our waste. A country's footprint is calculated on an annual basis as the sum
of the cropland, grazing land, forest and fishing grounds required to produce the
food, fibre and timber it consumes, to sequester the carbon dioxide it emits from
energy use, and to provide space for its infrastructure.

The Ecological Footprint can be compared to *biocapacity,* a measure of the
capacity of ecosystems, including agro-ecosystems, to produce useful biological
materials and to absorb waste products in a given year. In 2005, the global Ecological Footprint was 17.5 billion global hectares (gha), or 2.7 gha per person (a global
hectare is a hectare with world-average capacity to produce resources and absorb
wastes, Ewing *et al.,* 2008). On the supply side, the total biocapacity was 13.6 billion gha, or 2.1 gha per person, made up of cropland, grazing land, forest and fishing grounds. Our demands thus exceeded the planet's regenerative capacity by over
30% (compared to 25% in 2003). The growth of Ecological Footprint over time is
shown in Figure 1a, where one planet represents the biocapacity of the planet based
on contemporaneous management schemes and extraction technologies.

The second measure is the Living Planet Index (LPI), which tracks the populations
of 1686 vertebrate species across all regions of the world (Collen *et al.,* 2009). It indicates that global biodiversity has declined by nearly 30% over just the past 35
years (see Fig. 1b). The LPI shows that wild species and natural ecosystems are
under pressure across all biomes and regions. As human appropriation of the planet's resources increases, so we can expect increased impacts on the living organisms whose abundance in ecosystems is critical in maintaining habitat stability and
in providing the ecosystem services that underpin human well-being.

So what does the future hold? Figure 2 projects the growth in the Ecological
Footprint up to 2050 based on a set of moderate scenarios for future demands on
renewable resources. Based on this, our annual demands on the planet's regenerative capacity will exceed that capacity by 100% by the mid 2030s, or, in other words,

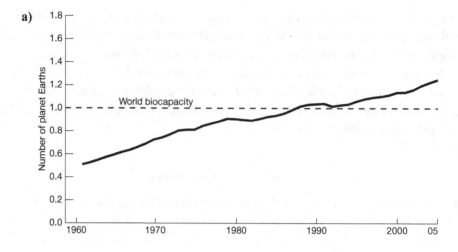

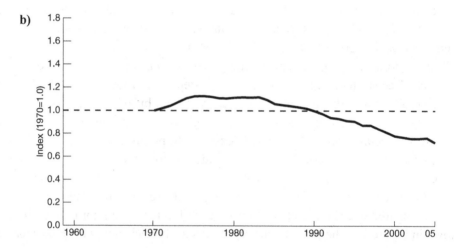

**Fig. 1.** Evolution in (a) humanity's Ecological Footprint and (b) the Living Planet Index. Changes in footprint are expressed in number of planet Earths where one planet represents the biocapacity available in a given year. (*Source:* WWF *et al.,* 2008)

we would need two planets to keep up with our demands for natural resources and waste assimilation.

While appealing in its simplicity, this business-as-usual scenario is conservative in that it assumes only very limited feedback between anthropogenic pressures and future bio-productivity. In practice, excessive demands on natural systems – measured as overshoot in footprint terms and shown as accumulated ecological debt in Figure 2 – are already compromising and will continue to compromise the planet's regenerative capacity.

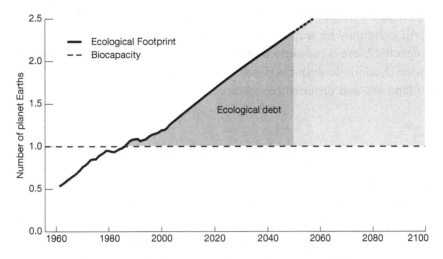

**Fig. 2.** 2050 scenario for ecological overshoot based on projected carbon emissions (Nacicenovic and Swart, 2000), moderate population growth (United Nations Population Division, 2006), utilization of food, fibre and forest products (FAO, 2002; FAO, 2006), and fisheries (Worm *et al.*, 2006). Changes in footprint are expressed in number of planet Earths where one planet represents the biocapacity available in a given year. (*Source:* WWF *et al.*, 2008)

We can already see how the direct impacts of resource over-extraction, ranging from fisheries collapse (e.g. Worm *et al.*, 2006) to deforestation, are undermining ecosystem services. A simple analogy could be pulled up between drawing on natural capital and drawing on capital in the bank, but in reality things are not so straightforward. For example, there are significant time lags between cause and effect for many pressures, the effects of changing atmospheric chemistry on ocean chemistry being a classic example. In addition, responses to environmental pressures are frequently synergistic and non-linear; scientists talk of thresholds, tipping points and discontinuities (e.g. Folke *et al.*, 2002). Extensive changes in marine ecosystem structure as a result of fisheries pressure (e.g. Sherman, 1994; Worm *et al.*, 2006) and arctic amplification with near-surface temperature rises in the region nearly two times the global average (e.g. Graversen *et al.*, 2008) are examples of such complexity.

Furthermore, not all anthropogenic pressures are readily measured in Ecological Footprint terms. One notable omission is the discharge of pollutants other than carbon dioxide, including other greenhouse gases, toxic chemicals and radioactivity. Similarly, conventional footprint accounting does not take measure of the now pervasive but indirect environmental effects of agricultural production, ranging from soil erosion to hydrological changes and biodiversity loss, nor of factors which may limit biocapacity, such as water availability, an issue of growing concern in the face of climate change (IPCC, 2007). The contribution of such factors to over-shoot will eventually be seen in national footprint accounts as results of declines in biocapacity. More sophisticated projections of overshoot are now being developed using dynamic footprint accounting which attempts to incorporate the influence of land use and disturbance, species diversity, and pollution (Lenzen *et al.*, 2007).

### An interconnected world

Just as conventional trade statistics describe the growth and changing patterns of international trade, ecological and water footprint analyses are revealing the way in which we draw on the environmental assets of other countries and regions to support our consumption patterns. Ecological Footprint accounts show that countries are increasingly relying on one another's biocapacity to support their preferred patterns of consumption. In 1961 the total footprint of goods and services traded internationally was 8% of humanity's total footprint. By 2005, this had risen to more than 40% of a much greater footprint. The imports of high-income countries averaged 61% of their total consumption footprint.

We are also increasingly relying on the water supplies of other countries to support our lifestyles. The water footprint of a country is the total volume of water used globally to produce the goods and services consumed by its inhabitants (Chapagain and Hoekstra, 2004; Hoekstra and Chapagain, 2007). Part of this footprint, the external water footprint, results from consumption of imported goods, or in other words, water that is used in the country which produces these goods. Worldwide, the external water footprint accounts for 16% of the average person's water footprint, though this varies enormously within and between countries. Twenty-seven countries have an external water footprint which accounts for more than half of their total water use.

As we externalize our water footprint and Ecological Footprint we also externalize the environmental impact associated with the goods and services we consume. A significant part of the Ecological Footprint is made up of carbon emissions that enter the global atmosphere, but food and fibre imports represent direct pressures on the ecological assets of other countries. The impact of the water footprint depends on where and when water is extracted. Water use in an area where water is plentiful is unlikely to have an adverse effect on people or the environment, but the same level of water use in an area experiencing water shortages may result in the drying up of rivers and the destruction of ecosystems, with associated loss of ecosystem services, biodiversity and livelihoods.

The global commodity markets and agricultural policies that sustain our consumption patterns generally overlook the environmental, economic and social costs to producer countries and the global environment. Production of palm oil for margarines and biscuits, soy production for animal fodder, shrimp farming, timber trade and biofuels are driving the destruction of some of the world's most valuable and biodiverse ecosystems.

### Footprint, income and development

Sustainable development has been defined as 'improving the quality of human life while living within the carrying capacity of supporting ecosystems' (IUCN *et al.*, 1991). One can see the difficulty of this challenge by mapping development progress against growth in footprint (WWF *et al.*, 2006).

Countries' progress towards sustainability can be assessed using the Human Development Index (HDI) as a measure of quality of life and Ecological Footprint as a measure of demand on supporting ecosystems (see Fig. 3). An HDI value of more than 0.8 is considered to be 'high-human development'. A footprint to global biocapacity per capita ratio of less than one is sustainable insofar as that it is replicable at a global level.

Figure 3 illustrates that as regions develop their footprint quickly becomes unsustainable. In fact, no region meets both criteria for sustainable development. Asia Pacific and Africa have been successful in achieving significant increases in HDI while still living within the available biocapacity per capita, but neither region meets the criterion for human well-being. North America and Western Europe have continued to achieve gains in human development, but their footprints soared disproportionately over the same period and are now several times greater than sustainable levels. In 2003, just one country met both criteria for sustainable development (Moran *et al.*, 2008).

Figure 4 shows how relative contribution of population and per capita footprint in driving overall national footprint has evolved in countries in different income categories (based on the World Bank's 2005 categorization). On a global scale, both population and average per capita footprint have increased since 1961. Since around 1970, however, population growth has been the principal driver in the growth of total footprint. Despite advances in agricultural productivity, the more than doubling of world population between 1961 and 2005 has its corollary in the halving of the average available biocapacity per capita.

The principal driver of increased footprint in high-income countries has been the growth in per person footprint, which grew by 76% from 1961 to 2005. The 15% of the world's population that live in high-income countries account for 36% of humanity's 2005 total footprint. In contrast, the principal driver of footprint in low and medium-income countries, as well as at a global scale, has been population. Population in low-income countries nearly trebled between 1961 and 2005 while in

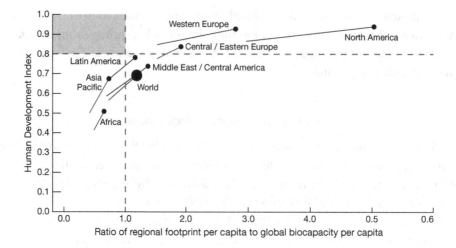

**Fig. 3.** Human Development Index (HDI) and Ecological Footprint. Points indicate values for 2003, and grey trailing lines show trends from 1975 to 2003. The shaded box represents a domain where both criteria for 'sustainable development' are met. (*Source:* Moran *et al.,* 2008)

middle-income countries it more than doubled. The per capita footprint of low-income countries actually decreased over this period while middle-income countries saw a 21% increase.

These data speak for themselves. Neither the rapid population growth nor the reckless consumption seen in different parts of the world are sustainable and both issues deserve our attention.

Clearly, a key challenge for this century is that faced by emerging economies such as China. China's per capita footprint and population roughly doubled between 1961 and 2005 producing more than a four-fold increase in its total Ecological Footprint. While population growth has remained steady, growth in per capita footprint has escalated in recent years and has overtaken population as the principal factor driving national footprint growth. China's HDI grew from 0.53 in 1975 to 0.77, at the threshold of high human development, in 2005. Will China now join the ranks of countries like Korea whose footprint growth has accelerated relative to its gains in HDI or will it find a 'third way' (see Potsdam Memorandum, this volume)? The challenge here is that of decoupling human development from footprint: how do we enable development without costing the Earth?

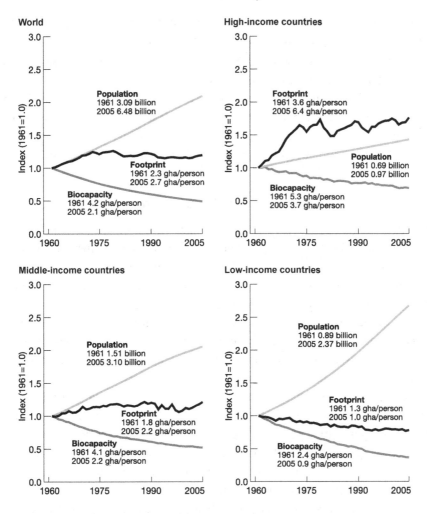

**Fig. 4.** Evolution of per capita footprint, biocapacity and population between 1961 and 2005 for the world and for high-income, middle-income and low-income countries. (*Source:* WWF *et al.,* 2008)

## Turning the tide: towards sustainability

### *An end to overshoot*

The above paragraphs have highlighted some aspects of the multi-faceted challenge we face in finding the 'third way' between environmental destabilization and persisting underdevelopment that the Potsdam Nobel Laureate Symposium was concerned with. The fundamental imperative for achieving sustainability is to ensure that humanity's global footprint stays within the Earth's capacity to sustain life, while achieving an acceptable standard of living for all.

Figure 5 charts a conceptual return to sustainability where humanity's footprint is reduced over the next three decades to fall within the planet's biocapacity. Instead of accumulating ecological debt we would maintain an ecological reserve, providing a buffer against environmental variability and shocks.

In the following paragraphs we present a two-pronged approach to maintaining and restoring the ecosystem services on which humanity depends, building on this conceptual framework of Ecological Footprint, biocapacity and overshoot but extending practical action beyond the metrics included in national footprint accounts.

### Turning the tide on humanity's footprint

Humanity's footprint is a product of population, consumption per capita and resource use and waste production intensity. Managing our footprint to sustain our natural capital requires re-examining the nature of the pressures exerted by each of the production sectors that meet our basic food, fibre and timber requirements: forestry, grazing, agriculture and fisheries. It means reconsidering the way we convert some of the world's richest ecosystems into built-up land and redefining the resource-intensive lifestyles that come with city living. And it means curbing the pollution that is overwhelming the assimilative capacity of natural systems and building up a toxic legacy for future generations.

We need to tackle all aspects of our footprint in order to sustain sustainable lifestyles, but one area deserves particular attention. In 2005, energy demands in our homes, industry and transportation represented the largest component of our footprint, with energy production from fossil fuels accounting for nearly 45% of the global footprint. High-income countries saw a nine-fold growth in the carbon component of their footprint between 1961 and 2005: this is a development pathway the global community cannot afford to see replicated at a global scale. WWF developed a 'Climate Solutions Model' to illustrate how it is technically possible to dramatically reduce climate-threatening emissions from energy services while meeting the needs of both the developing and developed countries in the twenty-first century (see Box).

### Building resilience

While we reduce our footprint, we must also find ways to restore the Earth's ability to support us, its biocapacity. Biocapacity can at least theoretically be increased by enhancing either the area of land or water available, or the productivity of those lands or waters.

The major challenge to maintaining biocapacity is the ongoing destabilization of ecosystems and attrition of ecosystem services. A recent study suggests that by

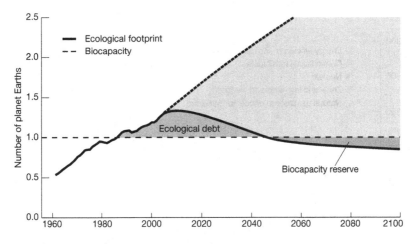

**Fig. 5.** A conceptual return to sustainability. (*Source:* WWF *et al.,* 2008)

### 'Climate Solutions' – Meeting the Carbon Challenge

Inspired by Pacala and Socolow's (2004) energy wedges, the WWF Climate Solutions Model explores whether it is possible to meet the projected 2050 demand for global energy services while achieving significant reductions in global greenhouse gas emissions through a concerted shift to already-available and more sustainable energy resources and technologies (Mallon *et al.,* 2007).

Figure 6 shows an output of the model which achieves reductions of 60–80% in carbon dioxide emissions by 2050 yet meets the three-fold increase in energy services projected in the IPCC's A1B scenario (Nakicenovic and Swart, 2000). The model embraces three parallel strategies:

- Breaking the link between energy services and primary energy production by expansion of energy efficiency in industry, buildings, and all forms of transport to stabilize the overall energy demand by 2025;
- Concurrent growth of low- to zero-emissions technolzogies through the use of renewable energies such as wind, hydro, solar and thermal, and bio-energy;
- An expansion of carbon capture and storage to phase out remaining emissions from conventional fossil fuels used for power and industrial processes.

In addition, an increase in the use of natural gas is proposed as an interim measure, creating a gas bubble which extends from 2010 to 2040.

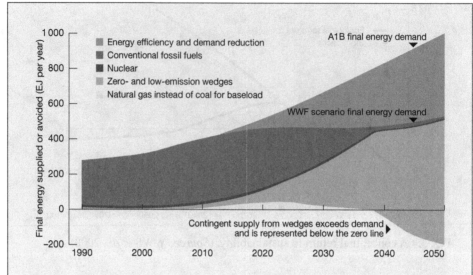

**Fig. 6.** Output of the WWF Climate Solutions Model. Energy efficiency and demand reduction (middle grey) allow the rising demand for energy services to be met by a more or less level supply of energy by 2025. This is complemented by introduction of zero- and low-emission energy supply technologies (light grey). By 2040, fossil-fuel use (dark grey) is reduced to a residual level for applications which are hard to substitute. The scenario provides spare capacity as a contingency, represented by energy supply in light grey shown below the x-axis. (*Source:* Mallon *et al.,* 2007)

2050 biodiversity decline will be the strongest negative influence on biocapacity as a result of the associated impacts on ecosystem functioning (Lenzen *et al.,* 2007). Yet, the global community is not on track to meet even the modest goal of the Convention on Biological Diversity, to reduce by 2010 the rate at which global biodiversity is being lost. Biodiversity conservation and ecosystem restoration can thus be seen as crucial management approaches in the face of growing pressures on ecosystems.

Practical measures to maintain and build resilience include putting in place effective protected area systems, integrated into surrounding landscapes, with the effective participation of local communities. But critical ecosystem services cannot be maintained simply through allocating specific areas to biodiversity conservation. Management and restoration of ecosystems, and measures to reduce direct and indirect pressures on biodiversity all have a role to play, and such efforts need to encompass highly modified landscapes as well as relatively pristine areas.

## A global agenda

In the face of growing human populations, uneven distribution of biocapacity and water resources, and the effects of climate change now being felt, the rising oil and food prices experienced in 2008 provided a glimpse of some of the stark choices that may face decision makers in the decades to come as they try to improve the quality of human life while remaining within the capacity of supporting ecosystems. The Earth simply cannot support the growing demands we are placing on its ecosystems; we are threatening our future prosperity and security.

In the following conclusive paragraphs we will set out four cross-cutting elements of a global agenda to reduce humanity's footprint and build ecosystem resilience.

### *Global action*

We will only meet the challenge of sustainability if we find a way to mobilize global action. The most pressing need is for action to curb humanity's emissions of carbon into the atmosphere – which accounts for nearly half of our Ecological Footprint. Specifically, the transformations of technology and infrastructure needed to achieve the climate solutions outlined above depend on three policy imperatives. These are *strong leadership* to agree on targets, strategies and investments in energy development; *a global effort*, with every country acting in accordance with its local challenges and capacity; and *urgency,* to address the real-world constraints on industrial transition and the risks of becoming locked in to energy-intensive technologies (Mallon *et al,* 2007).

While the specific challenges faced by developed and developing countries differ, the scale and ubiquitous nature of environmental challenges – from global warming to resource depletion – call for a global response for political as well as practical reasons. Looking ahead to 2050, the world's leaders and society as a whole will need to face some thornier issues surrounding global sustainability that go to the very root of our identities and industrial economy, namely population growth and burgeoning individual consumption. This implies a fundamental transition, at a global scale.

### *Market transformation*

Success will also require that we find ways to harness global markets for the cause of sustainability. The pioneering efforts of the Forest Stewardship Council (for wood products) and the Marine Stewardship Council (for fisheries) are paving the way for a wide range of initiatives to create markets for companies who commit themselves

to producing goods sustainably. A powerful blend of best practice, innovative partnerships and market opportunities offers a formula for transforming the production
of commodities such as timber, pulp and paper, cotton, palm oil, soy and wild-
caught and farmed seafood.

Further efforts are needed to increase the market share of ecologically and socially sustainable goods and services. These include developing positive incentives
for provisioning and trade of these goods and services, removing trade-distorting
and environmentally harmful subsidies, and establishing disincentives for providing goods and services that impede the long-term goal of achieving sustainability.

In the long run, we will need to develop more sophisticated tools to better account for externalities and resource scarcity in the pricing of goods and services.
The concept of virtual water, originally developed in the 1990s to show how the
import of water-intensive commodities can be an effective strategy for a country
experiencing internal water shortages (Hoekstra and Chapagain, 2007), provides a
basis for rethinking comparative advantage in trade of environmental assets.

### *Ecosystem-based management*

On the ground, or in the water, sustainability requires that we learn to manage natural systems on nature's terms – shifting from management of individual resources
like timber or water to management of whole ecosystems. We need an ambitious
effort to secure biodiversity and ecosystem services that takes us beyond traditional habitat and species protection measures to an integrated and system-wide
approach to conservation. Ecosystem-based management is an adaptive approach
that aims to achieve sustainable use of natural resources by balancing the social
and economic needs of human communities with the maintenance of healthy ecosystems.[1] Implementing ecosystem-based management requires mainstreaming
environmental protection and conservation action into decision making from local
to regional levels, and requires new ways of working across sectors and between
state and non-state actors.

### *Strategic alliance with the scientific community*

The final element is to reinforce science-based decision making. Environmental
science has come of age in recent decades. Panels and processes such as the *Intergovernmental Panel on Climate Change* and *Millennium Ecosystem Assessment*
have brought scientific understanding into the heart of the public debate and the

---

[1] Ecosystem-based management can be seen as the practical application of the ecosystem approach that has been
adopted by the Parties to the Convention on Biological Diversity

policy process, and have succeeded in transcending the disciplinary silos that still characterize so much scientific endeavour. Their findings have had profound impacts on the way we think about our impacts on the planet and on the nature of our responses.

There is little doubt that investment in strengthening scientific capacity both in the developed and developing world could further inform our choices and broaden our options towards a more sustainable future. To be effective, a new global contract between science and society (see Part V, this volume) will entail a broadening of the dialogue between policy makers, NGOs and the media on the one hand and the scientific community on the other. This will assure that scientists are able to communicate in clear and compelling terms the actions that need to be taken to sustain human well-being and that policy makers and society as a whole are able to respond with confidence.

The level of economic and social transformation required to put humanity on the pathway to sustainability may look daunting, but we only need to look back a few decades to see just how fast our societies and lifestyle can change. Some of the ingredients of this transformation have been set out above, bringing with them new opportunities to harness technology and innovation, to rebuild our energy economy, to reinvigorate and reform our food production systems, and to build a future in which humans live in harmony with nature and the natural systems on which we depend. Climate change brings a fresh imperative and a renewed momentum to harness humankind's ingenuity: we believe the challenge can be met but our response in the next decade may determine whether we thrive or decline as a species.

## References

Chapagain, A. K. and Hoekstra, A. Y. (2004). Water footprints of nations. *Value of Water Research Report Series,* **16.** Delft.

Collen, B., Loh, J., Whitmee, S. *et al.* (2009). Monitoring change in vertebrate abundance: the Living Planet Index. *Conservation Biology,* **23**(2), 317–27.

Diamond, J. (2006). *Collapse: How Societies Choose to Fail or Survive.* London.

Ewing, B., Goldfinger, S., Wackernagel, M. *et al.* (2008). *The Ecological Footprint Atlas 2008.* Oakland.

FAO – Food and Agriculture Organization (2002). *World Agriculture: Towards 2015/2030.* Summary report. Rome.

FAO – Food and Agriculture Organization (2006). *World Agriculture: Towards 2030/2050.* Interim report. Rome. Available at http://www.fao.org/ES/esd/AT 2050web.pdf.

Folke, C., Carpenter, S., Elmqvist, T. *et al.* (2002). Resilience and sustainable development: building adaptive capacity in a world of transformations. *Ambio,* **31**(5), 337–40.

Graversen, R. G., Mauritsen, T., Tjernström, M., Källén, E. and Svensson, G. (2008). Vertical structure of recent Arctic warming. *Nature,* **451**(7174), 53–6.

Hoekstra, A.Y. and Chapagain, A. K. (2007). Water footprints of nations: water use by people as a function of their consumption pattern. *Water Resources Management,* **21**(1), 35–48.

IPCC (2007). *Climate change 2007: Synthesis report. Summary for Policymakers.* Geneva. Available at http://www.ipcc.ch/pdf/assessment-report/ar4/syr/ar4_syr_spm. pdf.

IUCN, UNEP and WWF – The World Conservation Union, United Nations Environment Programme and World Wide Fund for Nature (1991). *Caring for the Earth: A Strategy for Sustainable Living.* Gland.

Lenzen, M., Wiedmann, T., Foran, B. *et al.* (2007). *Forecasting the Ecological Footprint of Nations: A Blueprint for a Dynamic Approach.* ISA (Centre for Integrated Sustainability Analysis) research report 07–01. Sydney.

Mallon, K., Bourne, G. and Mott, R. (2007). *Climate Solutions: WWF Vision for 2050.* Gland. Available at http://www.panda.org/what_we_do/footprint/climate_deal/our_ solutions.

Moran, D., Wackernagel, M., Kitzes, J., Goldfinger, S. and Boutaud, A. (2008). Measuring sustainable development – nation by nation. *Ecological Economics,* **64**(3), 470–74.

Nakicenovic, N., Swart, R., eds. (2000). *Special Report on Emissions Scenarios.* A Special Report of Working Group III of the IPCC. Cambridge.

Pacala, S. and Socolow, R. (2004). Stabilization wedges: solving the climate problem for the next 50 years with current technologies. *Science,* **305**(5686), 968–72.

Sherman, K. (1994). Sustainability, biomass yields, and health of coastal ecosystems: an ecological perspective. *Marine Ecology – Progress Series,* **112**(3), 277–301.

United Nations Population Division (2007). *World Population Prospects: the 2006 Revision.* New York.

Worm, B., Barbier, E. B., Beaumont, N. *et al.* (2006). Impacts of biodiversity loss on ocean ecosystem services. *Science,* **314**(5800), 787–90.

WWF, Global Footprint Network and Zoological Society of London (2006). *Living Planet Report 2006.* Gland.

WWF, Global Footprint Network and Zoological Society of London (2008). *Living Planet Report 2008.* Gland.

# Part II

Climate stabilization and sustainable development

# Chapter 6

# Scientific understanding of climate change and consequences for a global deal

———————

## Stefan Rahmstorf, Jennifer Morgan, Anders Levermann, and Karsten Sach

Stefan Rahmstorf obtained his PhD in oceanography from Victoria University of Wellington, New Zealand in 1990. After working at the New Zealand Oceanographic Institute and at the Institute of Marine Science in Kiel, he joined the Potsdam Institute for Climate Impact Research in 1996, and was appointed professor at Potsdam University in 2000. His work focuses on the role of the oceans in climate change. Rahmstorf is a member of the German Advisory Council on Global Change, and is one of the lead authors of the Fourth IPCC Assessment Report.

*Note:* Photos and biographies of co-authors can be found in the appendix.

## The scientific basis

Several important findings of climate research have been confirmed in recent decades and are now generally accepted as fact by the scientific community. These include the rapid increase in carbon dioxide concentrations in the atmosphere during the last 150 years, from 280 ppm (a value typical for warm periods during at least the past 700 000 years), to the 2007 level of 383 ppm (Global Carbon Project, 2008). This increase is entirely caused by humans and is primarily due to the burning of fossil fuels, with a smaller contribution from deforestation. Carbon dioxide is a gas that affects the Earth's climate by changing its radiation budget: an increase in its concentration leads to a rise in near-surface temperature. If the concentration doubles, the resulting global mean warming will likely be between 2 °C and 4 °C (the most probable value is approximately 3 °C according to the IPCC – UN Intergovernmental Panel on Climate Change (Solomon *et al.*, 2007)). Since 1900, the global climate has warmed by approximately 0.8 °C. Temperatures in the past ten years have been the highest since measured records began in the nineteenth century and, as shown by other climate indicators, for many centuries before that (see Fig. 1).

Most of this warming is due to the rising concentration of carbon dioxide and other anthropogenic gases (Solomon *et al.*, 2007). It follows that a further increase in carbon dioxide concentration must lead to a further rise in global mean temperature (see Fig. 2). Considering a range of plausible assumptions about future emissions, this rise will be in a range from approximately 2 °C to approximately 7 °C above preindustrial levels.

By comparison, the last major period of global warming occurred at the end of the last great ice age (about 15 000 years ago), and involved global warming of approximately 5 °C over a time span of 5000 years (Schneider von Deimling *et al.*, 2006). Unchecked anthropogenic warming could reach a similar magnitude over a fraction of this time – and, of course, starting from an already warm climate.

## Impacts and risks

Whether this warming constitutes 'dangerous' change cannot, of course, be determined by scientists alone, as such an assessment depends on societal value judgments about what is dangerous. However, science can help to state and clarify the risks that arise from such unprecedented warming. Among the most important risks are the following:

- **Increase in sea level and loss of ice sheets.** In the twentieth century global sea level rose by 15–20 cm. Currently, sea level is rising at a rate of over three

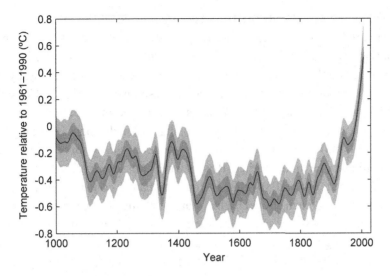

**Fig. 1.** Global temperature over land and ocean during the past millennium, based on a variety of proxies including ice cores, tree rings, corals and sediment data. The grey bands show the 25–75 and 5–95 % uncertainty ranges. (*Source:* Mann *et al.,* 2008)

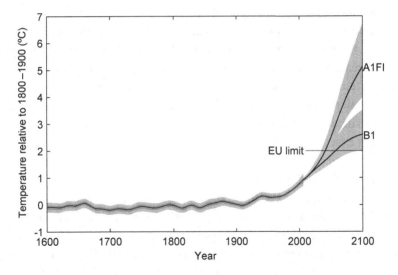

**Fig. 2.** IPCC projections for global mean temperature in the twenty-first century in comparison to past variability as shown in Fig. 1. The lowest (B1) and highest (A1FI) emission scenarios are shown with their respective projection uncertainties: for B1 emissions, warming will be between 2 °C and 3 °C, for A1FI emissions between 4 °C and 7 °C. The 2 °C limit adopted by the EU and many countries is also shown. (*Sources:* Mann *et al.,* 2008; Solomon *et al.,* 2007)

centimetres per decade, about 50% faster than projected in the scenarios of the IPCC Third Assessment Report (Rahmstorf *et al.*, 2007). If warming is not limited, a rise of around one metre by 2100 is not unlikely (Rahmstorf, 2007). Even if warming is halted at 3 °C, the sea level will probably keep rising by several metres in subsequent centuries as a delayed response (see Fig. 3). Coastal cities and low-lying islands are at risk. What is now a once-in-a-century occurrence of extreme flooding in New York City (causing major damage, including flooded subway stations) would happen on average about every three years if the sea level were just one metre higher (Rosenzweig and Solecki, 2001).

- **Loss of ecosystems and species.** If climate change continues unabated global temperatures will reach a level higher than for millions of years and this increase will be much too fast for many species to adapt to. A large fraction of species – some studies suggest up to one third of all species – could be doomed to extinction by the year 2050 (Thomas *et al.*, 2004). Life in the oceans is not only threatened by climate change but by the equally serious problem of ongoing global ocean acidification, which is a direct chemical result of our carbon dioxide emissions independent of the warming effect.

- **Risk of extreme events.** In a warmer climate, the risk of extreme flooding events will increase, as warmer air can hold more water (approximately 7% more for each degree Celsius of warming). Hurricanes are expected to become more destructive. Both physical considerations and data suggest an increase in the force of hurricanes in response to rising sea surface temperatures (see Fig. 4).

- **Risk to water and food supplies.** While total global agricultural production may increase with moderate global warming due to temperature gains in colder regions, many poorer and warmer countries may experience reductions in yields due to water shortages and weather extremes. Agricultural productivity is expected to decline globally in the event of warming between 2 °C and 4 °C. Should warming exceed 4 °C major losses are to be feared (Parry *et al.*, 2007). The water supply of major cities (such as Lima) and of agricultural lands (such as those fed by rivers draining the Tibetan Plateau) is threatened when mountain glaciers and snow packs disappear (WBGU, 2007).

- **Non-linear responses – tipping elements.** Positive feedbacks have been identified for a number of climatic subsystems, and these feedbacks may self-amplify the response to external disturbances (Lenton *et al.*, 2008). For example, the Arctic sea ice cover has shown a drastic reduction in recent years (see real-time sea ice data at the National Snow and Ice Data Center, http://nsidc.org). While, for instance, the melting of the Himalayan glaciers is a likely result of increased temperatures, other tipping elements, like global monsoon systems, represent a risk due to their extraordinary impact (Auffhammer *et al.*, 2006; Zhang *et al.*, 2008), but it remains difficult to assess the probability of their occurrence. The

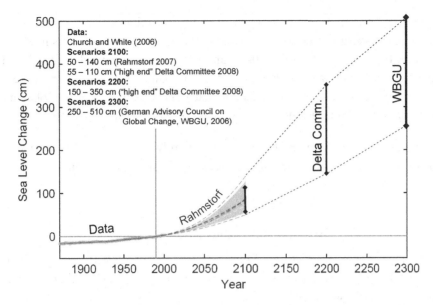

**Fig. 3.** Observed sea level up to 2000 and several recent projections up to the year 2300. (*Sources:* Church and White, 2006; Rahmstorf, 2007; WBGU, 2006; Vellinga *et al.,* 2008)

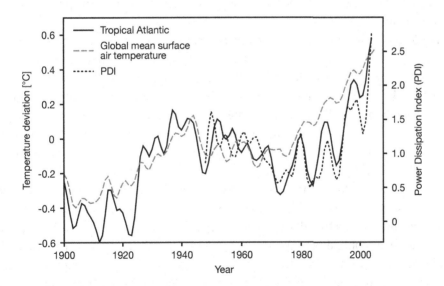

**Fig. 4.** Changes in the force of tropical storms over time (Power Dissipation Index – PDI, dotted) and the average sea-surface temperature in the tropical Atlantic from August to October (solid). For comparison the evolution of globally averaged near-surface air temperature is also shown (broken grey line). (*Source:* Emanuel, 2005)

triggering of some tipping elements, such as the major ocean circulation in the Atlantic and the great ice sheets in Greenland and West Antarctica, is likely to be irreversible (Toniazzo *et al.,* 2004; Rahmstorf *et al.,* 2005; Schoof, 2007). The release of methane from a thawing of the Siberian and North American perma-frost is an additional tipping element that may directly increase global mean temperature and thus accelerate the process of global warming.

It is important to remember that these are merely examples. The exact conse-quences of such a major change in climate are difficult to predict, and surprises are likely.

## Avoiding dangerous climate change

Following the United Nations Framework Convention on Climate Change (UNFCCC, 1992), in which states committed themselves to preventing 'dangerous interference' with the climate system, the European Union went one step further in pledging to limit the increase in global average temperature to 2 °C above prein-dustrial levels (EU limit, see Fig. 2). This means that global carbon dioxide emis-sions must be reduced by 50–80 % of the 1990 level by 2050 (Meinshausen *et al.,* 2009). This range of necessary emission reductions arises from uncertainty within the carbon cycle and the physical climate system, and from different possible emis-sion pathways up to 2050.

While the exact emission reduction pathways are uncertain, one crucial fact clearly follows from a decision to stabilize temperatures: ultimately, carbon dioxide emissions must be reduced to practically zero. The reasons are that few permanent natural sinks exist, and that carbon dioxide, once released to the atmosphere, is removed only on a millennial time scale (Solomon *et al.,* 2009). Thus, the stabili-zation target of global mean temperature is determined by cumulative emissions (Allen *et al.,* 2009), and delayed emissions reduction results in the necessity to reduce more rapidly.

On the issue of stabilization targets, two further issues need to be kept in mind. First, for some major climatic subsystems, such as the great ice sheets in Green-land and Antarctica, a two-degree target might not be sufficient to avoid dangerous interference. Second, due to the slow pace and delayed reaction of the global cli-matic system, even after phasing out anthropogenic carbon dioxide emissions tem-peratures will not drop for several centuries and, additionally, sea level will continue to rise (Solomon *et al.,* 2009).

**The global deal**

A global deal is required to tackle the climate-related challenges outlined in the previous sections (see also Stern and Garbett-Shiels, this volume). If successful, such a far-ranging international effort will not only help avoid the worst impacts of climate change and protect the world's most vulnerable people, but will also initiate the process of restructuring the global economy towards ecologically oriented growth, focused on the creation of new jobs in low-carbon industries. New economic instruments and business models will be developed to put a price on carbon, and to ensure that opportunities for efficiency and renewable energies are generated. The best way to counteract the insecurity driven by high energy and commodity prices is to demonstrate the practical opportunities available in the creation of efficient and resilient societies. Currently, countries are focused on short-term crises and are not able to plan proactively for a low-carbon economy.

The UNFCCC and the Kyoto Protocol have been guided by the findings of climate science as documented by the Intergovernmental Panel on Climate Change (IPCC). In order to avoid operational disruption after the first commitment period of the Kyoto Protocol (ending in 2012), the next round of focused negotiations are scheduled to be completed by 2009 in Copenhagen. A post-2012 climate agreement that would represent a major step towards accomplishing a global deal must include the following key elements:

### 1. Developing a vision for international climate protection: low-carbon development and the 2 °C objective

The agreement should aim for all countries to achieve their national economic and development goals within the framework of a low-carbon strategy that safeguards the environment, strengthens their ability to adapt to the changes already underway, and allows for sustained economic welfare. Industrialized countries should demonstrate their intention to lead the way on the low-carbon transition, and should agree to support developing countries in their transitions. This vision acknowledges the fundamental need for ambitious adaptation support, particularly in the world's poorest countries, and reiterates the commitment to achieving the UN Millennium Development Goals. It also reaffirms that all peoples, nations and cultures have the right to survive.

The vision should include the 2 °C objective, which means countries agree to a long-term goal to at least halve global emissions of greenhouse gases by 2050 compared to 1990 levels, and to bring about a peak in global emissions in the next 10–15 years. If countries want a higher probability of staying below 2 °C, then a more ambitious target should be agreed. In addition to a vision for 2050, countries should agree on emissions pathways for industrialized and developing countries with benchmarks in 2020 and 2030 that lead the way to the almost complete decarbonisation required by 2050. The goals should be regularly assessed in light of the latest scientific findings to avoid the risk of triggering critical climate tipping points.

## 2. Creating a global carbon market

Establishing a reliable and long-term price signal for carbon dioxide creates effective incentives for worldwide mitigation of greenhouse gas emissions (see Edenhofer *et al.,* this volume). A carbon market generates this price signal, while also creating the flexibility that participating companies need with regard to the timing and location of their required emissions reductions. Alongside emissions trading, the Kyoto mechanisms should be scaled up and the European Emissions Trading Scheme linked up to comparable systems in other regions (for example, in North America, Australia, Japan and other countries, including emerging economies).

Some of the Kyoto mechanisms should be reviewed, however. In particular, new sector-based crediting mechanisms for larger emerging economies and reform of the project-based Clean Development Mechanism for smaller developing countries are needed (see Liverman, this volume).

## 3. Agreeing on ambitious emissions reduction commitments for industrialized countries

A stable and sufficiently high price level on the international carbon market presupposes ambitious, absolute and binding emissions reduction targets for industrialized countries. Such binding targets also represent a politically necessary signal by those countries primarily responsible for the currently observed levels of climate change. By 2020 the industrialized countries should have reduced their emissions by around 30% compared to 1990. By 2050 the emissions of this group of countries must be reduced by approximately 80% (or even 90 or 95% if a higher probability of risk reduction is desired). In order to build confidence in their intention to decarbonize their societies and encourage long-term planning, industrialized countries should put forward low-carbon action plans. These plans need to outline the process of economic transformation that will be undertaken to address unsustainable patterns of consumption and production (see Gell-Mann, this volume, for a more detailed discussion of transformation to sustainability). This process must promote a low-carbon economy and ensure deep emissions reduction targets in line with commitments.

## 4. Taking action in developing countries

One of the greatest challenges for a global deal is to build the confidence and capacity of developing countries to provide their populations with development, energy services and food security, while they start to decarbonize their societies. There is no model for developing countries to follow. The mitigation agreement should therefore provide significant external incentives for developing countries to move beyond business-as-usual (BAU) pathways, but should also expect them to undertake 'no-regrets' and low-cost measures based on their own means as well.

Overall, actions undertaken in developing countries should result in a substantial reduction of emissions below BAU. Collectively, these actions would be enough to ensure that developing country emissions peak no later than 2020–2025, a requirement if global warming is to remain below 2 °C.

Guided by the commitment to a substantial deviation from BAU, each major developing economy should submit a set of actions that can be incorporated into a low-carbon action plan. This would include measures that the country will implement unilaterally in defining its national baseline. The action plan would also outline what other measures can be taken conditional upon greater access to the global carbon market and technological and financial support.

What could such a low-carbon action plan look like? Some countries could assume sectoral obligations (for example, in the electricity sector); others could adopt more ambitious national policies and measures (for example, on renewables targets and housing standards); and yet other countries could adopt national efficiency targets (for example, on energy consumption in relation to GDP). Each plan should describe in a quantitatively verifiable way the substantial deviation from BAU.

## 5. Reducing emissions from deforestation and degradation (REDD) and other land use sectors

Climate change cannot be solved by addressing the energy system alone. Improved ecosystem management would avoid a substantial amount of emissions and restore many of the carbon sinks that once existed. In particular, about 17% of global emissions are caused by deforestation and forest degradation (Metz *et al.,* 2007, Fig. TS.1b). Therefore, the new climate change agreement needs to include enhanced actions for this sector. Each country could submit a national deforestation plan. This plan would outline the country's commitment and strategy to reduce deforestation emissions from an agreed national baseline, provided that financial support from industrialized countries is guaranteed. The agreement must also include the most efficient mechanism to provide this funding within the framework of a carbon market: a commitment by industrialized countries to auction a percentage of their national allocation and designate it to REDD. In addition, a credible monitoring and review mechanism should be included to assess if, when, and how deforestation credits might be permitted to enter the global carbon market without jeopardizing market stability or causing the carbon price to fall dramatically.

In addition to tropical forests, peatlands and agriculture are priority areas where action is needed. A recent study of the United Nations Environment Programme (UNEP) suggests that the agricultural sector could be broadly carbon-neutral by 2030 if best management practices were widely adopted (Trumper *et al.,* 2009). Other societal goals could be achieved alongside carbon storage, such as improved soil fertility, new employment and income-generating opportunities, and biodiversity conservation.

## 6. Promoting technology:
### investment, innovation and transfer

The agreement should include a scaled-up technology cooperation mechanism, one that strikes a balance between building the capacity of all major economies to become innovation leaders, and supporting the needs of some countries for technology transfer.

Support for technology cooperation and diffusion needs to be rapidly expanded in order to meet the mitigation and adaptation challenges posed by climate change. A robust and comprehensive approach is needed to correct market failures and provide support along the entire technology innovation chain. This approach should leverage public and private finance to spur innovation and technology cooperation, with substantial focus on the international agreement but even greater focus on bringing bi-lateral and private capital in line with low-carbon action plans and strategies.

To address the need for rapid technology development and diffusion in the near-term the agreement should include a 'technology development objective' to at least double current levels of research, development and demonstration by 2012 and quadruple those budgets by 2020. In order to ensure focused investment, the agreement should contain a commitment from all countries to jointly develop a set of strategically important adaptation and mitigation technologies incorporated in 'technology action programmes'. The agreement should also include a new fund with two distinct functions: to increase investment in research, development & demonstration (RD & D); and to increase diffusion of new technologies in developing countries. The fund, through matching grants and other blended financing, would leverage public financing to catalyze a shift of private investment into low-carbon technologies. Private companies and developing countries would bid into this new fund.

Furthermore, to protect the interests of the innovator while also promoting diffusion of low-carbon technologies, the agreement should contain a 'protect and share' framework for managing intellectual property rights (IPR). This would facilitate joint ventures and public-private partnerships, and define systems for enhanced access, conditional on strengthened IPR protection. Countries failing to robustly protect low-carbon IPR would run the risk of losing their access to the proposed technology fund.

## 7. Supporting adaptation: climate-proof investments and risk management

The post-2012 agreement should send a clear signal to the poorest and most vulnerable countries of the world that they will not be left alone to deal with the increasing impacts of climate change. A new 'global adaptation framework' will need to be created to provide the vision, strategy and coordination to respond to catastrophes and climate-change impacts as they occur.

This framework needs to incorporate the key institutions with relevant expertise (Food and Agriculture Organization, World Health Organization, World Bank, UNDP, Red Cross, etc.) and should join up efforts inside and outside the Convention. It should base those strategies on input from regional adaptation centres, regional information systems on climate risks in developing countries, and national plans. The confidence of the donor community could be enhanced by assuring that the billions raised will be applied to the most urgent and critical needs.

In support of this framework, industrialized countries should give a firm undertaking both to honour their existing official development assistance (ODA) commitments and to provide additional resources for adaptation to climate change. A substantial share of the new resources should be channelled through an Adaptation Fund in order to promote predictability and transparency. In addition, donors should scale up their investment in disaster prevention and response, and should develop a global reinsurance scheme to provide a safety net for poor people exposed to climate change risk. Also, donors should agree to incorporate adaptation strategies and measures into their existing bilateral and multilateral aid programmes, and 'climate-proof' their investments without diverting funds from existing aid budgets.

## 8. Financing

A major element of the agreement will be *measurable, reportable* and *verifiable* financing. It is estimated that by 2015 the annual costs of action in developing countries will be approximately USD 70 billion for mitigation efforts in the power and transport sectors (Anderson, 2006), and USD 90 billion for adaptation (UNDP, 2007).

This compares with the 2008 ODA levels of approximately USD 120 billion (OECD, 2009). Developing countries will have to meet some of these costs themselves but will also expect substantial international support. The mix of financing responsibility between industrialized and developing countries will determine the 'fairness' of abatement commitments.

The expansion of the international carbon market will generate significant additional financing for mitigation programmes in developing countries.[1] These will be concentrated in industrializing middle-income countries such as China, India, Brazil and Mexico. As noted above, additional funding is needed in the areas of adaptation, reduction of emissions from deforestation, and technology cooperation. While the technology fund can likely leverage private sector funding, adaptation and deforestation will be more dependent on public funding. A number of potential mechanisms exist to generate the needed revenue: a) industrialized countries could pledge to contribute a share of their auction revenues from domestic emissions trading to mitigation and adaptation in developing countries; b) a share of each country's 'assigned amount' for the next commitment period could be monetized and invested in a set of international funds; c) a tax could be introduced on international bunker fuels to generate revenue (as well as to include the aviation and maritime sectors in national commitments); and d) countries could pledge direct budgetary support based on a set of agreed criteria.

In order to leverage external funding such as bi-lateral funds, measurable, reportable and verifiable criteria need to be agreed upon. This would enable donors to get 'credit' for contributions to low-carbon development plans not only through UNFCCC-related funds but also through other multilateral and bilateral initiatives.

### 9. Including international air and maritime transport

The sector with the most rapidly increasing emissions worldwide is international aviation and maritime transport. Up to now these sources have been exempt from emissions restrictions. The post-2012 regime should include targets to reduce emissions from these sectors.

In the midst of a financial crisis it may be difficult to imagine that a global deal such as that outlined above is possible. Countries are focusing on national economic priorities and job creation, dealing with a recession that is raising fears and could lead to greater isolationism. If a post-2012 agreement is not reached and the focus continues to rest on national-level activities, it is highly unlikely that the

---

[1] The World Bank (Capoor and Ambrosi, 2007) estimated CDM flows at USD 5 billion per year in 2006, and the UNFCCC (2006) estimates substantial future growth, generating USD 12 billion per year by 2012.

scientific challenge of climate change mitigation will be met. International agreements are created to raise the level of ambition, to generate a shared vision for a common endeavour, and to stimulate action at a faster pace than countries would normally pursue. Such agreements provide not only motivation but also the security that other major economies are also taking significant investment decisions to move in a new direction. A global deal on climate change is needed to build trust between industrialized and developing countries, trust that will be hard to rebuild if a deal is not struck. Exceptional and determined leadership is likewise needed to ensure that global transformation happens in time.

## References

Allen, M. R., Frame, D. J., Huntingford, C., *et al.* (2009). Warming caused by cumulative carbon emissions towards the trillionth tonne. *Nature,* **458**(7242), 1163–6.

Anderson, D. (2006). *Costs and Finance of Abating Carbon Emissions in the Energy Sector.* Stern Review supporting research paper. London.

Auffhammer, M., Ramanathan V. and Vincent, J. R. (2006). Integrated model shows that atmospheric brown clouds and greenhouse gases have reduced rice harvests in India. *Proceedings of the National Academy of Sciences of the United States of America,* **103**(52), 19668–72.

Capoor, K. and Ambrosi, P. (2007). *State and Trends of the Carbon Market 2007.* Washington, D. C. Available at http://web.worldbank.org/WBSITE/EXTERNAL/NE WS/0,,contentMDK:21319781~pagePK:64257043~piPK:437376~theSiteP K:4607,00.html.

Church, J. A. and White, N. J. (2006). A 20th century acceleration in global sea-level rise. *Geophysical Research Letters,* **33**(1), L01602.

Emanuel, K. (2005). Increasing destructiveness of tropical cyclones over the past 30 years. *Nature,* **436**(7051), 686–8.

Global Carbon Project (2008). Carbon budget and trends 2007. Available at http://www. globalcarbonproject.org.

Lenton, T. M., Held, H., Kriegler, E. *et al.* (2008). Tipping elements in the Earth's climate system. *Proceedings of the National Academy of Sciences of the United States of America,* **105**(6), 1786–93.

Mann, M. E., Zhang, Z. H., Hughes, M. K. *et al.* (2008). Proxy-based reconstructions of hemispheric and global surface temperature variations over the past two millennia. *Proceedings of the National Academy of Sciences of the United States of America,* **105**(36), 13252–7.

Meinshausen, M., Meinshausen, N., Hare, W. *et al.* (2009). Greenhouse-gas emission targets for limiting global warming to 2 °C. *Nature,* **458**(7242), 1158–63.

Metz, B., Davidson, O. R., Bosch, P. R., Dave, R., Meyer, L. A., eds. (2007). *Climate Change 2007: Mitigation of Climate Change. Contribution of Working Group III to the Fourth Assessment Report of the Intergovernmental Panel on Climate Change.* Cambridge.

OECD (2009). Development aid is at its highest level ever in 2008. *Development Co-operation Directorate (DCD-DAC) Statistics,* 30/03/2009. Paris. Available at http://www.oecd.org/document/35/0,3343,en_2649_34487_42458595_1_1_1_1,00. html.

Parry, M. L., Canziani, O. F., Palutikof, J. P. *et al.*, eds. (2007). *Climate Change 2007: Impacts, Adaptation and Vulnerability. Contribution of Working Group II to the Fourth Assessment Report of the Intergovernmental Panel on Climate Change.* Cambridge.

Rahmstorf, S., Crucifix, M., Ganopolski, A., *et al.* (2005). Thermohaline circulation hysteresis: a model intercomparison. *Geophysical Research Letters,* **32,** L23605.

Rahmstorf, S. (2007). A semi-empirical approach to projecting future sea-level rise. *Science,* **315**(5810), 368–70.

Rahmstorf, S., Cazenave, A., Church, J. A. *et al.* (2007). Recent climate observations compared to projections. *Science,* **316**(5825), 709.

Rosenzweig, C. and Solecki, W. D. (2001). *Climate Change and a Global City – The Potential Consequences of Climate Variability and Change.* New York.

Schneider von Deimling, T., Ganopolski, A., Held, H. and Rahmstorf, S. (2006). How cold was the last glacial maximum? *Geophysical Research Letters,* **33**(14), L14709.

Schoof, C. (2007). Ice sheet grounding line dynamics: steady states, stability, and hysteresis. *Journal of Geophysical Research,* **112,** F03S28.

Solomon, S., Qin, D., Manning, M. *et al.*, eds. (2007). *Climate Change 2007: The Physical Science Basis. Contribution of Working Group I to the Fourth Assessment Report of the Intergovernmental Panel on Climate Change.* Cambridge.

Solomon, S., Plattner, G. K., Knutti, R. and Friedlingstein, P. (2009). Irreversible climate change due to carbon dioxide emissions. *Proceedings of the National Academy of Sciences of the United States of America,* **106**(6), 1704–9.

Thomas, C., Cameron, A., Green, R. E. *et al.* (2004). Extinction risk from climate change. *Nature,* **427**(6970), 145–8.

Toniazzo, T., Gregory, J. M. and Huybrechts, P. (2004). Climatic impact of a Greenland deglaciation and its possible irreversibility. *Journal of Climate,* **17,** 21–33.

Trumper, K., Bertzky, M., Dickson, B. *et al.* (2009). *The Natural Fix? The Role of Ecosystems in Climate Mitigation.* A UNEP rapid response assessment. Cambridge.

UNDP – United Nations Development Programme (2007). *Human Development Report 2007/2008. Fighting Climate Change: Human Solidarity in a Divided World.* Houndmills.

UNFCCC (1992). *United Nations Framework Convention on Climate Change.* New York. Available at http://www.unfccc.org.

UNFCCC (2006). *Annual Green Investment Flow of Some 100 Billion Dollars Possible as Part of Fight Against Global Warming.* Press release of the UNFCCC Secretariat, 19 September. Bonn. Available at http://unfccc.int/files/press/news_room/press_re leases_and_advisories/application/pdf/20060919_riyadh_press_release_vs5.pdf.

Vellinga, P., Katsman, C., Sterl, A. *et al.* (2008). *Exploring High-end Climate Change Scenarios for Flood Protection of the Netherlands.* International scientific assessment carried out at request of the Delta Committee. Wageningen.

WBGU – German Advisory Council on Global Change: Schubert, R., Schellnhuber, H.-J., Buchmann, N. *et al.*, eds. (2006). *The Future Oceans – Warming Up, Rising High, Turning Sour. Special report.* Berlin.

WBGU – German Advisory Council on Global Change: Schubert, R., Schellnhuber, H.-J., Buchmann, N. *et al.*, eds. (2007). *World in Transition – Climate Change as a Security Risk.* London.

Zhang P., Cheng, H., Edwards, R. L. *et al.* (2008). A test of climate, sun, and culture relationships from an 1810-year chinese cave record. *Science,* **322**(5903), 940–42.

# Chapter 7

# Towards a global deal on climate change

---

## Nicholas Stern and
## Su-Lin Garbett-Shiels

Nicholas Stern is IG Patel Professor of Economics and Government at the London School of Economics and Political Science, and Chairman of the Grantham Research Institute on Climate Change and the Environment. Stern was head of the UK Government Economic Service from 2003–7, and Chief Economist of the World Bank from 2000–2003 and of the European Bank for Reconstruction and Development from 1994–9. He authored the Stern Review on the Economics of Climate Change, reporting to the UK Prime Minister and the Chancellor of the Exchequer from 2005–7. He was knighted for services to economics in 2004, and was appointed to the UK House of Lords as Lord Stern of Brentford in 2007.

Su-Lin Garbett-Shiels is an economic advisor working in the Stern Team at the UK's Department for Energy and Climate Change, leading studies on the regional economics of climate change. She was private secretary to Lord Stern during the Stern Review on the Economics of Climate Change. Previously she worked in macroeconomics analysis teams at the UK Treasury Department.

Greenhouse gas (GHG) emissions represent the biggest market failure the world has ever seen. GHGs cause damage and without specific policies nobody pays for this damage. We all contribute to producing them; and many people around the world are already suffering from the effects of past emissions. Moreover, current emissions have the potential to cause catastrophic damage in the future. Due to the global nature of the link between emissions and damage, we need a global response to this problem. Failure to analyse the problem in terms of the great risks, and the long-term and global co-operation required, will produce (and has produced) approaches to policy that are misleading and dangerous. The arguments for strong and timely action are overwhelming. The costs of inaction, which means continuing with current paths and practices, or 'business as usual' (BAU), should be measured in terms of the possible outcomes and damages compared to a global strategy that sets sensible targets.

The world must create and implement a global deal that is effective, efficient, and equitable.[1] The world must create this deal quickly – indeed, the meeting of the United Nations Framework Convention on Climate Change in Copenhagen in December 2009 should be our deadline. Importantly, this deal must be implemented with real commitment by all countries of the world.

We do not wish to pretend that reaching agreement will be easy; on the contrary, the road to Copenhagen and beyond will be very tough and full of obstacles, principally in the form of resentment, particularly towards rich countries for their historical responsibility for high-carbon growth, and narrow perspectives of self-interest. Central amongst these obstacles will be an argument that a first priority should be to deal with the current economic crisis and that action on climate change can be postponed. Often this argument comes from those who are, in any case, not keen on taking action and who use the economic crisis as an excuse. This argument is erroneous and must be confronted (see Rockström *et al.,* this volume). There is no doubt that the economic crisis is extremely serious, and requires strong, co-ordinated action, both nationally and internationally. The error lies in seeing responses to the economic crisis and to climate change as being in conflict. They are not: the economic crisis becomes an obstacle to urgent action on climate change only if we allow it to do so by failing to put the arguments clearly.

There are two important lessons from the economic crisis that are relevant for action on climate change. First, by ignoring the dangers and delaying action we risk greatly magnifying the ensuing damages. Second, our reaction to the current crisis should not sow the seeds of the next economic bubble, as was the case after the dot-com bubble of the 1990s, when economic policies helped to create the housing bubble of the 2000s, which was a prime cause of the economic crisis of

---

[1] A more extensive discussion of a global deal is provided in the book *A Blueprint for a Safer Planet* (Stern, 2009)

2008–9. Investments to manage climate change are less costly during a slow-down. Furthermore, the foundations for growth in the next two or three decades (i.e. investments in low-carbon technologies) can be created now.

In order to construct and implement a deal on climate change the peoples of the world and their leaders require a clear understanding; not only of the huge risks we face, and thus why such a deal is necessary, but also of the whole range of technologies and policies that are available to us to make effective action possible. The purpose of this book and this essay is to contribute to this understanding.

The price of failure will be a world that is subject to devastating physical change, mass movement of people, and conflict. The prize of success will be sustainable growth, a significant reduction in world poverty, and a cleaner, safer, quieter, more diverse, and more prosperous future for all.

## Risks, targets and costs

### *Targets*

The relation between the stock of GHGs in the atmosphere and the resulting temperature increase is at the heart of any risk analysis. It is the clearest way to begin and anchors most of the discussion. There are many models that estimate these links: running a model many times for different parameter choices yields probability distributions of outcomes – in other words, it allows us to take into account the uncertainties in the link between emissions and temperature changes (see Table 1).

Current concentrations of GHGs are around 430 parts per million (ppm) of carbon dioxide equivalent ($CO_2$-eq – which aggregates carbon dioxide and other GHGs). We are currently adding about 2.5 ppm $CO_2$-eq per annum. The rate of emissions growth appears to be accelerating, as a result of continued rapid growth in the developing world. There seems little doubt that, in the absence of any restraining policy, the annual increase in the overall quantity of GHGs will average somewhere above 3 ppm $CO_2$-eq – potentially 4 ppm $CO_2$-eq or more – over the next 100 years. That is likely to take us beyond 750 ppm $CO_2$-eq by the end of this century.

This level of concentration would give us, if we were to stabilize there by 2100, a fifty-fifty chance of a temperature increase above 5 °C. We do not really know what the world would look like with a climate 5 °C warmer than in preindustrial times. The most recent warm period was around three million years ago when the world experienced temperatures 2 °C or 3 °C higher than today (Jansen *et al.*, 2007). Humans have not experienced anything that high (Hansen *et al.*, 2006). During the last *glacial maximum* (around 21 000 years ago) global temperatures were around 4–7 °C cooler than today (Solomon *et al.*, 2007), and ice sheets extended to latitudes

**Table 1.** *Probabilities of exceeding a temperature increase at equilibrium (%). (Source:* based on Stern, 2007, p. 220, using Hadley Centre modelling (Murphy *et al.,* 2004)).

| Stabilization level (in ppm $CO_2$-eq) | 2 °C | 3 °C | 4 °C | 5 °C | 6 °C | 7 °C |
|---|---|---|---|---|---|---|
| 450 | 78 | 18 | 3 | 1 | 0 | 0 |
| 500 | 96 | 44 | 11 | 3 | 1 | 0 |
| 550 | 99 | 69 | 24 | 7 | 2 | 1 |
| 600 | 100 | 94 | 58 | 24 | 9 | 4 |
| 750 | 100 | 99 | 82 | 47 | 22 | 9 |

just north of London and just south of New York.[2] As the ice melted and sea levels rose, and taking into account the changed topography, Britain separated from the European continent and there was major re-routing of much of the global river flow. Such magnitudes of temperature change can transform the planet.

The last time the Earth's temperature lay in the region of 5 °C above the preindustrial level was in the Eocene period around 35 – 55 million years ago.[3] Much of the world was covered by swampy forests and there were alligators near the North Pole. The point is not particularly about alligators, it is about the transformation of the world; these kinds of variations would bring very radical changes to where and how different types of species, including humans, could live. Many of the changes would take place over 100 or 200 years rather than thousands or millions of years. At a temperature increase of 5 °C most of the world's ice and snow would disappear, most likely including the Arctic and Antarctic ice sheets and the snows and glaciers of the Himalayas. According to the IPCC Fourth Assessment Report, the former effect would – taking the two ice sheets together – eventually lead to a sealevel rise of over 10 metres, and possibly much higher. The latter effect would thoroughly disrupt the flows of the major rivers from the Himalayas, which serve countries containing around half of the world's population. There would be severe torrents in the rainy season and dry rivers in the dry season. The world would probably lose more than half its species. The intensity of storms, floods and droughts is likely to be much higher than at present.

Whilst we cannot be precise about the magnitude of the effects associated with temperature increases of such size, it does seem reasonable to suppose that they would be, or are at least likely to be, disastrous. They would probably involve very

[2] http://math.ucr.edu/home/baez/temperature
[3] See footnote 2.

large movements of population from regions where human life would become extremely difficult or impossible. History tells us that large movements of population are likely to bring major conflict, and this movement would probably be on a huge scale.

If we fail to act there is a high probability that these devastating impacts and conflicts will become reality. As Table 1 shows, we can cut the probability of temperature change above 5 °C from 50 % to 3 % by stabilizing emissions at 500 ppm $CO_2$-eq. We cannot be very precise about these probabilities (the ones we have used here, from the Hadley Centre, are probably cautious)[4], however, the point is that the reduction in risk is huge.

By using extremely simple models one can try to quantify the avoided damages although our description of the risks should make it clear that it is very hard to attach convincing figures to the potential losses. Even from a very narrow perspective, world wars seem to involve losses of 15 % or more of GDP and the conflicts we are discussing are likely to be on a greater scale, lasting longer and, of course, affecting much more than GDP. The Stern Review (Stern, 2007), which looks at damages up to the year 2200 and extrapolated thereafter, concluded that such costs can be estimated as being equivalent to a 5–20 % loss of global GDP averaged over space, time and possible outcomes. Such models can provide useful insights but we warn strongly against taking them too literally.

## *Recent developments on the risk and potential damages of climate change*

There are a number of factors that climate change scientists and economists have raised recently which point to a worsening of the prospects on climate risk. First, recent data – particularly from developing countries – indicates that emissions are growing more quickly than we thought. For example, a recent study by Max Auffhammer, University of California Berkeley, and Richard Carson, University of California San Diego, indicates that carbon dioxide emissions in China over the period 2004–10 will have grown at 11% per annum (Auffhammer and Carson, 2008). BAU assumptions used by the IPCC (Solomon *et al.*, 2007) projected a growth of only 2.5–5 % per annum. At this pace, by 2010 China will have increased its carbon emissions from 2000 by around 1.5–3 billion metric tons of carbon dioxide. To put it another way, the projected annual increase in China over the next several years alone is greater than the current emissions produced by Germany. If

[4] Work by the Hadley Centre and the IPCC (Murphy *et al.,* 2004 and Wigley and Raper, 2001) suggests that 550 ppm $CO_2$-eq is associated with a 24 % probability of exceeding 4 °C, a level at which it is projected that significant and irreversible changes would occur. Stabilization below 500 ppm $CO_2$-eq would be significantly less risky (11% probability of exceeding 4 °C). For details see Stern (2007), p. 220.

indeed emissions are growing more quickly than we thought, then the dangerous concentration levels associated with higher probabilities of disastrous temperature increases will be reached much more quickly.

Second, the key feedbacks of the carbon cycle, such as the reduction in the absorptive capacity of the oceans (and thus the reduced effectiveness of a key carbon sink) and the release of methane from the permafrost have not been taken into consideration in the projected concentration increases quoted here. If these factors are considered it is likely that stabilizing GHG concentrations at stocks associated with lower probabilities of disastrous temperature increases could be even more difficult.

Third, it is increasingly clear that we know little about what would happen to the planet if we were to see very high concentrations of GHGs. However, given the nature of feedback mechanisms scientists agree that the damages associated with very high GHG concentrations could be enormous. Most of the current research on damages makes conservative assumptions about the implications of high levels of concentrations. As the Harvard economist Martin Weitzman, (Weitzman, 2008; Weitzman, 2007a; Weitzman, 2007b), among others, has convincingly shown in his research, considering the risk of very high GHG concentrations escalates the estimations of climate change impact – and its potential cost to the economy.

The balance of the evidence implies that the level of risk suggested by the IPCC Fourth Assessment Report (Solomon *et al.*, 2007) and the Review (Stern, 2007) may be underestimated. Therefore, the opinion expressed by some commentators – that the Stern Review was alarmist – is simply wrong.

## *Costs of abatement*

Up to this point our discussion of targets has focused on those for the stabilization of stocks of GHGs in the atmosphere. We must now ask about the implications for emissions pathways and how much, with good policy, GHG abatement would cost. A broad answer was given in the Stern Review (Stern, 2007) – around 1–2% of world GDP per annum to get below 550 ppm $CO_2$-eq – but we must look at the argument in a little more detail.

Figure 1 illustrates possible paths for stabilization at 550 ppm $CO_2$-eq (long-dashed line), 500 ppm $CO_2$-eq (dotted line), and 450 ppm $CO_2$-eq (dot-dashed line); the solid line represents BAU. There are many paths for stabilization at a given level – see, for example, Stern Review, Fig. 8.2 (Stern, 2007, p. 226) – but all of them form a similar pattern to those shown (if a path peaks later it must fall faster). And if the carbon cycle weakens, the cuts would have to be larger to achieve stabilization at a given level – see Stern Review Fig. 8.1 (Stern, 2007, p. 222). Broadly speaking, however, a path stabilizing at 550 ppm $CO_2$-eq or below will have to

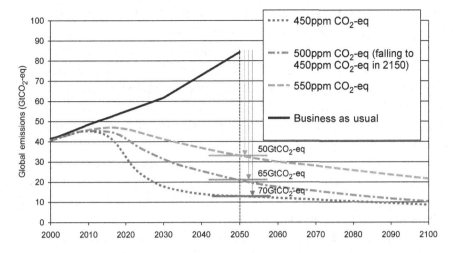

**Fig. 1.** Business as usual (BAU) and stabilization trajectories for 450–550 ppm $CO_2$-eq. (*Source:* Stern, 2007, p. 233)

show emissions peaking in the next 20 years. For lower stabilization levels, the peak will have to occur sooner. The magnitudes of the implied emission reductions between 2000 and 2050 are around 30 % for 550 ppm $CO_2$-eq, 50 % for 500 ppm $CO_2$-eq, and 70 % for 450 ppm $CO_2$-eq. Cuts relative to BAU are indicated in the figure. The stabilization pathway includes different options for cutting emissions that would be more or less prominent at different times. In the earlier periods there would be greater scope for energy efficiency and halting deforestation. With technical progress different technologies in the power and transport sector would play an increasingly strong role.

Both the bottom-up and the top-down studies in the Stern Review (Stern, 2007) produced figures in similar ranges – around 1 % of world GDP per annum for stabilization below 550 ppm $CO_2$-eq. We would now argue, given the growing evidence on the magnitude of the risks, that holding concentrations below 500 ppm $CO_2$-eq and then attempting to reduce from there to below 450 ppm $CO_2$-eq would be an appropriate target to limit temperature increases to not more than 2 °C which many climate scientists believe is the threshold beyond which serious impacts would occur. The costs involved might be of the order of 1–2 % of world GDP per annum.

The calculated order of magnitude may be understood as follows. The reductions required to keep concentrations below 500 ppm $CO_2$-eq in 2050 may be around 65 gigatonnes $CO_2$-eq compared to business-as-usual emissions (see Fig. 1). An average cost of USD 30 per tonne would produce an overall cost of around USD 2 trillion. If global GDP doubled by then, to USD 100 trillion, the overall cost would equate to around 2 % of global GDP.

As we learn more about new technologies, methods and economic policies these costs may fall sharply and so the above projections may apply only over a few decades. They also ignore the many co-benefits of action, including less pollution, greater energy security, and increased biodiversity. There is, of course, considerable uncertainty over cost estimates. Bad policy or delayed decisions could produce higher figures. Greater technological progress could also result in lower figures. Assumptions about substitutability between different goods and options also matter.

Since the Stern Review (Stern, 2007) was published there have been a number of new studies, both bottom-up and top-down. Significant examples of the former include those from McKinsey (Enkvist *et al.,* 2007) and the IEA (2007), both of which indicated costs for given targets either in the range we suggested in the Stern Review, or somewhat lower. Similar conclusions were drawn in the IPCC Fourth Assessment Report (Metz *et al.,* 2007).

In summary, looking back at the Review, we would suggest that subsequent evidence and analysis have confirmed, or at least indicated, that the range of our cost estimates for stabilization of GHG concentrations may be on the high side. Good policy and timely decision-making are, however, crucial to keeping costs down. Merely adopting a 'wait and see' approach, or a 'climate policy ramp', risks not only excessive and dangerous levels of GHG stocks but also much more costly abatement if, as is likely, we later realize that the response was delayed and inadequate.

## A structure for a global deal

The balance of scientific evidence clearly demands that all countries plan credible emissions reduction policies now. If mankind is to avoid substantial damages to future generations, large-scale and urgent international action is required. Market mechanisms should be central in this, with both economic instruments and discretionary policy being used to provide incentives for behavioural change. The UN Conference of the Parties in Copenhagen in late 2009 will be decisive in determining the post-2012 policy frameworks, and designing an effective institutional architecture. It is important that the text of any deal agreed in Copenhagen is guided by clear principles based on rigorous analytic foundations and a common understanding of the key challenges.

The challenges are far-reaching, comprehensive, and global; but they are also manageable. The activities and technologies necessary to eliminate the bulk of the risks associated with climate change are already available, or can be developed through appropriate policies to support innovation. Policies must be designed and applied carefully. Badly implemented policies can create additional market distortions, introduce perverse incentives, and foster protectionism. Care must be taken to ensure that additional policies are not simply layered on top of existing bad policies, such as distortionary energy market subsidies, trade restrictions, or inadequate agricultural policies. Where possible, policies must encourage market-based solutions, minimize transactions costs, and stimulate reform of existing distortion mechanisms. For markets and entrepreneurship to work, the policy framework must be credible, durable, and predictable, while allowing the necessary flexibility.

The following is an attempt to describe the outline of a possible global deal (under six broad headings), based on the preceding analysis and on personal involvement in public discussion over the last two years. This work is described in full in a paper entitled *Key Elements of a Global Deal* (Stern, 2008), which was published in March 2008 at the London School of Economics. The purpose of this paper was to put forward a coherent set of proposals on global policy that satisfy three basic principles:

- Effectiveness – it must lead to cuts in GHG emissions on the scale required to keep the risks from climate change at acceptable levels.
- Efficiency – it must be implemented in the most cost-effective way, with mitigation being undertaken where it is cheapest.
- Equity – it must take account of the fact that it is poor countries that are often hit earliest and hardest, while rich countries have a particular responsibility for past emissions.

Different technologies and different policy instruments can be applied to different sectors and countries. Indeed, the more differentiated the global strategy, the greater the scope for learning, so it is important not to be unduly prescriptive on the details of policy action. However, it is also important that the various initiatives address the overall objective. A global treaty needs to be agreed by 2009 and translated into national policy and action plans by 2012, when the current Kyoto agreement ends.

### Emissions targets

Total anthropogenic greenhouse gas emissions in 1990 are estimated to have been 41 gigatonnes compared to approximately 45 gigatonnes in 2005, with significant shifts in the international distribution of these emissions over that period (CAIT, 2008). The scale of the emissions reductions required, and the welcome rapid economic growth in populous parts of the developing world, means it is necessary for developing countries to play an active role if the deep cuts in emissions, suggested at the G8 conference in Heiligendamm, Germany in June 2007 and confirmed at Hokkaido, Japan in June 2008 and L'Aquila, Italy in 2009, are to be implemented (G8, 2007; G8, 2008; G8, 2009). By 2050, eight billion out of a world population of around nine billion will live in what is currently termed the developing world (United Nations Secretariat, 2006). It is not in these countries' national interests to wait and allow developed countries to take the lead. Countries with strong emissions growth such as China and India will need to plan to limit and reduce emissions within the next ten to twenty years. For this they will require global co-operation, and they are unlikely to be able or willing to achieve these ambitious reductions without substantial technological and financial support and opportunities to develop, and ultimately export, low-carbon technologies.

Effective action must produce the following outcomes:

- World targets for global annual emissions of no more than around 30 gigatonnes $CO_2$-eq by 2030 and around 16 gigatonnes $CO_2$-eq by 2050 from the present level of 47 gigatonnes $CO_2$-eq. These reductions are required to have a reasonable chance of containing temperature increases to no more than 2 °C and to limit the grave risks associated with severe climate change.
- Average per-capita global emissions will – as a matter of basic arithmetic – need to be around 2 tonnes $CO_2$-eq by 2050 (20 gigatonnes divided by 9 billion people).
- The developed world must lead in committing to strong mid- and long-term targets.
- By 2020, developed countries need to demonstrate that they can deliver credible reductions, without threatening growth, and that they can design mechanisms and institutions to transfer funds and technologies to developing countries.
- Subject to the above, a formal declaration is needed stating that developing countries will also be expected to take on binding national targets of their own by 2020, but benefit from one-sided selling of emissions credits in the interim.
- Fast-growing middle-income developing countries will need to take immediate action in order to stabilize and reverse emissions growth, adopting sectoral targets immediately and possibly even national targets before 2020.
- Irrespective of targets, all countries need to commit to developing the institutions, data and monitoring capabilities and policies to avoid high-GHG infrastructural lock-in.

Only sound, measured and co-ordinated policy, and timely international collaboration can deliver strong and clean growth for all at reasonable cost. It is important to weigh up the competitiveness risks and opportunities for firms, countries, and sectors, especially where some countries or sectors apply climate policies earlier and more ambitiously than others do.

There will be losers, and the impacts of transition will need to be managed. However, transition to a GHG-constrained world will create opportunities for companies and sectors that anticipate new markets. Moreover, the evidence to date suggests that few firms are likely to relocate activities to less restrictive jurisdictions. Overstating the problems relative to the opportunities carries the risk of encouraging involved parties to wait for others to act before taking action themselves. By contrast, the expectation of a credible global agreement would heighten the incentives for companies and governments to act quickly and effectively.

### The role of developing countries

Emission reductions on the required scale cannot be achieved without contributions from all countries, both rich and poor. Already, developing countries account for about 50% of energy-related carbon emissions, and their share is expected to rise to 70% by 2030 in the absence of appropriate policies (IEA, 2006).

The arithmetic of global climate change abatement is such that, under the Kyoto successor treaty, the role of developing countries will have to be scaled up substantially. China, for example, currently emits about 6 tonnes $CO_2$-eq per person, and India is approaching 2 tonnes $CO_2$-eq. As a matter of pure arithmetic, climate stabilization will require all countries to reduce and stabilize their emissions at around 2 tonnes $CO_2$-eq by 2050. This target for per-capita emissions by mid-century is so low that there is little scope for any major group to go significantly above or below it. If one or two large countries were to merely reduce emissions to, say, three or four tonnes per capita, it is unlikely that other major countries or groups of countries would be able to offset this by reducing emissions close to zero; as a result the global target would most likely be missed. Indeed, all emissions trajectories should be designed with the target of two tonnes in mind. This is a pragmatic approach and not a strongly equitable one. It takes little account of the greater per-capita contributions of the developed countries to historical and future GHG emissions.

Achieving growth and fighting poverty are key objectives for all countries, but particularly for the developing countries. The world community must recognize that the poorer countries will see emissions grow for some time. But it is not slower growth that will allow developing countries to achieve this fall in emissions. It will be low-carbon growth using technologies demonstrated and shared by the rich countries augmented by developing countries' own technological advances and drives for energy efficiency.

In principle, the best way of achieving this would be to assign emissions targets to most countries, including major 'emerging emitters' such as Brazil, China, India, and Indonesia. However, it may not be feasible politically for these countries to make firm commitments at Copenhagen in December 2009. Developing countries may not be ready to adopt such targets until there is substantial evidence through actions in the developed countries that:

1. Low-carbon economic growth is possible;
2. Financial flows to countries with cheap opportunities to abate GHGs can be substantial; and
3. Low-carbon technologies will be available and shared, allowing developing countries to innovate, develop, and ultimately export their own low-GHG technologies.

If developed countries meet these conditions, developing countries might be willing to discuss binding caps from 2020. These caps cannot be decided upon now, but would be subject to the experience and performance of all countries over the next decade, and would both differ according to local circumstances and reflect countries' 'common but differentiated responsibilities', including historical contributions to emissions. A global deal could embody the presumption that once countries meet certain baseline criteria – for example in terms of GDP per capita or other metrics of economic development – they would be expected to adopt emission caps.

In any case, developing countries should start planning on this basis now, setting out credible action plans to achieve ambitious stabilization targets in the long term. Development plans have to place climate change – both mitigation and adaptation – at their core. Deforestation, in particular, must be a key element.

### *International emission trading*

Putting a price on greenhouse gas emissions should be a central pillar of mitigation policy (see Edenhofer *et al.* and Mirrlees, this volume). It is crucial in making polluters pay for the damages they cause in order to change behaviour on the massive, widespread, and cross-cutting scale necessary to tackle climate change. If there was a clear price to pay for every tonne of carbon dioxide emitted, then consumers and producers across the economy would think hard about whether there were less carbon-intensive products they could buy, or produce. In order to provide the most effective marginal incentive, the price needs to be credible, long-term, and applied across the whole economy.

International cap-and-trade means, first and foremost, that an upper limit is placed on emissions of GHGs. Imposing a fixed quantity target on the world reduces the risk of dangerous climate change impacts and tipping points. A fixed quantity target is therefore a direct link between the science and the policy instrument, thus ensuring that policy is effective. Trading in turn allows the required reductions in emissions to be achieved as cost-effectively as possible.

Currently there are several regional and national emissions trading schemes in existence. The EU emissions trading scheme (ETS) achieved sales of around

USD 50 billion in 2007 (Phase I and II), while sales under the UNFCCC's clean development mechanism (CDM) were approximately USD 13 billion (up from USD 24 billion for the EU ETS and USD 6 billion for the CDM in 2006). It is a major challenge to link, improve and expand these schemes, designing the right institutional frameworks, laws, accrediting and monitoring systems. But the world possesses the resources and the experience of successful cap-and-trade schemes to do it, and the potential rewards are huge.

The vision of the international emissions trading regime outlined in this essay is of a full cap-and-trade scheme covering all gases, sectors and including more advanced developing countries by 2020. In the transition to this goal, most of the effort (and demand for credits) will come from developed countries, while developing countries will receive finance for low-carbon development through selling credits. Prices high enough to generate a strong response will depend on ambitious and binding national targets. This underpins the 'demand side' and will also ensure strong action domestically. Efficiency requires that the supply side works smoothly and effectively.

### *Supporting emissions reductions from deforestation*

Addressing forestry as part of a global climate change deal – and in particular deforestation and forest degradation in tropical rainforest countries – is essential if overall targets for stabilizing carbon emissions are to be met. A total of 13 million hectares of forests are destroyed every year (FAO, 2005) – an area half the size of the United Kingdom, or one third of the size of Japan. According to the International Panel on Climate Change, 'forestry' currently contributes 17.4% of global annual GHG emissions, the overwhelming majority of which comes from burning or decomposition of tropical forests. These emissions include around 5.9 gigatonnes of carbon dioxide, approximately equivalent to the total annual carbon dioxide emissions from the USA (Solomon *et al.*, 2007; IEA, 2007).[5]

Forestry measures, in particular to reduce deforestation, have the potential to make a substantial and relatively immediate contribution to a low-cost global mitigation portfolio that combines synergies with adaptation and sustainable development. Standing forests also perform other significant environmental services, such as the regulation of water supplies and the conservation of biodiversity.

There are many causes of tropical deforestation. It will not be possible to reduce emissions effectively unless these drivers are addressed. Poor local communities are often blamed, but more often it is government incentives and the demand for

---

[5] World Energy Outlook 2007 estimate for US $CO_2$ emissions in 2005, not including land-use change, is 5.8 Gt $CO_2$.

internationally traded commodities such as timber, palm oil, and soy that drive deforestation. The issue of biofuels is just one example of a policy, pursued by developed and developing nations alike, that may (indirectly) play a role in incentivizing deforestation by increasing the demand for agricultural commodities and at the same time the profits to be made from converting forests to agricultural use (see Creutzig and Kammen, this volume).

Reducing deforestation will involve reversing this equation to make standing forests worth conserving. Consequently, any financing framework that successfully addresses the mitigation costs of reduced deforestation needs to be on a scale sufficient to cover these opportunity costs, as well as any transaction costs (including administration, implementation, and enforcement) and insurance.

Global estimates for the opportunity costs involved in halving deforestation have ranged from USD 3 billion[6] to USD 33 billion annually (Obersteiner, 2006), with a number of estimates in between. There is likely to be a large amount of deforestation that can be avoided at modest cost although marginal costs may rise substantially with amounts avoided. Much depends on assumptions concerning 'leakage', and leakage in turn depends on the scale and effectiveness of action.[7] Furthermore, the administration costs associated with achieving reduced deforestation through national payment schemes (one of a number of options) have been estimated to range from USD 250 million to USD 1 billion annually by the tenth year of operation (Grieg-Gran, 2006).

## *Technology*

Over the past 100 years, the global economy has developed largely on the back of the increasing application of carbon- and energy-intensive technologies in all major sectors. In recent years this trend has accelerated, driven by (a) surging growth in the developing world (especially China), (b) relatively low energy prices until 2005, and (c) increasing use of coal as the primary energy source for the power sector. The underlying rate of decrease in carbon intensity, defined as tonnes of carbon per GDP, is 1% per annum. Hence, given that the world economy continues to grow by 3–4% per annum, carbon emissions will continue to grow at 2–3% per annum under a business-as-usual scenario.

The challenge of significantly reducing emissions while maintaining economic growth requires a dramatic shift in the technologies that determine the carbon intensity of the economy. A number of studies indicate that the required GHG abatement can be achieved through the deployment of existing and near-commercial

---

[6] The lowest cost in the Grieg-Gran (2006) range, which does not take into account returns to selective logging before deforestation takes place.
[7] For example, Blaser *et al.* (2007) estimate costs of USD 12.2 billion annually to reduce emissions to zero by 2030.

technologies. New technologies will further lower the costs of transitioning to a low-carbon economy and are thus highly desirable. But in order for existing technologies to be fully diffused and adopted, and for new innovations to occur, three forms of market failure must be overcome. First is the general failure to internalize the costs of GHG emissions. This can be addressed by an appropriately determined carbon price. Second are market failures that have restricted the deployment of many existing energy-efficient technologies despite rising energy prices, and that cannot easily be addressed with a price on carbon. These include principal-agent problems (e.g., property owners not having incentives to deploy energy-saving technologies in commercial buildings), overly high consumer discount rates, lack of information, government energy subsidies that encourage energy consumption, and energy or carbon costs that are low in terms of individual purchase decisions but high in aggregate terms. Third and finally, are market failures specific to the nature of technology itself. These include lock-in of high-carbon technologies due to infrastructure or increasing return effects, risk aversion in the face of technological or carbon price uncertainty, spillovers of investment in research and development that benefit competitors, and learning-curve effects that create high prices for early adopters, thus discouraging demand.

Thus, the key message is that while a clear, appropriately determined and institutionally stable market price for carbon is necessary to stimulate the required technology response, it is not sufficient. An effective, efficient, and equitable policy response in this area must not only motivate market forces, but also overcome market imperfections.

### *Adaptation*

In addition to a fair distribution of the burden of emissions reduction, a further policy response is required to assist those facing the impact of emissions for which they were not responsible. This requires support for adaptation in those countries hardest hit by climate change. The most effective form of adaptation to a changing climate is robust, climate-resilient development. Adaptation assistance needs to be integrated into development spending to deliver development goals in a climate-resilient manner, rather than being earmarked for climate-specific projects.

Just as adaptation planning needs to be integrated into development plans and strategies, so adaptation funding should be integrated into development spending at regional, national and local levels, ideally by delivery through the same multilateral channels, and not by setting up parallel processes. Money should be spent through national development plans, reflecting overall national priorities, with delivery following the principles of the Paris Declaration (OECD, 2005): ownership, alignment, harmonization, managing for results, and mutual accountability.

Allocation of funding between countries will need to reflect a combination of several factors: impacts of climate change, vulnerability to those impacts, capacity for internal investment, and the commitment and ability of local governments to deliver appropriate outcomes.

Money and other assistance will be best used if national governments are responsible for using funds to deliver broad contracts on issues such as poverty, health and climate vulnerability. Delivery of these goals will need to be monitored and evaluated. Based on this evaluation, recipient governments will in turn need to be held accountable by their citizens – who stand to lose most from a changing climate – and by the international community. International financial institutions, including the World Bank and International Monetary Fund, should monitor, report on and, where necessary, facilitate non-financial aid such as access to insurance, technology and information, as well as other market-based facilities.

### A blueprint for a safer planet

We have described how we can offer a 'blueprint for a safer planet'. If we follow the route we have tried to chart, or something similar, as we believe we can, we will not only protect the planet for our grandchildren but we will also reduce dramatically the severe threat of global conflict that unmanaged climate change would eventually cause.

It is crystal clear, however, that this is a global challenge and can be confronted effectively only by concerted action across the world. It will require international collaboration on an unprecedented scale; that is the only way it can work. While there are different forms of mutual understandings and institutions that can support such action, a spirit of internationalism, mutual dependence and shared destiny is fundamental. If we cannot create this collaboration we will have failed future generations and ourselves.

The meeting of the United Nations Framework Convention on Climate Change at Copenhagen in December 2009 is the most important international gathering since the Second World War. The world set itself the task, in Bali in December 2007, of reaching an agreement on a successor to the Kyoto Protocol by the end of 2009. If we fail to construct a strong global deal in Copenhagen we risk years of dangerous delay. Delay means higher concentrations and growing emissions; it means that the starting point for both stocks and flows will make the required emission reductions greater and more difficult to achieve; and furthermore, it means that the confidence in future policy of the investors, those who will take the practical measures, will be severely damaged. The emerging carbon markets, crucial to necessary incentives, will be undermined. We cannot postpone the construction, agreement and action on a global deal.

We can and must now handle the short-term economic crisis, foster sound economic growth in the medium term, and protect the planet from devastating climate change in the long term. All three can be done in unison and all three are urgent. To try to set them against each other as a three-horse race is as confused analytically as it is dangerous economically and environmentally. The current economic crisis certainly requires an urgent response, but so does the climate crisis.

We need political leadership that is not only thoughtful and measured but also courageous and inspirational. That leadership must set out the compelling scientific and economic case for strong action. It must show not only the severe risks posed by climate change, but also that if we act sensibly and strongly starting now, we can dramatically reduce those risks at reasonable cost. That leadership must be courageous too in confronting the short-term, narrow and confused interests that will make a lot of noise and argue for postponement of action, or in some cases for little or no action. It is a time for clarity and strength in both vision and action.

Strong action on climate change will not only protect the lives and livelihoods of our children and grandchildren, it will allow them to experience the wonder of the natural environment which we still enjoy. Low-carbon growth will deliver much more than this. It will also create an industrial revolution that will drive growth in the coming decades. But still more important, it will create a world that is much freer from conflict over scarce resources, including water and hydrocarbons; a world that will be more secure in its energy supplies; a world that will be quieter and cleaner; a world with greater biodiversity, less pollution, and more beautiful in the physical and natural environment. It will also be a more co-operative world where we have a much better chance of dealing with the many global problems, above all entrenched poverty, that we face and will face as citizens of one planet.

This is indeed an inspirational story. But it is also a practical story, indeed the only practical story. We have a short window of opportunity to turn it into a reality. Whilst it is time for leadership, we must all contribute to the creation of this reality; from my own world of the university and of policy analysis from those who will invest in the new opportunities, and from those who will change the way they consume. We know what we have to do, and the prize is enormous. The people and politicians of the world, community by community, nation by nation, will now determine whether we can create and sustain the international vision, commitment and collaboration that will allow us to take this special opportunity and to rise to the challenge of a planet in peril.

*Acknowledgements:* We are very grateful for the advice and support of Elisa Fenzi and James Rydge.

# References

Auffhammer, M. and Carson, R. T. (2008). Forecasting the path of China's $CO_2$ emissions using province level information. *Journal of Environmental Economics and Management,* **55**(3), 229–47. Available at http://repositories.cdlib.org/are_ucb/971.

Blaser, J. and Robledo, C. (2007). *Initial Analysis on the Mitigation Potential in the Forestry Sector.* Prepared for the UNFCCC Secretariat, Bonn. Available at http://unfccc.int/cooperation_and_support/financial_mechanism/financial_mechanism_gef/items/4054.php.

CAIT – Climate Analysis Indicators Tool, Version 5.0. (2008). Washington, D. C.

Enkvist, P.-A., Nauclér, T. and Rosander, J. (2007). A cost curve for greenhouse gas reduction. *The McKinsey Quarterly,* **1,** 35–45.

FAO – Food and Agricultural Organization (2005). *Global Forest Resources Assessment 2005: Progress Towards Sustainable Forest Management.* FAO Forestry Paper 147. Rome.

G 8 (2007). *Heiligendamm Chair's Summary.* Heiligendamm. Available at http://www.g-8.de/Content/EN/Artikel/__g8-summit/anlagen/chairs-summary,templateId=raw,property=publicationFile.pdf/chairs-summary.pdf.

G 8 (2008). *Hokkaido Toyako Summit Leaders Declaration.* Hokkaido. Available at http://www.g8summit.go.jp/eng/doc/doc080714__en.html.

G 8 (2009). L'Aquila Chair's Summary. L'Aquila. Available at www.g8italia2009.it/static/G8_Allegato/Chair_Summary,1.pdf.

Grieg-Gran, M. (2006). *The Cost of Avoiding Deforestation.* Report prepared for Stern Review. London.

Hansen, J., Sato, M., Ruedy, R. *et al.* (2006): Global temperature change. *Proceedings of the National Academy of Sciences of the United States of America,* **103**(39), 14288–93.

IEA – International Energy Agency (2006). *World Energy Outlook 2006.* Paris.

IEA – International Energy Agency (2007). *World Energy Outlook 2007.* Paris.

Jansen, E., Overpeck, J., Briffa, K. R. (2007). Palaeoclimate. In Solomon, S., Qin, D., Manning, M. *et al.,* eds., *Climate Change 2007: The Physical Science Basis. Contribution of Working Group I to the Fourth Assessment Report of the Intergovernmental Panel on Climate Change.* Cambridge, pp. 433–97.

Metz, B., Davidson, O. R., Bosch, P. R. *et al.,* eds. (2007). *Climate Change 2007: Mitigation of Climate Change. Contribution of Working Group III to the Fourth Assessment Report of the Intergovernmental Panel on Climate Change.* Cambridge.

Murphy, J., Sexton, D. M. H., Barnett, D. N. (2004): Quantification of modelling uncertainties in a large ensemble of climate change simulations. *Nature,* **430**(7001), 768–72.

Obersteiner, M. (2006). *Economics of Avoiding Deforestation.* International Institute for Applied Analysis, Austria.

OECD – Organisation for Economic Co-operation and Development (2005). *The Paris Declaration.* Paris. Available at http://www.oecd.org/document/18/0,2340,en_2649_3236398_35401554_1_1_1_1,00.html.

Solomon, S., Qin, D., Manning, M. *et al.,* eds. (2007). *Climate Change 2007: The Physical Science Basis. Contribution of Working Group I to the Fourth Assessment Report of the Intergovernmental Panel on Climate Change.* Cambridge.

Stern, N. H. (2007). *The Economics of Climate Change: The Stern Review.* Cambridge.

Stern, N. H. (2008). *Key Elements of a Global Deal on Climate Change.* London.

Stern, N. H. (2009). *A Blueprint for a Safer Planet: How to Manage Climate Change and Create a New Era of Progress and Prosperity.* London.

United Nations Secretariat (2006). *World Population Prospects: The 2006 Revision.* New York.

Weitzman, M. L. (2008. *On Modeling and Interpreting the Economics of Catastrophic Climate Change.* Harvard.

Weitzman, M. L. (2007a). A review of the Stern review of the economics of climate change. *Journal of Economic Literature,* **45**(3), 703–24.

Weitzman, M. L. (2007b). Subjective expectations and asset-return puzzles. *American Economic Review,* **97**(4), 1102–30.

Wigley, T. M. L. and Raper, S. C. B. (2001). Interpretation of high projections for global-mean warming. *Science,* **293**(5529), 451–4.

# Chapter 8

# The German contribution to a global deal

———————

## Sigmar Gabriel

Sigmar Gabriel was born in 1959 in Goslar, Germany. He began his political career in his home town as local president of Die Falken, the youth group of the Social Democratic Party of Germany (SPD). He joined the SPD in 1977. He studied at the University of Göttingen to become a grammar-school teacher. Gabriel was elected to the Landtag (state parliament) of Lower Saxony in 1990, and served there until 2005, becoming SPD parliamentary leader in 1998. During this time he was a member of the Landtag's environment committee, was a local councillor of the town of Goslar, and was spokesman for domestic policy of the SPD parliamentary fraction. In December 1999 he became Minister President of Lower Saxony and subsequently chairman of the SPD parliamentary fraction. From 2005 to 2009 Sigmar Gabriel was the German Federal Minister for the Environment, Nature Conservation, and Nuclear Safety.

*Note:* This chapter is a commentary on chapter 7.

We live in turbulent times. With the world economic and financial system in turmoil, and growing concerns about competitiveness and job losses, times have been better for proponents of ambitious climate policies. However, these policies are more necessary than ever.

They are necessary not only to avoid the worst impacts of climate change and protect the world's most vulnerable, but also in order to begin the restructuring of the global economy towards ecologically oriented growth. They are necessary to create new economic instruments and business models to ensure that the opportunities for efficiency and renewable energies are fulfilled and result in the creation of new jobs in low-carbon industries. Hence, ambitious climate policies, which promote the practical opportunities that are already available, are the best way to counteract growing politics of insecurity being driven by high energy and commodity prices.

The Intergovernmental Panel on Climate Change (IPCC) has pointed out clearly that we are running out of time. At the 2007 UN climate conference in Bali, countries have agreed to negotiate a post-2012 agreement building on the UN Framework Convention on Climate Change (UNFCCC) and its Kyoto Protocol. They have also agreed that negotiations should be concluded in 2009 at the UN climate conference in Copenhagen.

This unprecedented global negotiation effort aims to bring North, South, East and West together in a solidarity pact to save the planet. To avoid triggering dangerous tipping points in the climate system these negotiations must succeed in placing the world on a path to a future with a global temperature increase of not more than $2\,^\circ$C. There is no time left for us to have a realistic second chance at agreement.

Suggestions like Lord Nicholas Stern's global deal are therefore most welcome and help to focus our thinking and introduce possible pathways to solutions. I agree with most of the elements that Lord Stern puts forward and also with his assumption that we can only reach a comprehensive agreement in Copenhagen if we design package solutions and reach a comprehensive global deal.

Germany is ready to contribute its fair share to this package. In fact, we are determined to continue our leadership role and to demonstrate in practice what Lord Stern has provided the economic analysis for: climate protection pays off.

## A shared vision for climate protection to guide our efforts

A global agreement will need to spell out a vision for all countries to achieve their national economic and development goals in a low-carbon fashion that safeguards the environment, strengthens countries' ability to adapt to the changes already underway and allows for sustained economic welfare. As Lord Stern describes, industrialized countries need to demonstrate their intent to lead the way on driving the low-carbon transition, and agree to support developing countries in their own transition.

The vision should include the objective of limiting global warming to 2 °C compared to preindustrial levels. This would imply a long-term goal to at least halve global emissions of greenhouse gases by 2050 compared with 1990 levels and to bring about a peak in global emissions in the next 10 to 15 years. This will be needed in order to ensure that no critical climate tipping points are at risk of being triggered.

## Mitigation of climate change: an environmental and economic imperative

It is an economic as well as environmental necessity to put our economies on the path to low-carbon intensity. The IPCC has provided the scientific analysis; the Stern Review has provided the economic arguments. If we take our 2 °C target seriously, developed countries need to adopt nationally binding caps ensuring that, collectively, they reduce their emissions by 25–40% below 1990 levels by 2020.

This may seem hard to reach. In Germany, however, we are already well underway to showing that reductions in this range are economically feasible: We have committed ourselves to reducing our 1990 emissions by 40% by 2020. In concert with the other EU countries that are willing to reduce their emissions by 30% as part of an international agreement, Germany and Europe are showing the way.

Nonetheless, no matter how ambitious our climate protection targets may be, developed countries cannot reach the 2 °C target on their own. We also need enhanced actions by major developing countries, which should result in a significant deviation below business-as-usual (BAU) greenhouse gas emission growth in the order of 15–30%. Collectively, these enhanced actions should be enough to ensure that developing country emissions peak no later than 2020–2025. I welcome Lord Stern's proposal that, guided by the overall commitment to a substantial deviation from BAU growth, each major developing economy should submit a low-carbon development plan that demonstrates what it will implement unilaterally domestically. I would further argue that these plans should also outline additional measures the country could undertake, given both greater access to the global carbon market and technological and financial support. This will indicate the degree to which the level of ambition of developing countries would be dependent on the level of support provided by developed countries.

## Financing and powering the transition to low-carbon economies

One of the major elements of a global deal is undoubtedly measurable, reportable and verifiable financing and support for technology transfer. The review by Lord

Stern and the analysis done by the UN Climate Secretariat on Climate Change and Investment Flows serve as a good basis to grasp the dimension of climate protection investments and the instruments required to achieve this. The figures are impressive: the Stern Review estimated that the overall global costs of unmitigated climate change may add up to 5–20% of our Gross Domestic Product (GDP) per year, whereas the costs of action – stabilizing greenhouse gas emissions at 550 ppm – can be limited to around 1% of global GDP each year.

According to the UN Climate Secretariat, additional investment and financial flows of 200–210 billion US dollars will be necessary for mitigation purposes in the year 2030 in order to return global greenhouse gas emissions to current levels. In 2030 overall additional investment and financial flows needed for adaptation amount to several tens of billions US dollars per year.

The numbers sound huge, but there is no reason to shy away from the challenge: Lord Stern's analysis reconfirms that the money and technologies needed are broadly available.

Compared to the global volume of investments, the additional investment costs will be rather low – in 2030 they will amount to a share of 1.1 to 1.7%. The private sector will play a decisive role with regard to these investments with a share of more than 80%. At the same time, public funds will play an important role as a catalyst, which makes it obvious that we need additional instruments to mobilize investments.

The International Energy Agency estimates that by 2030, global investments in energy infrastructure will have reached a level of around 20 trillion US dollars. Our main task as political decision makers is to set the political framework so that these trillions of dollars are spent in a climate-friendly way. Framework conditions such as the international carbon market or national climate and energy policies are the necessary prerequisites to make market forces work in favour of climate protection. Therefore, I strongly support the expansion of the carbon market, as promoted by Lord Stern, which will allow the flow of funds and technologies to be directed towards the markets of developing and newly industrialising countries.

In 2008 the international carbon market already reached a volume of 64 billion US dollars – and this trend is set to rise. Depending on the structure of international provisions, it is calculated that in 2025 international emissions trading may achieve a turnover of up to EUR 800 billion, according to the UN Climate Secretariat.

During the first period of the European Emission Trading Scheme (until 2012) Germany will auction nearly 10% of the allowances. The revenues – around EUR 400 million annually – will be used to finance additional climate protection measures at home and in developing countries as part of our Climate Initiative. EUR 120 million will be available for projects abroad, particularly in developing countries and economies in transition, on a yearly basis starting in 2008. We will focus on sustainable energy systems, adaptation and sustainable forest management projects in developing countries. With a 100 % auctioning in the future, even more financial flows will be generated – some billion euros per year in Germany alone. I do not see at this moment any approach other than the carbon market that can create investment flows comparable to these numbers.

Therefore, I fully support Lord Stern's pledge that an international carbon market based on internationally binding and absolute reduction targets by industrialized countries should serve as the basis for a post-2012 climate regime under the UNFCCC. The more ambitious our reduction targets are, the greater the returns will be. The expansion of the international carbon market will generate significant additional financing for developing country mitigation. To meet the huge challenges we face, we need to create self-financing mechanisms for climate protection in the long run. The German example shows how using revenues from emissions trading for climate protection projects can yield a double dividend for the climate and at the same time secure jobs and prosperity and make our economy fit for the future.

### Adapting to the inevitable impacts of climate change

At the same time, we must send a clear signal to the poorest and most vulnerable countries and people of the world that they will not be left alone to deal with the increasing impacts of climate change. I agree with Lord Stern that in a post-2012 climate deal, we need to pay more attention to supporting adaptation in developing countries.

As Lord Stern puts it, developed countries, whose emissions have been primarily responsible for climate change, have an obligation to pay at least part of the additional costs that arise from adapting to the impacts of climate change. This means that developed countries should give a firm undertaking both to honour their existing Official Development Assistance (ODA) commitments and at the same time to provide additional resources for adaptation to climate change.

Firstly, donors would need to agree to mainstream adaptation into their existing bilateral and multilateral aid programmes and 'climate proof' their investments without using funds from existing aid budgets. Secondly, we need additional resources to finance the additional support needed. In Germany we have committed new and additional money as part of our International Climate Initiative to supporting adaptation in developing countries. One option that we are exploring, which also Lord Stern points out as particularly promising, is to support disaster prevention and develop insurance schemes to provide a safety net for poor people exposed to climate change risk.

### 'Mission possible'

In German climate policies we are committed to practising what we preach: we have recently adopted an energy and climate package, which is unparalleled in the world and has brought us a large step closer to our goal of reducing our emissions by 40% compared to 1990 levels by 2020. This package of measures will bring new momentum to all $CO_2$-relevant sectors and advance climate protection in Germany.

The planned reductions are not only compatible with economic growth but offer new business opportunities and create a large number of jobs. We are convinced that Germany will play a pioneering role on the lead markets of the future. Successful energy and climate policy will have positive impacts for Germany as a location for business and innovation.

Germany's contribution to a global deal will be to lead by example. We are aware that the implementation of European and national climate protection targets up to 2020 and beyond requires nothing less than the radical restructuring of our industrial society – and we are willing to face up to this challenge.

# Chapter 9

# A 'just' climate agreement: the framework for an effective global deal

---

## Sunita Narain

Sunita Narain, born in 1961, joined the New-Delhi-based Centre for Science and Environment (CSE), one of India's leading environmental NGOs, after graduating high school in 1982. In her work, Narain has focused on the relationship between environment and development, and on the formation of a public consciousness regarding the need for sustainable development. Her research interests range from global democracy, with a special focus on climate change, to the need for local democracy, with a focus on forest-related resource management and community-based water issues. She is currently Director of the CSE and Director of the Society for Environmental Communications, which publishes a fortnightly magazine, Down To Earth.

I remember how I first learned about global warming. It was in the late 1980s. My colleague Anil Agarwal and I were searching for policies and practices to regenerate degraded common lands. We quickly learned to look beyond trees, at ways to deepen democracy, so that these commons – in India, forests are mostly owned by government agencies, but it is the poor who use them – could be regenerated. It became clear that without community participation, planting trees was not possible. For people to be involved, the rules for engagement had to be respected. To be respected, the rules had to be fair.

In the same period, data released by a prestigious US research institution, the World Resources Institute, convinced our then environment minister that it was the poor who contributed substantially to global warming – by doing 'unsustainable' things like growing rice or keeping animals. Anil and I were pulled into this debate when a flummoxed chief minister of a hill state called us. He had received a government circular that asked him to prevent people from keeping animals. 'How do I do this?' he asked us. 'Do the animals of the poor really disrupt the world's climate system?' We were equally perplexed. It seemed absurd. Our work told us that the poor were the victims of environmental degradation. Suddenly they were being presented as the villains. How was this possible?

With this question in mind we embarked on our climate research journey. We began to grasp climate change issues, and quickly learned that there was not much difference between managing a local forest and managing the global climate. Both are common property resources. What was needed most of all was a property rights framework that encouraged cooperation. We argued in the following way:

- First, the world needed to differentiate between the emissions of the poor (for example, from subsistence farming) and those of the rich (from, say, cars). Survival emissions were and could not be equivalent to luxury emissions.
- Second, managing a global common resource required cooperation between countries. Just as stray cattle or goats are likely to chew up saplings in the forest, any country could destroy a climate protection agreement if it emitted more than the atmosphere can take. Cooperation was only possible – and this was where our forest experience came in useful – if benefits were distributed equally. We then developed the concept of per-capita entitlements (each nation's share of the atmosphere), and used the property rights of entitlement to set up rules of engagement that were fair and equitable. We said that countries using less than their share of the atmosphere could trade their unused quota. This would give them an incentive to invest in technologies that would not increase their emissions. But within this process, we told climate negotiators, it was useful to think of the local forest and learn that the issue of equity is not a luxury; it is a prerequisite.

That was 20 years ago. Today, in 2009, we have come a long way, principally in our acceptance that climate change is the greatest existential crisis that human beings have ever faced. We remain weak in our commitment to bringing about change; we are big on words and small on action. In 2009, we have reached the point where we must commit to a very different future.

The framework would propose for this just and effective global climate deal is as follows.

### Climate change is all about the economy, stupid

It is important to note that industrialized countries have managed to de-couple sulphur dioxide emissions from economic growth. In other words, emissions have fallen even as national income has risen. But they have failed to do the same with carbon dioxide emissions. Per-capita carbon dioxide emissions remain closely related to a country's level of economic development and standard of living. It is evident that as long as the world economy is carbon-based – driven by energy from coal, oil, and natural gas – growth cannot be substantially de-coupled from carbon dioxide emissions.

The only way to avert environmental devastation is to reduce emissions dramatically. However, in a world where things are never quite so simple, the use of these fuels, and hence carbon dioxide emissions, are closely linked to economic growth and lifestyle. Every human being contributes to the carbon dioxide concentrations in the atmosphere, though the amount emitted depends on the person's lifestyle. The more prosperous a country's economy and the higher its per-capita income, the higher is its fossil fuel consumption for power generation and transport, and therefore the higher its greenhouse gas emissions.

Industrialized countries owe their current prosperity to 'historical' emissions, which have accumulated in the atmosphere since the start of the Industrial Revolution, as well as to high levels of current emissions. Developing countries, meanwhile, have only recently set out on the path of industrialization, and their per-capita emissions are still comparatively low.

Under these circumstances, any limit on carbon dioxide emissions amounts to a limit on economic growth, turning climate change mitigation into an intensely political issue. International negotiations under the UN Framework Convention on Climate Change – aimed at limiting greenhouse gas emissions – have turned into a tug of war, with rich countries unwilling to 'compromise their lifestyles', and poor countries unwilling to accept a premature cap on their right to basic development.

## Complexity is no excuse for inaction

Climate change is undoubtedly the greatest challenge of our century. Its sheer complexity and urgency seem overwhelming. For the past 18 years – the first intergovernmental negotiation took place in Washington DC in early 1991 – the world has been haggling about what it knows but does not want to accept. It has been desperately seeking every excuse not to act, even as science has confirmed and reconfirmed that climate change is real, that it is related to carbon dioxide and other emissions, and that these emissions are related to economic growth and wealth. In other words, it is man-made and can destroy the world as we know it.

The scientific community is not just certain but *unequivocal* that climate change and its devastating consequences are now inevitable. But along with understanding the still obtuse science we must begin to put a human face on the effects of climate change that are becoming evident all around us. We must see climate change in the faces of the millions who have lost their homes in the Sidr and Nargis cyclones that ripped through Bangladesh and then Myanmar. After all, science has clearly established that the intensity and frequency of tropical cyclones will increase as the Earth heats up (Solomon *et al.,* 2007). We need to see climate change in the faces of those who lost everything in the floods caused by intense rainfall events. We need to understand that the thousands of people who died in these disasters did so because the rich have failed to contain the emissions upon which their growth has been built.

## Inaction of the rich world

As the call for action has become more strident and urgent (as it must), the world has looked for small answers and petty responses. On the one hand, there is a well-orchestrated media and civil society campaign to paint the Chinese and Indians as the villains of the piece. If they 'cry' about their need to develop, the response is to tell them that they are most vulnerable. Rich countries seem to be saying: 'We cannot afford to waste time in the blame game. Even if, in the past, the Western world created the problem, *you* must, in *your* interest, take the lead in reparations.'

This hysteria is growing. But unfortunately, action is not keeping pace.

In late 1997, after years of protracted negotiations, the Kyoto Protocol was established. Under this agreement, the industrialized world agreed to cut its emissions by just 5.2% of 1990 levels by 2008–2012. It is important to realize that the world is nowhere close to achieving even this reduction. Not only has the world's largest polluter – the United States – walked out of the global agreement, even Europe is finding it difficult to reach this modest target. A review by the secretariat of the UN Framework Convention on Climate Change (UNFCCC, 2007) has found that between 1990 and 2006, while carbon dioxide emissions of all industrialized countries (classified as Annex I under the convention) declined by 1.3%, this reduction was primarily due to the countries whose economies are in transition. The carbon dioxide emissions of the Annex I countries, excluding countries in transition, actually increased by 14.5% (see Fig. 1).

During the same period, the carbon dioxide emissions of key polluters increased – in the case of the US by 18%, and by a whopping 40.5% in Australia. Even most European countries have seen an increase in their emissions. The only countries that have cut carbon dioxide emissions are Sweden, the UK and Germany. But it is important to note that emissions in the UK and Germany are beginning to increase again. The reason is simple. The UK partly gained its emissions reduction by switching from coal to natural gas, a transition that is now predominantly completed. Germany reduced its emissions greatly because of the reunification of the industrialized west with the economically depressed east. New answers must now be found. In other words, these emission cuts were nowhere close to what was needed, then or now, to avert catastrophic climate change. The industrialized countries have reneged on their commitment. They have let us all down.

So far, the rich world has found only small answers to existential problems. It not only wants to keep its coal-burning power plants (even as it points the finger at China and India), but wants to build new ones. It believes it can keep polluting while finding new 'fixes'. The latest solution it has come up with is 'Carbon Capture and Storage' – to pipe the emissions underground and hope the problem will simply go away. In this way, the rich world hopes it can have its cake and eat it too. Is it not ironic that in spite of science telling us that drastic reductions are needed, no country is talking seriously about limiting its energy consumption? Every analysis shows that while efficiency is part of the answer it is meaningless without sufficiency. Cars have become more fuel-efficient but people now drive more and own more cars. We have to realize that without a global cap on carbon emissions, any measures to improve energy efficiency will remain ineffective.

**Energy is the key**

It is the world's need for energy – to run everything from factories to cars – that is the principle cause of climate change. After years of talking about the problem no country has been able to de-couple its growth from the growth of carbon dioxide emissions. No country has yet shown how to build a low-carbon economy. No country has yet been able to re-invent its pathway to growth. This, then, is the challenge. After years of talk, the proportion of new renewable energy – wind, solar, geothermal, biofuels – comprised only about 1% of the world's primary energy supply in

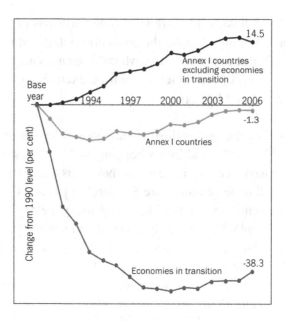

**Fig. 1.** Carbon dioxide emissions of Annex I countries under the UN Framework Convention on Climate Change, excluding land use, land-use change, and forestry (LULUCF). (*Source:* UNFCCC, 2007)

2006 (IEA, 2007). It is misleading to say that renewable sources add more electricity than nuclear power. It is 'old renewable' energy – hydroelectric power – that makes the world light up.

One of the tragedies of the climate change debate is that the world is hiding behind the poverty of its people to fudge its climate maths. Biomass combustion contributes greatly to the renewable sector – the firewood, cow dung or leaves and twigs used by the desperately poor in our world to cook their food and to light their homes. It is this that is providing the world its space to breathe.

## We are the change

What, then, is the way ahead? First, we must accept that the rich world must reduce emissions drastically. There is a stock of greenhouse gases in the atmosphere, built up over centuries in the process of creating nations' wealth. This has already made our climate unstable. Poorer nations will add to this stock through their desire for economic growth. But that is no excuse for the rich world to avoid adopting tough and binding emission reduction targets. The principle should be that the rich reduce so that the poor can grow. Second, any agreement must recognize that poor and emerging countries need to grow. Their engagement should therefore not be legally binding but based on national targets and programmes. The challenge is to find low-carbon growth strategies for emerging countries, without compromising their right to develop. This can be done. It is clear that countries such as India and China have the opportunity to 'avoid' additional emissions. The reason is that they are still in the process of building their energy, transport and industrial infrastructures. They can make investments in 'leapfrog' technologies so as to avoid pollution. In other words, they can build their cities based around public transport; their energy security based on local and distributed systems – from biofuels to renewables; and their industries using the most energy-efficient and pollution-free technologies.

We know it is in our interest not to first pollute, then clean up; or first to be inefficient, and then to save energy. But we also know that the existing 'green' technologies are costly. It is not as if China and India are bent on first investing in dirty and fuel-inefficient technologies. They invest in these, as the now rich world has done, based on the principle, 'first create emissions, then make money, then invest in efficiency'.

### The just deal: what does it mean?

If we know that the emerging world can leapfrog to cleaner technologies, the question is, why is this not happening? Why is it that the world talks big yet makes only small changes?

As part of the Kyoto Protocol the 'Clean Development Mechanism' (CDM) was invented to pay for the transition in the poorer world. But the mechanism was destined to fail (for ideas on CDM reform see Liverman, this volume). The rich countries were obsessed with obtaining the cheapest emission reduction options. As a result, the price of CERs – the certified emission reduction units used in this transaction – has never reflected the cost of renewable and other high-technology options. It is a cheap and increasingly corrupt development mechanism. It is also a convoluted mechanism, in which governments are prevented by rules from considering major change. In fact, CDM currently provides disincentives for governments in the South to drive policies for clean energy or production. Any such policy that is designed independently of the CDM framework does not meet the criterion of 'additionality', and does not qualify for funding.

The world must realize the bitter truth. Equity is a prerequisite for an effective climate agreement. Without cooperation this global agreement will not work. It is for this reason that the world must seriously consider the concept of equal per-capita emission entitlements so that the rich reduce and the poor do not go beyond their climate quota. We need effective and responsible action on climate change now.

## Conditions for action on climate

Warming of the global atmosphere is possibly the biggest and most difficult economic and political issue the world has ever needed to confront. First, as emissions of carbon dioxide are directly linked to economic growth, growth as we know it is at risk. We will have to reinvent what we do and how we do it. There will be costs associated with this change, but these costs will be a fraction of what we will need to spend if we do not change. Second, admissible growth has to be shared equitably among nations and people. The question now is: Will the world share its right to emit (or pollute), or will it freeze inequities? Will the rich world, which has accumulated a huge 'natural debt' – overdrawing on its share of the global commons – repay it so that the poorer world can grow and use the same ecological space. Third, climate change is about international cooperation. Climate change teaches us more than anything else that the world is one; if the rich world pumped excessive quantities of carbon dioxide into the atmosphere yesterday, then the emerging rich world will do so today. It also tells us that the only way to control emissions is to ensure that there is fairness and equity in the agreement, so that the greatest level of cooperation is possible.

There is clear understanding that the rich and the emerging rich worlds need to make the transition to low-carbon economies. There is also much better understanding that the way ahead involves technologies that we already possess. The answer will lie in increasing efficiency in both the generation of energy and in its use for the manufacture of other products. It will also lie in changes to how we do things – from transportation in our cities to everything else. Fact is that we know how to change.

## The imperative of energy transformation

It is increasingly understood that the de-carbonization of economies is imperative if the world wants to tackle climate change. It will require substantial investments to move towards a zero-carbon-energy-based economy, eliminating the use of fossil fuels altogether. It is also clear that the existing and growing use of fossil fuels has the potential to 'lock in' this energy source for a much longer time than desired, and 'lock out' renewable energy sources. The question is, how will the world accomplish this energy transition? And is it even possible?

## The shift to renewable energy sources

How can the world make this rapid shift towards renewable energy technologies? If the world waits for most of its oil, gas, and coal resources to be exhausted before making this transition – something that probably will not occur before the end of this century – then the risk of serious climate change will be inordinately high. It is important to understand the nature of this challenge. The twentieth century saw a major transition away from renewable energy towards a fossil fuel-based global economy. Between 1900 and 2000, world energy use grew more than ten-fold. Even though the energy from renewable sources increased nearly five-fold during the century, its share in total energy use dropped from 42% to 19% (IEA, 2007).

This trend has continued. The January 2007 report of the International Energy Agency (IEA, 2007), estimates that in 2006 the share of renewable energy in the total primary energy supply was just 13%. Significantly, the bulk of the renewable energy budget was made up of biomass burning and hydro-electric power. For instance, the share of renewable energy in India is estimated to be 39%, because of the use of biomass by the poor to cook food. The contribution of new renewables – wind, solar, tidal and geothermal energy – was as little as 0.5% of the world's total energy consumption. The challenge now is to reverse this trend.

It is clear that the market for renewable energy technologies is growing. According to the IEA, wind energy saw growth of 50% per annum and solar energy 28% per annum between 1971 and end of 2006. Modern biomass energy, including new technologies that produce ethanol from agricultural waste, also contains immense potential. Technological advances are also taking place in the use of hydrogen fuel cells. The cost of these technologies has also fallen; but not enough to make them competitive with conventional energy options.

We know that the more the world gets locked into fossil-fuel-based systems, especially efficient and low-cost fossil fuel systems, the longer it will take to get out of them. If the huge energy investments that will be made by developing countries in the next three to four decades lock them into a carbon energy economy like that of the industrialized countries, this will result in an enormous build-up of greenhouse gases. The governments of the world will therefore have to play a key role in 'reinventing the energy system', just as they have played a key role in determining the modern carbon-based energy supply structure since the nineteenth century.

Today the biggest obstacles in the way of renewable technologies are low prices for fossil fuels and subsidies on fossil fuels in many countries. In addition, the renewable energy sector is facing problems of declining public- and private-sector research and development. Rapid expansion in the use of zero-carbon technologies will come only with proactive official policies aimed at increasing research investment and creating favourable economic conditions, allowing mass production to bring costs down even further.

If the solution lies in creating large markets for zero-carbon energy technologies, the advantage lies with the countries of the South, the low carbon emitters. These countries have for the most part not yet invested in the electricity grid; they are not yet locked into fossil-fuel-based energy systems. However, these countries require huge investment if they are to supply energy to their millions of households.

## Toward a framework for equitable entitlements

The tragedy of the atmospheric commons has been the lack of rights to this global ecological space. As a result, industrialized countries have borrowed or drawn heavily from it – and without any control. They have emitted greenhouse gases far in excess of what the Earth can withstand. This was because they were not bound by limits or quotas, and enjoyed 'free use' of this natural capital. Some researchers have called this the 'natural debt' of the North, as opposed to the financial debt of the South. In this context, curtailing emissions can only be achieved through the creation of rights and entitlements of each nation to the atmosphere so that future responsibilities are clearly demarcated. This allocation of the common space has to be made on the basis of each nation's past, present, and projected future contributions to the global warming crisis. The world needs to adopt the concept of equal per-capita entitlements to greenhouse gas emissions.

Solving the climate crisis is about sharing growth among nations and people. And clearly this has not yet happened. Between 1980 and 2005, the total emissions of just one country (the United States) were almost double those of China, and more than seven times those of India (see Fig. 2).

In per-capita terms, the injustice is even more unacceptable and immoral (see Fig. 3). Historical emissions – between 1890 and 2005 – for example, amount to about 1100 tonnes of carbon dioxide per capita for the UK and the USA, compared to 66 tonnes per capita for China, and 23 tonnes per capita for India (CSE, 2008). As yet, the world has seen no real change in this situation. No change it can believe in.

## Net versus gross emissions: sharing the world's common sinks

In 1990, the Washington-based World Resources Institute (WRI) published a report which showed that annual greenhouse gas emissions in the developing world almost equalled those in the industrialized world, and predicted that the emissions of the developing world would overtake those of the industrialized world in the near future (WRI, 1990). However, the critique of this report by the Delhi-based Centre for Science and Environment (CSE) found that the methodology used by WRI to compute the responsibility of each nation favoured the polluter (Agarwal and Narain, 1991).

Under the WRI methodology, each nation was assigned a share of the Earth's ecological sinks, but the assignment was proportional to the nation's contribution to the Earth's emissions. The sinks are natural systems – principally the oceans and

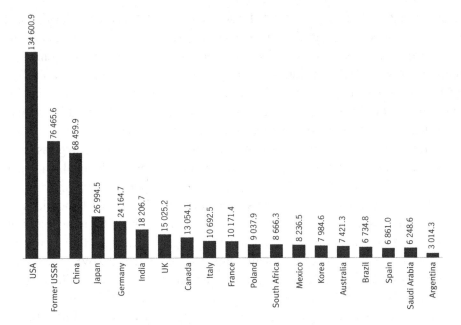

**Fig. 2.** Cumulative emissions of carbon dioxide 1890–2005 in million tonnes. Rich countries are still the major emitters of total carbon dioxide, with just 15% of the world's population they account for 45% of carbon dioxide emissions. (*Source:* CSE, 2008, calculated from the carbon dioxide information of the U.S. Department of Energy)

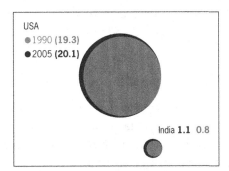

**Fig. 3.** The per-capita increase in annual carbon dioxide emissions between 1990 and 2005 in the USA is equal to three-quarters of India's total per-capita emissions in 2005. Current annual per-capita emissions in the USA are almost 20 times higher than in India. (*Source:* UNDP, 2008)

forests – that absorb emissions. Global warming occurs because emissions exceed the capacity of these sinks to absorb greenhouse gases. The WRI estimated that the world produces 31 billion tonnes of carbon dioxide and 255 million tonnes of methane every year. It then estimated that the Earth's sinks naturally assimilate 17.5 billion tonnes of carbon dioxide and 212 million tonnes of methane annually. On this basis, it calculated the 'net' emissions of each nation, by allocating a share of the sinks to each nation, based on its gross emissions contribution.

CSE in its critique argued that while terrestrial sinks, such as forests and grasslands, may be considered national property, oceanic sinks belong to humankind. They can be regarded as common global property. CSE then apportioned the sinks on the basis of a country's share of the world's population, arguing that each individual in the world has equal entitlement to the global commons. This allocation, based on individual rights to the Earth's natural cleansing capacity, changed the calculation of the nation's responsibility drastically. For instance, under the WRI methodology, the USA contributed 17% of the net emissions of the world, while the CSE methodology calculated that it actually contributed roughly 27.4% of net annual emissions. Similarly, the contribution of China decreased from the WRI estimate of 6.4% of net annual emissions to 0.57%, and India's from 3.9% to just 0.013% of net annual emissions.

This allocation of the Earth's global sinks to each nation, based on population, creates a system of per-capita emissions entitlements, which taken together form the 'permissible' emissions level of each country. This, according to CSE, could form a framework for trading between nations, as countries that exceed their annual quotas of carbon dioxide could trade with other countries that do not use up their 'permissible' emissions. This would create financial incentives for countries to keep their emissions as low as possible and to invest in zero-carbon trajectories.

### Chinese proposal for burden sharing

The Chinese Academy of Social Sciences has also presented its own model for a carbon budget based on equity and sustainability. Interestingly, the Academy says that, while there have been a variety of proposals with different interpretations of the equity principle for burden sharing and emissions entitlement, there is an 'imbalance' in the debate: only three of the 43-odd proposals for equity have come from researchers in the South.

The Chinese proposal is based on two concepts (Pan *et al.*, 2008): first, if we want to ensure that basic needs are fulfilled for all citizens of the world, then there is no space in the world for luxurious and wasteful emissions; second, emissions need to stay within the geophysical limits of the planet. That means, if emissions exceed the Earth's geophysical limits, then human society must reduce its emissions to adjust to what the planet can withstand. The proposal accepts that the global human community has to reduce emissions by 50% by 2050. This level then constitutes the global carbon budget that is to be shared between every individual in the world. The budget is allocated to every individual for meeting basic needs, and adjusted in terms of geographical, climatic and resource endowment. For the period 1900–2050, the total global carbon budget was 2272.5 billion tonnes of carbon dioxide, or 352.5 tonnes of accumulative carbon dioxide emissions per capita or 2.33 tonnes of carbon dioxide per capita per year.

The Chinese proposal includes two transfers – one of the budget and one of financial resources. As the developed countries have already exhausted their full share – until 2050 – the proposal calculates the price of this 'gift' to developed countries by the developing countries. It also transfers emissions budget to developed countries to meet basic needs. The total carbon budget acquired by developed countries would be in excess of the global average level. For instance, the US carbon budget would increase to 7.71 tonnes of carbon dioxide per capita, as compared to 2.33 tonnes per capita in the initial budget allocation. For Non-Annex I countries, the budget would decline from its initial allocation of 80.5% to 58.9%. This basic entitlement scheme would form the basis of the trading scheme. As a result, countries like the USA, Canada and Australia would be required to purchase 70% of their future emissions budgets.

It also proposes a progressive carbon tax that increases the rate of taxation on the basis of the amount of excessive emissions (from limited to moderate to severe). This tax should not be higher than the cost of renewable energy introduction, so that the framework supports the transition to cleaner energy.

## Ad-hoc emissions budgets and entitlements

Another possible approach would be to decide upon future atmospheric concentration limits for various dates on an ad-hoc basis, allowing for some build-up of greenhouse gases in the atmosphere. The targeted atmospheric concentrations could then be translated into a global emissions budget that can be distributed among nations in the form of equal per-capita entitlements. Both the targets and the emissions caps needed to meet them would be subject to periodic scientific review, and therefore per-capita entitlements based on this approach would be subject to review as well. A country that does not use its budget during a particular year could again have the right to trade its unused share. In this case, nations could also simply agree on an ad-hoc per-capita entitlement towards which all countries eventually will converge. This target could be more or less ambitious, but again it would be subject to periodic review, allowing for changes based on new scientific information.

### Entitlements within countries

As much as the world needs to design a system of equity between nations, the nations of the world need to design a system of equity within each nation. It is not the rich in India who emit less than their share of the global quota. It is the poor in India, who do not have access to energy, who provide us the breathing space. India had per-capita carbon dioxide emissions of 1–1.5 tonnes per year based on different estimates in 2005. Yet this figure hides huge disparities. The urban-industrial sector is energy-intensive and wasteful, while the rural subsistence sector is energy-poor and frugal. Currently it is estimated that only 31% of rural households use electricity. Connecting all of India's villages to grid-based electricity will be expensive and difficult. It is here that the option of leapfrogging to off-grid solutions based on renewable energy technologies becomes most economically viable. If India's entitlements were assigned on an equal, per-capita basis, so that the country's richer citizens pay the poor for excess energy use, this would provide both the resources and the incentives for current low-energy users to adopt zero-emissions technologies. In this way, too, a rights-based framework would stimulate a powerful demand for investment in new renewable energy technologies.

Let us be clear. The challenge of climate change is a make-or-break situation for the world. It forces us, perhaps for the very first time in our history, to realize that we live together on one Earth. It tells us that there are limits to carbon-based growth; and more importantly that growth will have to be shared between all. Ultimately, we cannot share a vision for how the world will combat climate change unless we are prepared to share the common atmospheric resources of the world. The big question is whether we will meet this challenge. The most convincing answer is that we have no choice. There is no other way.

### References

Agarwal, A. and Narain, S. (1991). *Global Warming in an Unequal World: A Case of Environmental Colonialism.* New Delhi.
CSE – Centre for Science and Environment (2008). *The Just Framework.* New Delhi.
IEA – International Energy Agency (2007). *Renewables Information 2007.* Paris.
Pan, J., Chen, Y., Wang, W., Li, C. (2008). *Carbon Budget Proposal, Global Emissions under Carbon Budget Constraint on an Individual Basis for an Equitable and Sustainable Post 2012 International Climate Regime.* Beijing.
Solomon, S., Qin, D., Manning, M. *et al.,* eds. (2007). *Climate Change 2007: The Physical Science Basis. Contribution of Working Group I to the Fourth Assessment Report of the Intergovernmental Panel on Climate Change.* Cambridge.
UNDP (2008). *Human Development Report 2007/2008: Fighting Climate Change. Human Solidarity in a Divided World.* New York.

UNFCCC (2007). Policies, measures, past and projected future greenhouse gas emissions. In *Compilation and Synthesis Report on Fifth National Communication* (Addendum 1). Bonn.

WRI – World Resources Institute (1990). *World Resources 1990–1991: A Guide to the Global Environment*. New York.

# Chapter 10

# Carbon justice and forestation – the African perspective

---

## Wangari Maathai

© Martin Rowe

Wangari Muta Maathai was born in Nyeri, Kenya, in 1940, the daughter of farmers in the highlands of Mount Kenya. The first woman in East and Central Africa to earn a doctoral degree, she subsequently became an associate Professor of Veterinary Anatomy in 1977 at the University of Nairobi. In the same year, she founded the Green Belt Movement, a grassroots environmental organization which has assisted women and their families in planting more than 40 million trees across Kenya to protect the environment and promote sustainable livelihoods. Since that time, Wangari Maathai has campaigned tirelessly for democracy, human rights and environmental conservation. She played a key role in the campaign to cancel debt in Africa, and has fought for the protection of public forests. In 2004, Wangari Maathai was awarded the Nobel Peace Prize, recognizing that for peace to be maintained there needs to be sustainable and equitable distribution of resources.

*Note:* This chapter is a commentary on chapter 9.

Climate stabilization will require that developing nations adopt a carbon-neutral energy system, and have robust adaptation plans. However, many developing nations today lack the financial resources to embrace climate-friendly technologies and protect their people from the impacts of climate change. It should therefore stand to reason that developed countries, which bear the greatest responsibility for past greenhouse gas emissions, must be the first to take action on climate change, and support the Global South in this process. As described by Sunita Narain, it is indeed the responsibility of industrialized countries to help developing countries start the transition towards green technologies. At the same time, all developing regions, including Africa, need to focus on the options that are available to them right now.

Although Africa has so far contributed little to global warming, as a region it will be one of the hardest hit by climate change. Many parts of Africa are already seeing the effects of climate change that science describes. The ice and snow on Mount Kilimanjaro and Mount Kenya are melting rapidly. Many of the rivers that flow from these mountains have either run dry or the volume of water has been greatly reduced. Droughts are prolonged and rains are coming at the wrong times. Poor land use practices are contributing to the expansion of African deserts – such as the Sahara in the north and the Kalahari in the south – as well as to forest and land degradation across the whole continent. As most Africans rely on the primary resources of their environment for their livelihoods (soil and land to grow food crops, water from rivers for domestic use, and forests for fuel and fodder), these changes greatly affect the livelihoods of the African people.

According to the most recent assessment report of the Intergovernmental Panel on Climate Change, deforestation and forest degradation account for up to 20% of global greenhouse gas emissions. The loss of healthy, stable forests therefore represents a significant factor in anthropogenic global warming. As in many parts of the world, deforestation is an issue of major concern in Africa. In Kenya, for example, the proportion of national territory covered by forest has been reduced from an original cover of about 30% to less than 2%. The UN recommends Kenya have at least 10% of its land under forestry to deliver essential ecosystem services such as water and climate control. Reforestation programmes, combined with the protection of standing forests, riverine systems and wetlands, are one of the many ways in which Africa can help face the huge challenge of climate change. By planting an appropriate number of trees, and protecting those that are already there, developing countries can help nature to regulate global temperatures.

It was in this spirit that the United Nations Environment Programme (UNEP),[1]

---

[1] http://www.unep.org

together with the organization which I founded, the Green Belt Movement,[2] and several other partners worldwide, including the World Agroforestry Centre (ICRAF), launched the Billion Tree Campaign.[3] As well as encouraging the planting of trees and taking action particularly at the individual level, this project, which has received a tremendous response across the world, aims to educate people about the very serious environmental risks humanity is facing.

In many developing countries we have found that environmental concerns are sidelined by other seemingly more urgent issues. However, we cannot survive without clean drinking water, food, and clean air. Environmental concerns are not a luxurious indulgence in Africa. When rivers dry up and soil erosion takes place, the land loses its fertility and the people who rely on the land for food and fuel lose their source of livelihood. These are some of the issues that governments in developing countries should stress in order to raise awareness among their own people, to highlight the serious risks the planet is facing, and to mobilize participation to tackle these challenges.

Forests play a major role as carbon sinks. We all have a moral duty to assist people and governments to rehabilitate and protect standing trees and vegetation. We need incentives in forestry to ensure indigenous forests are restored to promote the essential ecosystem services they deliver. Of course, financial mechanisms – both national and international – rely on principles that ensure accountability and responsible utilization of resources. However, excessive bureaucracy may block funds from reaching those who need them most. This especially concerns local communities and indigenous peoples who will need to both adapt to and mitigate climate change at grassroots level. Lack of access to financial resources and information constitute considerable barriers for pursuing the issues of justice, rights to sustainable development, and equity.

In addition to intensified reforestation efforts, existing forests must be protected. The Congo forest ecosystem, along with the Amazon and the forests of Southeast Asia can make an enormous contribution to sequestering carbon. It is therefore important to support countries that are willing to preserve their forests and which do not encourage logging. Initiatives such as the Congo Basin Forest Fund,[4] of which I am co-chair, are extremely valuable in this effort. The Fund aims to develop the capacity of the people and institutions of this region to manage their own forests, help local communities find livelihoods that are consistent with forest conservation, and reduce deforestation. Hopefully, such initiatives will contribute to a collective partnership with many of the countries that want to support forest-rich regions to retain these ecosystems, ensuring that they continue to contribute not

[2] http://www.greenbeltmovement.org
[3] http://www.unep.org/billiontreecampaign
[4] http://www.cbf-fund.org

only to carbon sequestration but also to the protection of biodiversity, the water cycle, and the global climate.

Another decisive factor in climate protection is climate justice: A large number of countries will be negatively impacted by climate change even though they have contributed little or nothing to the problem. Accelerating climate change is leaving little room for developing countries to increase their emissions as part of their struggle to overcome endemic poverty. These issues need to be addressed so that the discussion does not revolve only around the question of who is responsible but also includes what is a fair and just response to climate change.

Sunita Narain has described various mechanisms that can help to finance zero-carbon technologies in developing countries, based on a just allocation of emission allowances among all people of the world. The transfer of financial resources to developing countries, helping us to leapfrog to zero-carbon technologies will be needed. However, any initiatives developed will require careful checks and balances to ensure energy consumption is capped in developed countries. Extensive reduction in carbon emissions needs to be achieved in developed countries before any carbon burden is shared through a fair and equitable mechanism with developing countries. Such a mechanism will necessarily include significant financial transfers from developed to developing countries, and if this is structured correctly, new funds for protecting forests in the developing world could be generated. However many approaches that have been used to prevent deforestation have failed in the past. Innovative forest protection schemes – an emergency fund for forests and an initiative to reduce emissions from forest degradation and deforestation (REDD) – offer promising new opportunities to conserve and restore our forests.

Countries in Africa could certainly benefit from such schemes – primarily, because they support initiatives that do not require extensive funds or heavy technology, but rather the mobilization of citizens to do the work, such as planting trees. The experience gained in the Green Belt Movement during the past thirty years shows that it is possible to mobilize millions of individual citizens in every country to plant trees, prevent soil loss, harvest rain water, and practice less destructive forms of agriculture. It is important to educate citizens about the need to protect trees, especially indigenous mountain forests, which are sources of water and biological diversity. Through the Green Belt Movement we have learned that when local communities understand the link between trees and their own livelihoods they are more likely to protect them.

Many simple actions can be taken all over the world to change our consumption patterns now. Individuals can choose to reduce, reuse, and recycle wherever they live. Many people are opting for hybrid cars, public transportation, and alternative sources of energy. The Green Belt Movement in Kenya is encouraging people to plant trees to create a sustainable future. These trees serve both as carbon sinks and biodiversity reservoirs, thereby also making people aware of the linkages between poverty and the environment. While political leadership is important, it is also essential to mobilize citizens. In the end, it will be citizens who move their governments to more tangible commitments. We know what needs to be done to address climate change, and now is the time to do it.

# Chapter 11

# Carbon offsets, the CDM, and sustainable development

---

## Diana M. Liverman

Diana M. Liverman is Professor of Environmental Science at Oxford University, where she directs the Environmental Change Institute. She is also a faculty member at the University of Arizona. She has been a programme leader for research on post-2012 climate policy at the British Tyndall Centre for Climate Change, and chairs the science advisory committee for the international programme Global Environmental Change and Food Systems of the ESSP. Her research has focused on the human dimensions of global environmental change, especially the vulnerability of food systems, climate and development, and the role of carbon offsets. She is an IPCC contributing author and is a member of the US National Academy of Sciences Committee on America's Climate Choices.

Carbon offsets comprise one of the international climate regime's core strategies for reducing greenhouse gas emissions in the developing world. Carbon offsetting involves purchasing 'credits' from projects that reduce greenhouse gas emissions. By investing in such projects, emitters can compensate for emissions that an individual, organization or country is unwilling or unable to reduce domestically. Offset projects include energy efficiency, renewable energy and forestry, and include the full range of greenhouse gases through projects such as capture and destruction of industrial gases and methane from landfills.[1] They are managed under the flexible carbon trading options of the Kyoto Protocol known as the Clean Development Mechanism (CDM) and through an emerging voluntary market. In many cases the projects are identified and developed by private sector companies who prepare a project development document (PDD) to begin the process of demonstrating potential greenhouse gas reductions, obtaining project financing, and, in the case of the CDM, getting formal approval from the international CDM executive board. The growing potential of offsets to emission reductions is indicated by the projection that the CDM will produce more than 600 million tonnes (Mt) of carbon credits (in terms of carbon dioxide equivalents) each year until 2012 (UNFCCC, 2008, accessed on 22 November 2008) – compared, for example, to overall annual carbon dioxide emissions from fossil fuels of about 8000 million tonnes per year (Raupach *et al.,* 2007).

Several other papers and commentaries in this volume refer to bargains between North and South for emission reduction and forest projects. The use of offsets in reducing emissions is controversial, but is an important component of the UN climate negotiations that in Bali in 2007 set out a 'roadmap' for a major new agreement in 2009 in Copenhagen that would include a reformed CDM. One of the main arguments in favour of offsets is that they can contribute to sustainable development in developing and transitional economies through promoting a variety of direct and indirect benefits that include cheaper and healthier energy, forest and biodiversity protection, and income and jobs for local people. But for offsets to contribute to greenhouse gas reduction they must fund projects that 1) would not otherwise have taken place and 2) must reduce emissions compared to what would have happened otherwise (called additionality).

What are the problems and possibilities of carbon offsets in reducing the risks of dangerous climate change and contributing to sustainable development? In the remainder of this paper I discuss the scientific, ethical, and economic debates over offsets and argue that their sustainable development benefits depend critically on

---

[1] The CDM sets out to reduce the 'basket' of six greenhouse gases included in the Kyoto Protocol (carbon dioxide, methane, nitrous oxide, sulphur hexafluoride, hydrofluorocarbons and perfluorocarbons). For comparative purposes these are often converted, using a weighting related to their global warming potential, to carbon dioxide equivalent.

the nature of the offset project, especially the type of technology and the governance mechanisms. I conclude by looking at the potential role of offsets within the Bali roadmap to Copenhagen, and at the current proposals to regulate and reform the offset market.

## The origins of offsetting

Most of the initial carbon offset projects were forest conservation and reforestation projects designed to compensate for corporate carbon emissions by sequestering carbon dioxide in tropical forests, and included projects in countries such as Bolivia, Ecuador and Guatemala. Environmental non-governmental organizations were often involved in these early voluntary projects (dating from around 1990) which mirrored other initiatives to put a price on nature and its environmental services in the international market place. Offsets were brought into the UNFCCC framework in 1995 through a pilot programme (Activities Implemented Jointly – AIJ) which was supposed to allow for emission reduction projects in other countries to generate carbon credits. The United States and Brazil were influential in the discussions to include the developing world in the international climate regime through offsets and carbon trading, with Brazil proposing a clean development fund and the USA seeing offsets as a cheaper way to achieve reductions and foster developing world participation.

The 1997 Kyoto Protocol formalized offsetting within the set of flexible mechanisms for achieving emission reductions. The Clean Development Mechanism (CDM) has been called the Kyoto 'surprise' in that it provides a benefit to the developing world through allowing for emission reduction projects in the South. Such projects would produce certified emission reductions (CERs) that could be purchased and used to meet emission commitments under the protocol. The CDM was proposed as a cost-effective way for the North to achieve emission reductions through sustainable development in the South.

The potential for emission reductions in the developing countries was estimated by the IPCC in 2007, and shows potential savings of many billions tonnes of carbon dioxide equivalents even at relatively low carbon prices across activities that include energy supply shifts to low carbon alternatives, energy efficiency in buildings, and forestry (see Fig. 1). The developing world (light grey in the graph) has large potential for carbon reductions including in buildings, energy supply, and forestry, and at higher carbon prices in industry and agriculture. Offsets are one way to achieve these reductions without the developing world taking on their own binding commitments.

In anticipation of the first Kyoto commitment period (2008–12), the World Bank catalysed the offset market through a Prototype Carbon Fund which has now grown

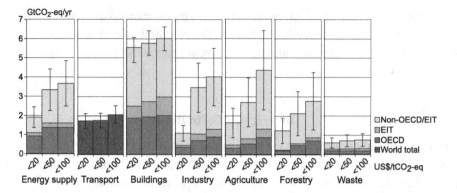

**Fig. 1.** Estimated potential for carbon reductions (in terms of carbon dioxide equivalents) per year as a function of carbon price in 2030. OECD represents the industrialized countries, EIT represents Economies in Transition – many in eastern Europe, and Non-OECD/EIT is the developing world. (*Source:* IPCC, 2007, p. 11)

into an investment worth almost USD two billion in ten different carbon funds supporting a wide variety of projects. A supplementary voluntary offset market also grew rapidly during the mid 1990s – mainly since 2005 with companies such as Climate Care and Future Forests – and developed carbon reduction projects which could provide carbon offset credits to consumers and businesses. As of 2007 about a billion tonnes per year of emission reductions (measured in carbon dioxide equivalents) had been contracted through the CDM and the voluntary market with carbon credits reaching a market value of almost USD 14 billion (World Bank, 2008, p. 1).

Although forestry dominated some of the early offset projects, they have now expanded to encompass a wide range of technologies and greenhouse gases. A large volume of credits has been generated by large-scale projects to capture and destroy HFC (hydrofluorocarbon) gases at refrigeration and other industrial plants and by projects that prevent emissions of methane from landfill sites (see Table 1). HFC and methane projects are popular with investors because the high global warming potential of these gases generates many more credits per unit of reduction than carbon dioxide (even though carbon dioxide has a much longer lifetime). Energy efficiency projects, ranging from manufacturing processes to light bulbs, and renewable energy projects including small-scale hydropower, biomass, solar and wind power, are also producing carbon credits in dozens of countries.

There are now several distinct markets for carbon offsets including the CDM and the voluntary sector. The CDM supplies credits for compliance with the Kyoto Protocol and with the European Union Emissions Trading Scheme (EUETS). The voluntary sector can operate in the United States (where the CDM is not yet used because of US failure to sign the Kyoto Protocol), serves individuals and firms, and often pilots methods and technologies not yet approved by the CDM.

**Table 1.** (*Source:* UNEP, 2008, accessed 1 Nov 2008)

| CDM Project Type | Number of projects | Total emission reduction credits per year (million) |
|---|---|---|
| HFC, PFC and $N_2O$ | 95 | 132 |
| Renewables (hydro, biomass, wind, solar, geothermal) | 2603 | 215 |
| CH4 (landfill, mines) and cement | 657 | 101 |
| Supply side energy efficiency | 425 | 70 |
| Fuel switching (e.g. coal to gas) | 135 | 44 |
| Demand-side energy efficiency | 194 | 8 |
| Forests | 34 | 2 |
| Transport | 8 | 1 |
| Total | 4151 | 572 |

## The arguments for offsets

Proposals that offsets should be a significant component of international, national, corporate and even individual responses to climate change are based on several key arguments:

1. The atmosphere is uniformly mixed and therefore emission reductions can occur anywhere, reduce the overall concentration of greenhouse gases and thus the risks of dangerous climate change. We can sequester and reduce greenhouse gases wherever it is easiest so long as there is a measurable impact on the carbon cycle.
2. It is often less costly to reduce emissions where energy use is less efficient and land, labour and other costs are cheaper (such as in many developing countries).
3. Including offset credits into emission reduction agreements can make it easier and cheaper for countries to join these agreements and to accept potentially more ambitious targets. In this way, the Kyoto agreement may only have been possible because it included flexible mechanisms and offsets.
4. Offsets provide multiple side benefits for sustainable development, especially when projects provide cheaper and healthier energy, jobs and incomes, and/or foster ecosystem conservation and restoration.
5. Emission reduction projects can initiate major shifts in attitudes and technologies in the developing world that set the stage for socio-technical and political transitions to low carbon futures and participation in international agreements.

6.  Carbon offsets are a way for individuals and firms to compensate for emissions that they are unable or unwilling to reduce. It is better than doing nothing and creates an internal price for carbon that can drive changes in behaviour and technology.
7.  Offsets are often small, locally based projects. While possibly inefficient from an economic standpoint, it is possible that the large quantity of projects will encourage a greater degree of experimentation and innovation than more top-down or sectoral approaches would engender.

These arguments have convinced a wide range of international institutions, countries, corporations, NGOs and individuals to include the CDM and voluntary offsets as part of their carbon management strategy. In addition to its role in meeting Kyoto emission reduction commitments the CDM is incorporated into EU climate policy, and into that of countries such as the United Kingdom and Japan. In addition, voluntary offsets are used or supported by organizations that include major banks (e.g. HSBC), airlines, and conservation groups (e.g. WWF).

## The arguments against offsets

A backlash against offsets has been led by activists and the media who argue that offsetting is unethical and ineffective (Smith, 2007) but offsets are also generating considerable discussion in the scientific and development community (Boyd *et al.,* 2007, Wara, 2008). The arguments against offsetting include the following:

1.  It is unethical to buy your way out of your carbon guilt by purchasing low-cost offsets to compensate for a high-consumption lifestyle (Smith, 2007). Offsets divert attention from the need to reduce consumption and to eliminate emissions from non-essential activities such as flying. Carbon trading has limited the overall potential of the Kyoto and other agreements because it reduced the need for domestic reductions.
2.  Some reductions (such as of HFCs) would be cheaper or more effective if achieved through direct payments or bans rather than through carbon finance, which may actually result in windfall profits that are many times the actual cost of reduction.
3.  Many offset projects may not provide verifiable emission reductions because of questions about
    *   measurement of GHG emissions and conversion to carbon equivalents
    *   the legitimacy and manipulation of baselines (emissions before the project started) and of projections of business as usual (it is difficult to establish a counterfactual scenario)

- proof that carbon finance was key to the project so that emission reductions are truly 'additional'. Some claim that most projects were already going to happen anyway
- permanence of reductions and risks of project failure, especially for voluntary forest offsets where some reforestation projects have failed
- the timing of delivery of reductions, especially in the voluntary sector where some object to selling offsets as forward contracts and not as reductions already achieved
- potential leakage as emissions are displaced outside the project boundary
- rebound as higher incomes or energy savings lead to other greenhouse-gas-emitting activities (which may be true of any efficiency savings, not just offsets).

4. Transaction costs associated with project development and verification are too high, especially for CDM projects (Michaelowa *et al.*, 2003).
5. The sustainable development benefits of offsets are often less than claimed because of the use of technologies that do not provide benefits to local people (e.g. capture of industrial gases like HFCs), the lack of participation in decisions, the unequal distribution of project benefits, lack of attention to customary land rights, the diversion of resources such as water to the projects and negative impacts on biodiversity (e.g. from large dams or forest monocultures).
6. The voluntary carbon market may include unscrupulous companies that resell the same credit several times ('double-counting'), confuse the consumer, and do not adhere to criteria for additionality and sustainable development.
7. Some of the most important potential offsets in terms of emission reductions and sustainable development are currently excluded from the CDM and ignored by the voluntary market. For example, the CDM allows credits for reforestation and afforestation but does not permit credits for reducing or avoiding deforestation (which may be responsible for 20% or more of carbon dioxide emissions). While some consider it a problem that technologies such as nuclear power and several geoengineering options are not eligible as offsets, others feel that these technologies do not meet criteria for sustainable development.
8. Project-based offsets are inefficient and too small in scale. It would be much more effective to fund emission reductions on a sectoral or programmatic/policy level, e.g. for the cement or electricity sector in a country or for a national forest or energy policy.

Many of these criticisms are based on case studies rather than a structured assessment of the risks and benefits of offset projects. There is a clear need for science and social science to undertake careful analysis of both the CDM and the voluntary market as well as for improved governance mechanisms that include self regulation and government oversight. But before we turn to potential reforms and improvements, it is helpful to focus in more detail on some sustainable development aspects of offsets.

## Sustainable development and the CDM

The Kyoto Protocol required the CDM to meet objectives of sustainable development but this has not been clearly defined and is currently implemented as a simple certification by the host country. This certification states that projects meet sustainability objectives. Countries vary in how strictly they define and implement the sustainability criteria and face a contradiction between the desire for investment and broader sustainable development objectives (Olsen and Fenhann, 2008). While comprehensive attention to sustainable development might examine the environmental, economic, and social benefits and costs of projects from the national to the local scale, it appears that for some governments and project developers, job creation or energy savings only are seen as enough to justify a sustainability check off (Brown and Corbera, 2004; Olsen, 2007).

The potential of the CDM to drive low-carbon energy transitions, to provide sustainable development benefits to the poor and to protect ecosystems has been less than promised, partly because of the inclusion of gases with high greenhouse-gas potential (HFCs, $N_2O$) which can be easily captured in large-scale projects at industrial plants (refrigerants, adipic acid, Teflon). These projects have dominated the carbon credits within the CDM to date, and until 2007 less than half of all carbon credits were associated with renewables and energy efficiency (although these project types are now expanding). It has also been argued that the large-scale industrial gas-capture projects receive far more money than needed to eliminate the targeted greenhouse gases. Wara (2008), for example, argues that to abate all HFC emission in the developing world would only cost USD 31 million per year whereas the CDM could pay up to 20 times that amount for eliminating these gases. He argues that companies earned three times more from reducing emissions from their operations in the CDM than from selling the products that are produced at the plants. He did also note a positive effect of this – both in Europe and North America HFC emissions have been reduced voluntarily or through regulation. However, there are also indications of somewhat perverse incentives to maintain production levels in installations where $N_2O$ emissions are captured as a CDM project whereas production elsewhere is scaled down in face of the current economic downturn.

Capturing industrial gases generates much fewer benefits to local people and ecosystems than energy efficiency, renewable energy, and forest projects. Unfortunately there are very few studies which assess these benefits in a consistent, comparative, long-term and carefully monitored fashion. Case studies suggest that the benefits to communities from projects involving wind or solar power or improved wood-stoves can include job creation, reduced indoor air pollution, lower energy costs, and direct carbon finance payments, while reforestation projects can protect watersheds and biodiversity.[2] A recent analysis of more than 700 project design documents found that the most likely overall benefits promised for CDM projects are jobs (66%), economic growth (46%), improved air quality (42%), cheaper energy (32%) and conservation (13%) (Olsen and Fenhann, 2008). This study also looked at different technologies and found that HFC and $N_2O$ projects generate the fewest sustainability benefits and projects on household efficiency, solar, hydro, and wind power, and cement the most. Both cement (switch from limestone to use of waste fly ash) and landfill/livestock methane-capture or fuel projects have environmental benefits in reduced pollution, cheaper energy and improved health.

One of the problems in linking offsets to sustainable development in very poor communities and countries is that emissions may be so low that savings are harder to achieve, such that projects on renewables are not alternatives to carbon-emitting activities but the first step towards greater (if low carbon) energy use. The cost of developing a project can also be prohibitive, both in terms of transaction costs and in finding investors willing to take risks with certain technologies and in weakly governed countries.

One added complication arises from factors external to the nature of offsets and carbon markets themselves. Many potential projects with very long-term emission reduction benefits and significant sustainable development benefits have not materialized under the CDM to date because of the prevailing relatively low carbon prices. This concerns projects such as renewable energy and small energy-efficiency projects. Low carbon prices are primarily due to the very modest emission-reduction targets agreed under the Kyoto Protocol which generate relatively low demand, and hence prices, for CDM offsets overall. This in fact has skewed investment into CDM projects towards cheap and technologically easy interventions such as industrial gas capture. One extremely important constraint is that the lack of a post-2012 international agreement, combined with a political debate in the EU, UK and USA about the future role of the CDM, has created considerable investor uncertainty and reluctance to make long-term investments.

## Sustainable development and the voluntary offset market

The voluntary offset market has tended to focus on projects with greater apparent sustainable development benefits because both individual and institutional purchasers

---

[2] Statement based on preliminary results from a number of student research projects in the Environmental Change Institute (www.eci.ox.ac.uk) and Corbera (2005).

see added value in offsets that help poor people and protect ecosystems (Lovell *et al.*, 2008). This is one of the reasons why many of the early voluntary offset projects focused on forests despite the technical challenges of securing forest carbon. Voluntary offset companies often highlight the sustainable development benefits of their carbon projects such as reduced indoor air pollution, lower energy costs or conservation values. But sustainable development goals have also been a source of criticism of offsets where activist and media investigations have suggested that forests have degraded after crediting, local people have not benefited from projects, or that profits are being made through neo-colonial practices that take advantage of low land and labour costs (Smith, 2007; Ma'anit, 2006).

## Responding to the challenge: reforming the CDM and setting voluntary standards

Improving the effectiveness and quality of both CDM and voluntary offsets has become a priority for both private-sector interests and governments in preparation for the 2009 Copenhagen climate negotiations and serious greenhouse gas reductions. Several standards have been proposed to ensure the quality of offsets, with stricter rules for both carbon and sustainable development in terms of proven additionality, appropriate technologies, and local participation and approval of projects (Kollmuss *et al.*, 2008).

The CDM has become a major focus of negotiations with proposals that include scaling up to sectoral, policy or programmatic CDM, providing incentives or extra credits for projects in poorer regions, streamlining and simplifying project approval, expanding the range of approved methods and technologies, discounting credits to account for risk of underperformance of projects and ensure real atmospheric benefits, and shifting to a model where industrial countries take on obligations to buy and retire CDM credits directly rather than as offsets (Boyd *et al.*, 2007; Cosbey *et al.*, 2005; Ellis *et al.*, 2004; Sterk and Wittneben, 2006). Sectoral and policy CDM has the potential to transform concentrated economic sectors representing a large share of emissions – such as cement, iron and steel, or electricity – to lower carbon futures using CDM-type carbon finance flows to countries rather than projects. Mechanisms might include large-scale sectoral investments linked to no-lose targets, negotiated binding sectoral intensity targets, commitments to use best technology and practices, or implementation of particular policies and measures (Höhne *et al.*, 2008; Schmidt *et al.*, 2008).

The other important new proposal on the roadmap from Bali to Copenhagen is to provide carbon credits to countries for avoided deforestation or reducing emissions from deforestation and degradation (REDD). Deforestation contributes up to 20% of carbon dioxide emissions yet forest protection was excluded from the CDM which only provides credit for new forests or reforestation (Ebeling and Yasué, 2008; Miles and Kapos, 2008; Schlamadinger *et al.,* 2007). Because REDD credits may parallel the CDM in that industrial countries may purchase them to meet emission reduction commitments, they could dramatically shift the offset market. Unlike the CDM which is project-based, REDD is likely to be negotiated at the country level, and may operate separately from the international carbon market because of concerns about cheap forest credits swamping markets and reducing carbon prices or discouraging domestic emission reductions. Several conservation NGOs are supportive of REDD because of the potential benefits to biodiversity and because REDD would allow exactly those countries to participate in carbon markets which have so far been largely excluded because they are poor and lack an industrial base to reduce emissions. However, there are certain sustainable development concerns about the participation of indigenous peoples, property rights and land tenure, about who will actually receive the funds, how benefits will be distributed to and at the local level, about the reduction of forest values solely to carbon, and about how countries will choose to enforce forest protection. These are added to technical concerns about the measurement of forest carbon and baselines, and the risk that climate change could reduce forest carbon benefits.

What is the future of carbon offsets in terms of sustainable development?

1. It must be noted that the CDM can be viewed as an interim mechanism pending the establishment of a broader or universal cap on carbon emissions. Offsets may only make sense up to the point where the cost of buying emission credits rises to a level where it would be cheaper to reduce carbon domestically.
2. The viability of both the CDM and voluntary offsets depends on assurances of additional, permanent and verifiable emission reductions, hopefully at a scale that produces rapid reductions, in order to address criticisms about the atmospheric benefit of offsets.
3. The sustainable-development value of offsets can be enhanced through a variety of reforms and incentives, including standards for sustainable development (e.g. the Gold Standard), although some new standards (such as the Voluntary Carbon Standard http://www.v-c-s.org) focus only on carbon and do not address sustainable development because of its complexity (see also Kollmuss *et al.,* 2008).
4. Finally, there is an urgent need for carefully designed empirical studies of the sustainable development benefits and risks of offsets in order to resolve some of the urgent questions about the value of offsets and the design of new mechanisms and agreements.

*Acknowledgements*: I would like to thank Emily Boyd, Harriet Bulkeley, Adam Bumpus, Esteve Corbera, John Cole, Chris Ellerman, Johannes Ebeling, Nate Hultman, Heather Lovell, Phil Mann, Mike Mason, and Timmons Roberts for their contribution to my understanding of offsets and sustainable development and comments on this chapter.

# References

Boyd, E., Hultman, N., Roberts, T. *et al.* (2007). *The Clean Development Mechanism: An Assessment of Current Practice and Future Approaches for Policy.* Tyndall working paper 114. Norwich. Available at http://www.tyndall.ac.uk/publications/working_papers/twp114.pdf.

Brown, K. and Corbera, E. (2003). Exploring equity and sustainable development in the new carbon economy. *Climate Policy,* **3**(Supplement 1), S41–56.

Corbera, E., (2005). Interrogating development in carbon forestry activities: a case study from Mexico. Unpublished Ph.D. thesis, University of East Anglia.

Cosbey, A., Parry, J., Browne, J. *et al.* (2005). *Realizing the Development Dividend: Making the CDM Work for Developing Countries (Phase I Report).* Report of the International Institute for Sustainable Development IISD, Canada. Available at http://www.iisd.org/publications/pub.aspx?id=694.

Ebeling, J. and Yasué, M. (2008). Generating carbon finance through avoided deforestation and its potential to create climatic, conservation and human development benefits. *Philosophical Transactions of the Royal Society B,* **363**(1498), 1917–24.

Ellis, J., Corfee-Morlot, J. and Winkler, H. (2004). *Taking Stock of Progress Under the Clean Development Mechanism (CDM).* Paris.

Höhne, N., Worrell, E., Ellerman, C., Vieweg, M. and Hagemann, M. (2008). *Sectoral Approach and Development.* Utrecht.

IPCC (2007). *Climate change 2007: Synthesis report. Summary for Policymakers.* Geneva. Available at www.ipcc.ch/pdf/assessment-report/ar4/syr/ar4_syr_spm.pdf.

Kollmuss, A., Zink, H. and Polycarp, C. (2008). *Making Sense of the Voluntary Carbon Market: A Comparison of Carbon Offset Standards.* Frankfurt/Main.

Lovell, H., Bulkeley, H. and Liverman, D.M. (2009). Carbon offsetting: sustaining consumption. *Environment and Planning A. Published online doi:10.1068/a40345.*

Ma'anit, A. (2006). $CO_2$nned: carbon offsets stripped bare. *New Internationalist,* **391**.

Michaelowa, A., Stronzik, M., Eckermann, F. and Hunt, A. (2003). Transaction costs of the Kyoto mechanisms. *Climate Policy,* **3**(3), 261–78.

Miles, L. and Kapos, V. (2008). Reducing greenhouse gas emissions from deforestation and forest degradation: global land-use implications. *Science,* **320**(5882), 1454–5.

Olsen, K.H. (2007). The clean development mechanism's contribution to sustainable development: a review of the literature. *Climatic Change,* **84**(1), 59–73.

Olsen, K.H. and Fenhann, J. (2008). Sustainable development benefits of clean development mechanism projects: a new methodology for sustainability assessment based on text analysis of the project design documents submitted for validation. *Energy Policy,* **36**(8), 2819–30.

Raupach, M.R., Marland, G., Ciais, P. *et al.* (2007). Global and regional drivers of accelerating $CO_2$ emissions. *Proceedings of the National Academy of Sciences,* **104**(24), 10288–93. Available at http://www.pnas.org/content/104/24/10288.abstract.

Schlamadinger, B., Johns, T., Ciccarese, L. *et al.* (2007). Options for including land use in a climate agreement post-2012: improving the Kyoto protocol approach. *Environmental Science and Policy,* **10**(4), 295–305.

Schmidt, J., Helme, N., Lee, J. and Houdashelt, M. (2008). Sector-based approach to the post-2012 climate change policy architecture. *Climate Policy,* **8**(5), 494–515.

Smith, K. (2007). *The Carbon Neutral Myth-Offset: Indulgences for your Climate Sins.* Amsterdam. Available at http://www.carbontradewatch.org/pubs/carbon_neutral_myth.pdf.

Sterk, W. and Wittneben, B. (2006). Enhancing the Clean Development Mechanism through sectoral approaches: definitions, applications and ways forward. *International Environmental Agreements: Politics, Law and Economics,* **6**(3), 271–87.

UNEP – UNEP RISOE Centre on Energy, Climate and Sustainable Development (2008). Risoe CDM/JI Pipeline Database. http://cdmpipeline.org/ accessed 2 December, 2008.

UNFCCC (2008). CDM statistics. Available at http://cdm.unfccc.int/Statistics/index.html.

Wara, M. (2008). Measuring the Clean Development Mechanism's performance and potential. *UCLA Law Review,* **55**(6), 1759–803.

World Bank (2008). *State of the Carbon Market.* Washington.

## Further Reading

Bumpus, A. G. and Liverman, D. M. (2008). Accumulation by decarbonization and the governance of carbon offsets. *Economic Geography,* **84**(2), 127–55.

ENDS – Environmental Data Services (2008). *The Essential Companion to Voluntary Carbon Markets.* London. Available at http://www.endscarbonoffsets.com.

Field, C. and Raupach, M., eds. (2004). *The Global Carbon Cycle: Integrating Humans, Climate, and the Natural World.* Washington, D. C.

Hultman, N. E., Boyd, E., Timmons, J. *et al.* (2009). How can the Clean Development Mechanism better contribute to sustainable development? *Ambio,* **38**(2), 120–22.

Muller, A. (2007). How to make the clean development mechanism sustainable: the potential of rent extraction. *Energy Policy,* **35**(6), 3203–12.

Yamin, F. (2005). *Climate Change And Carbon Markets: a Handbook of Emission Reduction Mechanisms.* London.

# Chapter 12

# Insights into the climate challenge

———

## Rajendra Pachauri

Rajendra Pachauri was born in Nainital, India, in 1940. He studied industrial engineering at North Carolina State University in Raleigh, USA, where he also obtained a PhD in industrial engineering and a PhD in economics. In 1982 he joined the Energy and Resources Institute (TERI), which conducts research in the fields of energy, environment, forestry, biotechnology, and the conservation of natural resources, providing professional support to governments, institutions, and corporate organizations worldwide. In 2002, Pachauri was elected Chairman of the Intergovernmental Panel on Climate Change (IPCC). Established by the World Meteorological Organization and the United Nations Environment Programme in 1988, the IPCC assesses scientific, technical and socioeconomic information relevant to the understanding of climate change, its potential impacts, and options for adaptation and mitigation.

Climate change has emerged as one of the most contentious and critical issues of our time, with far-reaching implications for the way the human race will live and develop, especially over this century. While skeptics continue to doubt the human contribution to the phenomenon, the majority of the scientific community has come to a clear conclusion regarding the reality of human-induced climate change. These studies, many of which are included in the reports of the Intergovernmental Panel on Climate Change (IPCC), clearly show that human activities are the main reason for altered climate patterns in the last century.

The findings of the IPCC's Fourth Assessment Report, released in 2007, indicate that the warming of the climate system is unequivocal (Solomon *et al.,* 2007). They also reveal several disturbing trends regarding levels of atmospheric greenhouse gases since the Industrial Revolution, and changes in climate over the same period. According to the report, continued emissions would lead to further warming of 1.1 °C to 6.4 °C over the twenty-first century, depending on different scenarios of economic growth, population projections, technological change, energy demand, structure of energy use, and other factors (see Rahmstorf *et al.,* this volume).

The impacts of climate change are widespread and complex, and are projected to vary according to the timing and magnitude of change, as well as according to adaptive capacity. It is clear, however, that climate change impacts have serious implications for the livelihoods of billions of people worldwide, and pose one of the greatest challenges to development in our time.

The ecological footprint, a sustainability indicator measuring the pressure exerted by human activity on the Earth's systems, indicates increasingly unsustainable global consumption trends (see Leape and Humphrey, this volume). The ecological footprint is an estimate of the amount of biologically productive land and sea area are needed to regenerate (if possible) the resources that a human consumes, and to absorb and neutralize the corresponding waste, given prevailing technology. According to the Global Footprint Network,[1] it currently takes one year and four months to regenerate the resources consumed globally in a year. The 'carbon footprint'[2] is by far the largest component of the overall ecological footprint, comprising half of the total. Climate change is clearly one of the most pressing sustainability challenges of the century, and one that urgently needs to be addressed as part of mainstream development policy.

---

[1] The *Global Footprint Network* is an international think tank working to advance sustainability.
[2] The amount of forests and other vegetated areas to sequester carbon dioxide emissions.

## Impacts of climate change on developing countries

The IPCC's Fourth Assessment Report projected that climate change will have a disproportionately high impact on developing countries, thereby exacerbating inequalities in health status and in access to adequate food, clean water, and other resources. In all countries, certain sections of the population, such as the elderly and poor, tend to be at a higher risk, thus also exacerbating inequalities within nations.

While industrialized countries bear the greatest responsibility for the changing climate, developing countries are already bearing the major burden of its effects. Between 1990 and 2005, nearly 3.5 billion people were affected by natural disasters, of which approximately 90% live in developing countries (LaFleur *et al.,* 2008). As if this inequality were not enough, developing countries also have far fewer resources to adapt to climate change than developed economies. Factors influencing vulnerability to climate change include dependence of communities on climate-sensitive resources, vitality of local communities, the integrity of key infrastructures, level of current preparedness and planning, the sophistication of public healthcare systems, and existing exposure to conflict.

The impacts of climate change on developing economies are projected to be severe in terms of several critical factors. These include not only access to key resources such as water, but also factors related to health and vulnerability to a rise in sea level. Climate change also has serious implications for the Millennium Development Goals, particularly those related to environmental sustainability and poverty reduction. Some of the projected impacts include:

**Access to food:** In the Sahel region of Africa, warmer and drier conditions have already led to a shorter growing season with detrimental effects on crops. In some countries of Africa, yields from rain-fed agriculture could be reduced by up to 50% by 2020 (Parry *et al.,* 2007, chapter 9). Local food supplies are projected to be negatively affected by decreasing fish populations in large lakes due to rising water temperatures, and the shortage may be exacerbated by continued over-fishing. These consequences would further adversely affect food security and increase malnutrition in Africa. By 2020, between 75 and 250 million people in Africa are projected to be exposed to increased water stress due to climate change (Parry *et al.,* 2007, chapter 9).

**Health problems:** In addition to malnutrition and consequent disorders in child growth and development, projected climate change is likely to affect the health status of millions of people – particularly those with low adaptive capacity – through increased deaths, disease, and injury due to heat waves, floods, storms, fires and droughts; the increased burden of diarrhoeal disease; the increased frequency of cardio-respiratory diseases due to higher concentrations of ground-level ozone related to climate change; and the altered geographical distribution of some infectious disease vectors (Parry *et al.,* 2007, chapter 8).

**Coastal risks:** By the end of the century, many millions more people than today are projected to experience floods every year due to the rise of sea levels (Parry *et al.,* 2007, chapter 6). The number of people affected by sea-level rise and by storm floods will be highest in the mega-deltas of Asia and Africa, while small islands are also especially vulnerable.

**Migration:** In addition to the existing migration due to resource scarcity, the numbers of environmental refugees could increase as coastal flooding, extreme weather events, famines, and conflicts that will arise due to these events become more frequent.

**Biodiversity:** By 2020, significant loss of biodiversity is projected to occur in some ecologically rich sites, including the Great Barrier Reef and Queensland Wet Tropics. There is also risk of significant biodiversity loss through species extinction in many areas of tropical Latin America. Increases in sea surface temperature of about $1-3\,°C$ are projected to result in more frequent coral bleaching events and widespread mortality, unless there is thermal adaptation or acclimatization by corals (Parry *et al.,* 2007, chapter 4).

These and the other projected vulnerabilities underscore the importance of promoting alternative, sustainable development paths for the 80 % of the world's population that lives in developing countries. This poses a significant, though achievable, challenge for the world economy, which has so far relied heavily on fossil fuels.

## Increasing emissions from developing and emerging nations

Whereas industrialized nations bear the greatest responsibility for the current situation, the contributions of emerging nations are becoming more and more problematic. Their rapidly growing economies will in the long run exacerbate the climate change problem. Clearly, incorporating sustainable patterns of consumption in countries at all levels of development is a critical component of a sustainable development path, and will be vital for ensuring the success of climate change policies.

At the same time, and in spite of rapid economic growth, per-capita emissions in the emerging countries are still a fraction of the per-capita emissions in most industrialized economies (see Narain, this volume). For instance, while China overtook the US in 2007 in terms of absolute emissions (see Fig. 1), the USA's per-capita emissions are still four to five times higher than China's. India's per-capita emissions are even lower, about one-twentieth of the US level.

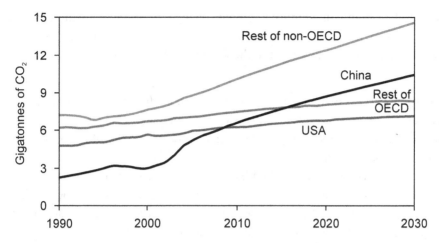

**Fig. 1.** Increase in carbon dioxide emissions in gigatonnes for different countries or groups of countries. (*Source:* IEA, 2006)

## Coping with climate change through adaptation and mitigation

Adaptation and mitigation measures both have the potential to minimize climate change impacts. Adaptation includes initiatives and measures to reduce the vulnerability of natural and human systems to actual or expected climate change effects. Examples include raising river or coastal dikes, and substituting more temperature- and shock-resistant plants for sensitive ones (Metz *et al.,* 2007). Several countries are already undertaking adaptation measures, including crop diversification, irrigation, water management, disaster risk management, and adjusted insurance rates for people in areas that are likely to be most severely affected. Mitigation involves technological change and substitution that reduce resource inputs and emissions per unit of output. While several social, economic and technological policies can indirectly enhance emissions reductions, in this context climate change mitigation refers to policies undertaken to directly reduce greenhouse gas emissions and to enhance carbon sinks such as forests (ecosystems that absorb carbon).

Adaptation requires a conscious reorientation of global priorities to ensure the availability of adequate resources. Even though several international funds have been established, there is still a lack of available adaptation funding, particularly in developing countries. Only USD 163.3 million and USD 57.1 million have been pledged to the UN's LDCF (Least Developed Country Fund) and SCCF (Special Climate Change Fund) funds, of which only USD 67.3 million and USD 49.3 million respectively have been received (GEF, 2007). By contrast, developed nations spent about USD 250 billion in 2005 supporting their own agriculture (WTO, 2006), and global military expenditure was about USD 1.2 trillion in 2006 (UNDP, 2007).

Given their high vulnerability to climate change, developing countries urgently need to increase their adaptive capacity. Adaptation mechanisms for developing and emerging nations that can decrease their vulnerability include (Parry *et al.*, 2007, chapter 10):

- Improving access to high quality information about the impacts of climate change and about best-response mechanisms to the anticipated effects;
- implementing early warning systems and information distribution systems to enhance disaster preparedness;
- reducing the vulnerability of livelihoods and infrastructure to climate change;
- promoting good governance, including responsible policy and decision making;
- empowering communities and other local stakeholders so that they actively participate in vulnerability assessment and adaptation;
- mainstreaming climate change into development planning at all scales, levels and sectors.

While adaptation measures are vital, particularly in the short term, sustainable solutions to climate change need to include a mix of adaptation and mitigation policies, suited to each country's vulnerabilities and level of development.

It has been well established that delaying emissions reduction leads to investments that lock in more emission-intensive infrastructure and development pathways. This significantly constrains the opportunities to achieve lower stabilization levels and increases the risk of more severe climate change impacts (Metz *et al.*, 2007, SPM). The fact that mitigation efforts have visible long-term impacts underscores the need to scale up current mitigation efforts. Even if the concentrations of all greenhouse gases and aerosols were kept constant at 2000 levels, further warming of about 0.1 °C per decade could be expected for the next two decades (Solomon *et al.*, 2007, SPM, p. 12). Energy system inertia adds a further dimension to the time scales involved in climate change. It has taken at least 50 years for each major energy source to move from a 1% penetration to a major position in global supplies. This inertia, as well as the even longer periods associated with interactions between systems, implies that abatement must begin as early as possible to ensure stabilization of greenhouse gases and temperature at targeted levels.

Figure 2 illustrates that the maximum projected cost of mitigation would not exceed 3 % of global GDP in 2030.

Several common drivers exist among policies addressing economic development, energy security, and health and climate change mitigation. Therefore, there are numerous co-benefits associated with mitigation, including health benefits and enhanced energy security. Mitigation measures present various opportunities for no-regrets policies, which should be integrated into the overall socio-economic policy framework.

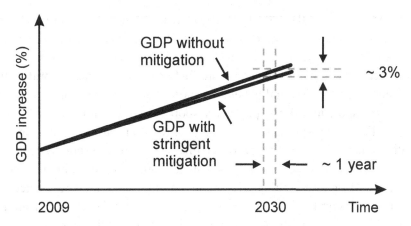

**Fig. 2.** Projected percentage increase of global GDP from today until 2030, with and without climate mitigation measures. The difference between climate mitigating efforts or the absence thereof would be relatively small: approximately 3 % of GDP. (*Source:* Pachauri, 2007)

Many technologies that have the potential to provide solutions for low-carbon development are already available, though several are not economically competitive at present. Investments in renewable technologies would contribute towards making them competitive with fossil fuels at an earlier stage, and this would enable greater energy security. Policies that divert unsustainable, distortive subsidies from fossil fuels to cleaner technologies would make resources available for increased investments in renewable energies, and thus facilitate the transition to low-carbon economies.

The four main sectors that require massive reductions in greenhouse gas emissions are energy supply, transportation, housing, and land use change:

**Energy supply.** The energy supply sector accounted for about 25.9 % of global greenhouse gas emissions in 2004 (IPCC SYR, 2007, Fig. 2.1). All assessed stabilization scenarios indicate that 60–80 % of reductions would come from energy supply and use and industrial processes, with energy efficiency playing a key role in many scenarios (IPCC SYR, 2007, p. 20). Mitigation technologies for this sector include improved supply and distribution efficiency, fuel switching from coal to gas, nuclear power, renewable heat and power (hydropower, solar, wind, geothermal and bioenergy), combined heat and power, and early application of carbon capture and storage (CCS) technology (see Bruckner *et al.,* this volume).

**Transport.** The transport sector accounted for about 13.2 % of global greenhouse gas emissions in 2004 (IPCC SYR, 2007, Fig. 2.1). Rapidly growing mobility demands from developing countries pose a significant challenge in terms of ensuring that mitigation efforts are not offset by increased transport activity. If current trends continue, by 2035 there will be around 250 million more cars and SUVs operating in China and India (USAID, 2007, p. 3). The increased demand for transportation will lead to a 2.6-fold increase in oil demand in developing Asia during this period, and a corresponding three-fold increase in carbon dioxide emissions. Currently available mitigation technologies include more fuel-efficient vehicles, hybrid vehicles, cleaner diesel vehicles, second-generation biofuels, electric vehicles, modal shifts from road transport to rail and public transport systems, non-motorized transport such as cycling or walking, and improved land-use and transport planning. Transport policies that enhance co-modality and efficient public transport systems would be crucial in supporting technological change to reduce emissions in this sector.

**Buildings.** Mitigation technologies in the building sector include efficient lighting and use of natural light, more efficient electrical appliances and heating and cooling devices, improved cooking stoves, improved insulation, passive and active solar building designs for heating and cooling, alternative refrigeration fluids, and the recovery and recycling of fluorinated gases. The building sector accounts for a sizeable share of overall emissions. The expansion of this sector in the rapidly growing transition economies provides the potential to integrate energy-efficient buildings into the infrastructural development process at an early stage, thereby providing co-benefits. It is vital, however, that energy-efficiency regulations are adaptable, suit local conditions, and draw on sustainable local building practices.

**Deforestation and land use change.** The IPCC estimates that the cutting down and degradation of forests currently account for close to 20% of all greenhouse gases entering the atmosphere (Metz *et al.,* 2007, TS, Fig.1b). Deforestation and forest degradation are significant causes of concern, particularly in the developing nations. Key mitigation initiatives and technologies in this sector include afforestation, reforestation, forest management, reduced deforestation, harvested wood product management, use of forestry products for bioenergy to replace fossil fuels, tree species improvement to increase biomass productivity and carbon sequestration, improved remote sensing technologies for analysis of vegetation/soil carbon sequestration potential, and mapping land use change. Policies incorporating financial incentives that value carbon sequestration and other ecosystem services provided by forests would represent potentially significant mitigation measures. To this end, the UN's REDD (Reduced Emissions from Deforestation and Degradation) programme has been initiated, one of the goals of which is to assess whether careful payment structures and capacity support can create the incentives to ensure actual, lasting, achievable, reliable, and measurable emission reductions, while maintaining and improving the other ecosystem services forests provide.

A vital component of market-based climate change policies is to put an accurate price on carbon that reflects the social costs of emissions. Policies that implement a real or implicit price on carbon could create incentives for producers and consumers to significantly invest in products, technologies and processes that produce low amounts of greenhouse gases. Such policies could include economic instruments, government funding and regulation. For stabilization at around 550 ppm carbon dioxide equivalent, carbon prices should reach USD 20–80 per tonne of carbon dioxide equivalent by 2030 (Metz *et al.,* 2007, SPM, p. 19). To limit global warming to the two-degree guardrail mandated by the Potsdam Memorandum (see pp. 369 ff.), deeper cuts in the short and medium term, leading to lower concentrations of greenhouse gases, will be necessary. This implies higher price ranges. However, it should be kept in mind that appropriate policies, such as those inducing technological development, have the potential for achieving the emissions reductions targets at generally lower price ranges.

## Common but differentiated responsibility

The United Nations Framework Convention on Climate Change (UNFCCC), which has been ratified by 192 countries, outlines the principle of common but differentiated responsibility (CISDL Legal Brief, 2002, see also Narain, this volume). This principle recognizes the need for concerted global action, while emphasizing the need for proportionate and appropriate action by nations, taking into account those nations' historical contributions to climate change. Developed nations need to reduce their per-capita emissions, and at the same time consider the requirements of developing nations to industrialize, with overall global per-capita emissions not exceeding acceptable agreed levels. In addition to taking the lead on mitigation, developed nations also need to transfer financial, technical and other resources to emerging and developing nations to facilitate adaptation and mitigation. In principle, one could anticipate that the share of global emissions from developing countries will initially grow in line with their social and development needs.

The contraction and convergence policy option proposes that equalizing global per-capita emissions across countries would ensure equity in the global climate change mitigation process. It supports climate change negotiations that aim to equalize per-capita emissions at a future date, with the levels of permissible global per-capita emissions and the different years by which the emissions have to be equalized varying according to several formulae. This would allow citizens of all countries, regardless of size or level of development, equal space in the atmosphere, and thus equal responsibility to mitigate. While there are concerns that contraction and convergence may provide incentives to high population growth rates, it is entirely feasible, and indeed widely proposed, to place a limit on population beyond which no further entitlements would be granted. Furthermore, countries with high population growth rates would still have to provide resources for their growing populations. Therefore, the economic incentive to encourage high population growth rates may not even exist.

### Adopting a sustainable development path

While there is immense potential for developing economies to integrate sustainable development initiatives into their economic and development policies, technology and capacity transfer is crucial to ensure widespread and effective mainstreaming of low-carbon technologies. An environment that is conducive to the transfer of low-carbon technologies would aid in implementing appropriate future policies for emerging economies, and would combine development policy with climate change mitigation. As mentioned earlier, there are numerous co-benefits associated with several mitigation measures, such as health benefits and enhanced energy security. Investing in sustainable infrastructure, planning cities with minimized environmental and ecological impacts, and conducting appropriate research and development (R&D) are some of the policy options that can re-orient an economy onto a sustainable path.

The assumption that economic growth is the panacea for all development problems, including climate change, may be worth discussing at this point. A narrow policy approach that solely promotes economic growth provides, at best, a partial solution to climate change by providing increased resources for adaptation while possibly worsening the overall problem. At worst, it will instigate a highly unsustainable development path that undercuts the foundations of future economic growth. The original inverted U-shaped Kuznets Curve suggests that with increasing economic growth income inequality will first increase, and then, after a point, decrease. Drawing on this concept, the Environmental Kuznets Curve suggests that economic growth would, after a point, lead to better environmental quality (Fig. 3). This carries the implication, at least to some degree, that there is potential for developing economies to 'grow out' of environmental degradation, since at a certain income level the population's preferences, or the increased resources due to development, would lead to better environmental quality, including decreased pollution and sustainable management of resources. The Environmental Kuznets Curve hypothesis is one of the most contentious empirical phenomena in environmental economics, at least in part due to its implications for economic and environmental policy in developing countries.

The Environmental Kuznets Curve for carbon dioxide emissions in particular, which predicts that as countries develop a certain level of wealth carbon dioxide emissions will fall, seems fraught with uncertainties on several grounds (Galeotti *et al.,* 2006; Stern, 2003). The most significant of these is that by the time most of the current high emitters have developed 'sufficiently' to reach the other side of the curve, it will be much too late to begin mitigation. Also, the very existence of the Environmental Kuznets Curve for global pollutants such as carbon dioxide is contentious in the first place. Clearly, climate change mitigation is not something a

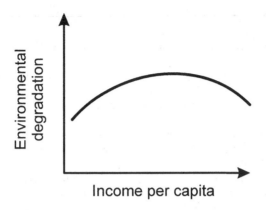

**Fig. 3.** Illustration of the Environmental Kuznets Curve.

developing country can simply 'grow into', but rather is a development path that needs to be agreed and acted upon quickly and effectively.

## Conclusion

Climate change is a critical global challenge, one that requires international and inter-sectoral collaboration on an unprecedented scale. In addition, the present direction of the global economy requires a re-orientation, both in terms of outlook and development priorities. No country will be unaffected by climate change, and the socio-economic links between countries may in some cases exacerbate these impacts. This poses a daunting challenge, but one – if acted upon quickly and effectively – that promises a more inclusive and less vulnerable planet for the global population as a whole.

## References

CISDL Legal Brief. (2002). *The Principle of Common But Differentiated Responsibilities: Origins and Scope.* Montreal. Available at http://www.cisdl.org/pdf/brief_common. pdf.

Galeotti, M., Lanza, A. and Pauli, F. (2006). Reassessing the Environmental Kuznets Curves for $CO_2$ emissions: a robustness exercise. *Ecological Economics,* **57**(1), 152–163.

GEF – Global Environment Facility (2007). *Special Climate Change Fund.* Washington, D. C. Available at http://www.gefweb.org/interior_right.aspx?id=192.

IEA – International Energy Agency (2006). *World Energy Outlook 2006.* Paris. Available at http://www.worldenergyoutlook.org/2006.asp.

IPCC SYR (2007). *Climate change 2007: Synthesis report: A Report of the Intergovernmental Panel on Climate Change.* Geneva. Available at http://www.ipcc.ch/publica tions_and_data/publications_ipcc_fourth_assessment_report_synthesis_report.htm.

LaFleur, V., Purvis, N. and Jones, A. (2009) *Double Jeopardy: What the Climate Crisis Means for the Poor.* Washington, D. C. Available at http://www.brookings.edu/~/media/Files/rc/reports/2009/02_climate_change_poverty/02_climate_change_poverty.pdf.

Metz, B., Davidson, O. R., Bosch, P. R., Dave, R., Meyer, L. A., eds. (2007). *Climate Change 2007: Mitigation of Climate Change. Contribution of Working Group III to the Fourth Assessment Report of the Intergovernmental Panel on Climate Change.* Cambridge.

Pachauri, R. K. (2007). *The IPCC Fourth Assessment Working Group Reports: Key findings.* New York. Available at http://www.ipcc.ch/pdf/presentations/pachauri-un_nyc_2007-09-07.pdf.

Parry, M. L., Canziani, O. F., Palutikof, J. P. *et al.,* eds. (2007). *Climate Change 2007: Impacts, Adaptation and Vulnerability. Contribution of Working Group II to the Fourth Assessment Report of the Intergovernmental Panel on Climate Change.* Cambridge.

Solomon, S., Qin, D., Manning, M. *et al.,* eds. (2007). *Climate Change 2007: The Physical Science Basis. Contribution of Working Group I to the Fourth Assessment Report of the Intergovernmental Panel on Climate Change.* Cambridge.

Stern, D. I. (2003). The Environmental Kuznets Curve. In *Internet Encyclopaedia of Ecological Economics.* Available at http://www.ecoeco.org/pdf/stern.pdf.

UNDP – United Nations Development Programme (2007). *Fighting Climate Change: Human Solidarity in a Divided World. Houndmills USA.* Available at http://hdr.undp.org/en/reports/global/hdr2007-2008.

USAID – United States Agency International Development (2007). From Ideas to Action: *Clean Energy Solutions for Asia to Address Climate Change.* Report of the International Resources Group (IRG) under the ECO-Asia Clean Development and Climate Program. Available at http://usaid.eco-asia.org/programs/cdcp/reports/Ideas-to-Action/Executive%20Summary.pdf.

WTO – World Trade Organization (2006). *World Trade Report 2006: Exploring the Links between Subsidies, Trade and the WTO.* Available at www.wto.org/english/res_e/booksp_e/anrep_e/world_trade_report06_e.pdf.

# Chapter 13

# Climate change – learning from the stratospheric ozone challenge

————

## Mario Molina

Mario Molina studied physical chemistry and obtained his PhD at the University of California, Berkeley. In 1974, well before the first measurements of the Antarctic ozone hole, he co-authored a paper that described how chlorofluorocarbon (CFC) gases, widely used in industry at that time, destroy the atmospheric ozone layer. In 1995, Molina was honoured with the Nobel Prize in Chemistry for his work on ozone depletion. As Professor of Chemistry and of Earth, Atmospheric and Planetary Sciences at the Massachusetts Institute of Technology, Molina continued his research on man-made changes of atmospheric chemistry. In 2004 he joined the faculty at the University of California in San Diego.

*Note:* This chapter is a commentary on chapter 12.

The most recent findings on climate change provide clear evidence that 'human activities, especially the combustion of fossil fuels, are influencing the climate in ways that threaten the well-being and continued development of human society' (Richardson *et al.,* 2009). At the same time, Pachauri (this volume) states that climate change 'poses a daunting challenge, but, if acted upon quickly and effectively, one that promises a more inclusive and less vulnerable planet for the global population as a whole'. We have, thus, an opportunity that we must not miss.

Although we know about the potentially disastrous consequences of destabilizing the climate, many people, organizations, and nations are still not responding adequately to the urgent call for action, and insist that more evidence is needed to warrant a global response. Such thinking is similar to that of a patient who asks for virtual certainty that a tumour is indeed malignant before agreeing to have it removed. Yet, most patients would surely agree to have surgery even if the probability of malignancy were merely ten or twenty percent. A similar attitude is still common when considering action on climate change. But we know that the risk of inaction, although difficult to quantify, is very significant, and we do not need more scientific evidence to conclude that drastic action is necessary.

If we continue to delay action we will miss a unique opportunity to make our world a more just and healthy place for all. It is time to remember that we have cooperated on solving global problems before, most notably the problem of ozone depletion in the stratosphere. It might be helpful to look back at previous successes and mistakes to increase the likelihood that our efforts to avoid climate destabilization will be successful.

## Stratospheric ozone – a short history

Ozone is found mainly in the stratosphere – the second layer of our atmosphere, at a height of about 10–50 km. An ozone molecule contains three atoms of oxygen instead of the two found in normal oxygen molecules; it is formed at high altitudes through the action of short-wavelength solar radiation on oxygen molecules. Stratospheric ozone has made it possible for life to evolve on our planet; it acts like a sunscreen, absorbing most of the harmful ultraviolet radiation that destroys the DNA molecule, which is essential for life as we know it.

In 1974, we discovered that chlorofluorocarbons (CFCs) – then commonly used in refrigeration and as propellants for spray cans – can have a detrimental effect on ozone (Molina and Rowland, 1974). In the stratosphere, CFCs decompose by the action of short wavelength solar radiation splitting off chlorine atoms, which in turn start chain reactions that break down ozone. We concluded that CFCs could cause a depletion of the ozone layer, potentially affecting human health and the environment.

Our theory was eventually confirmed by atmospheric observations, laboratory measurements and modelling studies. In 1985 the Antarctic ozone 'hole' was discovered (Farman *et al.*, 1985); in the middle of the stratosphere above Antarctica more than 95% of the ozone disappeared in the spring months, and subsequent measurements confirmed that the disappearance was caused by the CFCs. These discoveries initiated a political process that culminated in a multilateral agreement to phase out practically all substances that are responsible for stratospheric ozone depletion. This treaty, the Montreal Protocol, came into full force four years later, in January 1989, and has been amended several times since then. It can be regarded as one of the best examples of effective global collaboration on behalf of humanity and the environment; the amount of CFCs in the atmosphere has started to decline, and although the ozone hole still forms every year over Antarctica, the rest of the ozone layer has started to show signs of recovery from a less severe, but still noticeable thinning.

It turns out that the CFCs that affect stratospheric ozone are also powerful greenhouse gases, and thus the Montreal Protocol has also led to significant climate change mitigation. So far, it has been considerably more effective than the Kyoto Protocol, the treaty that was developed in 1997 to regulate greenhouse gases, and that is currently being reassessed (Velders *et al.*, 2007). Most of the compounds now replacing ozone depleting substances are hydrochlorofluorocarbons (HCFCs) and hydrofluorocarbons (HFCs); some of these chemicals are also strong greenhouse gases (Velders *et al.*, 2009), and for this reason recent amendments to the Montreal Protocol are now aimed at accelerating their phase-out.

## An example for global action: Montreal versus Kyoto treaties

The current global problem caused by greenhouse gas emissions has many similarities to the stratospheric ozone problem. In both cases it is crucial to exchange 'business as usual' for collaboration between nations as one global community. But the quick and effective implementation of the Montreal Protocol to protect the ozone layer stands in stark contrast to the Kyoto Protocol. Even though climate change is well documented by a large numbers of scientific studies, the Kyoto Treaty has not been successful on a global scale; global society has yet to find a way to agree on effective actions on climate change. Several important differences between the problems of ozone depletion and climate change are discussed below. They at least partially explain why the Montreal Protocol is more effective than the Kyoto Protocol. Recognizing these differences might enable us to find more effective solutions to climate change.

*The science behind ozone depletion is very well established.* Reproducible scientific data involving atmospheric observations, laboratory measurements and modelling studies clearly show that the chemical processes that are initiated by CFCs in the stratosphere result in the depletion of ozone. The basic science of climate change is also relatively straightforward; increased concentrations of greenhouse gases warm the surface of the planet. However, the Earth's climate system is quite complicated, and there are many feedbacks which affect the overall functioning of this system. The changes are gradual and occur in a dynamic and complex system; furthermore, at first sight these changes appear to be natural, and hence there is more room for scepticism.

*CFCs are clearly of human origin, and do not exist naturally in our atmosphere.* In contrast, carbon dioxide and methane, which are the main gases responsible for the greenhouse effect, have predominantly natural sources and play an important role in producing the benign climate that has facilitated the development of human civilization in the past 12 000 or so years. In fact, greenhouse gases make life on Earth possible; the average surface temperature of our planet would be about -15 °C without these gases, when in fact it is +15 °C as a consequence of the 'natural' greenhouse effect. The problem is that human activities are adding large amounts of these 'natural' gases to the atmosphere, causing an enhanced anthropogenic greenhouse effect that is significantly affecting the natural climate.

*The extent of change necessary to phase out CFCs was relatively small and relatively easy to monitor.* This is probably the most important difference to the climate change issue. The ozone-depleting chemicals were used mainly as refrigerants, solvents and as propellants for spray cans, and could be replaced with other compounds, most with very similar qualities. The Montreal Protocol called for a complete phase-out of the production of ozone-depleting chemicals, which has already been largely accomplished. Most people never even noticed the changes, as the required transition affected only a few industries.

In contrast, climate change is caused mainly by activities related to the production and consumption of fossil fuel energy, which has so far been essential for the functioning of our industrialized society. Effective action therefore requires a major transformation, not only involving a few industries, but affecting a great number of activities of society. Furthermore, the unwanted side-effects from these activities involve the generation of compounds that are naturally emitted to the environment; it is much harder to monitor not only who is responsible for the unwanted emissions but also if they are actually changing. Furthermore, it is not easy to establish the appropriate baseline to decide if the emissions in question are decreasing or increasing. It is harder still to monitor changes in the greenhouse gas emissions related to deforestation, agricultural and land-use practices, which contribute about one third of the total emissions responsible for climate change.

It is thus not surprising that efforts to mitigate climate change have been slow and difficult to implement compared to those of the Montreal Protocol. Today, action will only be effective if there is large-scale collaboration between politicians, industry and civil society, and between most nations. It is important to communicate the urgency of the problem to all these groups. It is also important to understand that its solution involves costs, but these are clearly smaller than the costs associated with inaction, as shown by recent economic studies. It is therefore essential not only to base any solutions on the best science available, but also to clearly communicate the short- and long-term benefits and challenges of the suggested solutions.

## Opportunities to act

Because effective climate action is more urgent than the scientific community had anticipated only a few years ago, it is imperative for society to find an effective way to move forward in an effort that will define the future of modern societies. Unfortunately, there is no 'silver bullet'; however, there are technologies currently available that could be implemented in the near future and would result in a significant reduction of greenhouse gas emissions at a relatively modest cost, namely a few percent of global GDP (Stern, 2006; Paltsev *et al.*, 2009; Stern and Garbett-Shiels, this volume). Some of these technologies involve significantly increasing the efficiency with which energy is consumed in a variety of sectors (industry, transportation, housing, etc.); some others involve the use of renewable energy sources (such as solar, wind and biomass); and yet others involve sequestering and capturing the carbon dioxide emitted in power plants consuming coal, oil or biomass.

## The role of developing nations in mitigating climate change

So far, developing nations have not been the major contributors to anthropogenic global climate change, but they are bearing the brunt of its effects (see Pachauri, this volume). This has led to the common perception that developing nations are the victims of unjust and ineffective policies, and that the industrialized nations have the responsibility to solve the problem they created. Along these lines, any changes to be carried out by the developing world to address the climate change issue would have to be paid for by the industrialized countries.

At the same time – and rightly so – developing nations are striving to achieve the same standard of living as the industrialized world, implying similar levels of energy consumption. The problem is that so far their economic growth is being achieved along the same path the industrialized countries followed in the past. Industrialized nations are thus reluctant to transfer the funds requested by developing nations as they believe that these funds might not be properly employed to significantly reduce the growth of greenhouse gas emissions.

As understandable as these attitudes might be, they do not help solve the problem.

There are not enough natural resources on our planet, and the atmosphere is not large enough to absorb the unwanted by-products of human activities without consequences. Clearly, economic development cannot continue along the same path it has followed in the past, and something has to change quite drastically. Developed nations have to understand, and most of them do, that for reasons of justice they must contribute to the solution of the problem by transferring economic resources and technology to developing nations. In fact, an important precedent was set by the Montreal Protocol: the creation of the 'Multilateral Fund'. This fund was instrumental in addressing the stratospheric ozone question by providing resources to developing nations to achieve a smooth transition to a CFC-free society. At the same time, developing nations have to realize that they can and must aim for a different system, one not heavily tied to the consumption of energy and the combustion of fossil fuels. They also have to acknowledge that these changes are very significant and should not occur only to the extent implied by a transfer of funds from developed nations. The climate change problem is truly global; all nations stand to benefit from an effective international treaty, and all nations stand to lose if no agreement is reached.

An example of a developing country with a positive attitude is Mexico; this country has already made a commitment to follow a low-carbon economic growth plan and to halve its greenhouse gas emissions by the year 2050. Furthermore, Mexico is proposing the establishment of a 'Green Fund' with contributions from both developed and developing countries to facilitate the global transition to low-carbon economies. Some of the proposed changes in Mexico will merely require new government regulations – for example, those that lead to more efficient energy use – while others will require economic assistance from abroad. The point is, however, that Mexico is already embarking on this new economic growth path with the expectation that a global agreement will be reached, that this new path will improve its competitiveness in the global economy, and that it will also end up facilitating the eradication of poverty. Fortunately, it appears that other nations, such as China and India, are also developing and beginning to implement similar plans. In the end, it is this type of positive attitude that might lead to a successful global treaty. The main problems that are currently being experienced in international negotiations result from excessive demands from some industrialized countries for 'binding commitments' by all developing nations, or excessive demands by some developing nations for economic contributions as a condition for change. Here again, the Montreal Protocol stands out as an example which demonstrates that an effective international agreement can indeed be negotiated.

## Air pollution and climate change

Air pollution continues to be a serious problem, particularly in many developing countries. The public health impacts of poor air quality are well documented, and thus the economic and quality-of-life benefits to society of air pollution controls provide ample justification for their implementation. Furthermore, it turns out that many of the measures required to address air pollution also provide important benefits in relation to climate change.

The most common components of air pollution include atmospheric ozone and aerosols. Although ozone is most abundant in the stratosphere, the lowest layer of the atmosphere, the troposphere, also contains ozone of natural origin. The concentration of this 'tropospheric ozone' has increased in recent years as a consequence of human activities, mainly the burning of fossil fuels and biomass. At high temperatures characteristic of a combustion process, small amounts of oxygen and nitrogen (the most abundant compounds in air) combine to form nitric oxide, which, together with carbon monoxide and unburned gaseous organic compounds, undergo a series of chemical reactions in the presence of sunlight to generate ozone. As a consequence of its detrimental health effects, ozone levels are controlled in many cities, but barely so in the background troposphere, where it acts as a powerful greenhouse gas. Thus, reducing emissions of ozone precursors (nitrogen oxides and gaseous organic compounds) leads not only to improved air quality, but also contributes to climate change mitigation.

Atmospheric aerosols are solid and/or liquid airborne particles. A large fraction of man-made aerosols come in the form of smoke from burning tropical forests, biomass, and fossil fuels. Black carbon is a component of smoke, and is generated in part by diesel engines not fitted with modern emission control devices. It turns out that black carbon emissions have not only serious public health impacts, but also contribute very significantly to climate change (Ramanathan and Carmichael, 2008). On the other hand, a major component of atmospheric aerosols of human origin comes in the form of sulphates, created by the burning of coal and oil. In contrast to black carbon, sulphate aerosols are white and reflect or scatter incoming solar radiation, and thus lead to climate cooling, compensating to some extent for the anthropogenic greenhouse effect. In fact, the true impact of greenhouse gases has been masked to some extent by this type of aerosol (Ramanathan and Feng, 2008). Nevertheless, air quality considerations alone justify the need to reduce emissions of these white aerosols, even if that means that stricter controls of greenhouse gases and black carbon will be required to properly reduce the climate change risk.

## The role of ethics in mitigating climate change

Even though we are moving dangerously close to reaching tipping points with nearly irreversible consequences for the Earth's climate system (Lenton *et al.*, 2008), the world as a whole is still debating what, if any, changes are needed to address the climate change crisis. Clearly, science and knowledge alone are not enough to move people to action. In addition to scientific communication, experts need to help decision-makers in society to truly understand what climate change is all about.

Global environmental problems have been caused so far predominantly by developed countries, which are home to about one fourth of the global population. The enormous challenge now facing society is to enable the economic development of the rest of the global population, so that they too can enjoy a satisfactory standard of living, without, however, degrading the natural environment. Our generation has the responsibility to address the climate change problem in such a way as to ensure that future generations have access to environment and natural resources suitable for the continued improvement of their economic well-being. Solving the climate change and air quality dilemmas is thus not just well justified from a purely economic point of view, but ethical considerations imply that it is a truly imperative endeavour for our generation.

## References

Farman, J. C., Gardiner, B. G. and Shanklin, J. D. (1985). Large losses of total ozone in Antarctica reveal seasonal ClOx/NOx interaction. *Nature,* **315,** 207–10. Available at http://www.ciesin.org/docs/011-430/011-430.html.

Lenton, T. M., Held, H., Kriegler, E. *et al.* (2008). Tipping elements in the Earth's climate system. *Proceedings of the National Academy of Sciences of the United States of America,* **105**(6), 1786–1793. Available at http://www.pnas.org/content/105/6/1786.full.pdf.

Molina, M. J. and Rowland, F. S. (1974). Stratospheric sink for chlorofluoromethanes: chlorine atom-catalysed destruction of ozone. *Nature,* **249,** 810–12.

Paltsev, S., Reilly, J. M., Jacoby, H. D. and Morris, J. F. (2009). *The Cost of Climate Policy in the United States.* MIT Joint Program Report Series, Report 173. Available at http://globalchange.mit.edu/pubs/abstract.php?publication_id=1965.

Ramanathan, V. and Carmichael, G. (2008). Global and regional climate changes due to black carbon. *Nature Geoscience,* **1,** 221–7.

Ramanathan, V. and Feng, Y. (2008). On avoiding dangerous anthropogenic interference with the climate system: formidable challenges ahead. *Proceedings of the National Academy of Sciences of the United States of America,* **105,** 14245–50.

Richardson, K., Steffen, W., Schellnhuber, H. J. *et al.* (2009). *Synthesis Report. Climate Change: Global Risks, Challenges & Decisions.* Available at http://climatecongress.ku.dk/pdf/synthesisreport.

Stern, N. (2006). *Stern Review on the Economics of Climate Change.* Available at http://www.hm-treasury.gov.uk/stern_review_report.htm.

Velders, G. J. M., Andersen, S. O., Daniel, J. S., Fahey, D. W. and McFarland, M. (2007). The importance of the Montreal Protocol in protecting climate. *Proceedings of the National Academy of Sciences of the United States of America,* **104,** 4814–19.

Velders, G. J. M., Fahey, D. W., Daniel, J. S. *et al.* (2009). The large contribution of projected HFC emissions to future climate forcing. *Proceedings of the National Academy of Sciences of the United States of America,* **106**(27), 10949–54.

# Chapter 14

# Climate change, poverty eradication, and sustainable development

———————

## Nitin Desai

Nitin Desai, a graduate of London School of Economics, taught economics at two British universities, worked briefly in the private sector, and had a long stint as a government official in India before joining the UN, where he rose to the position of Under-Secretary-General for Economic and Social Affairs. He retired from this position in 2003. His major work at the UN involved organizing a series of global summits, including the Rio Earth Summit (1992) and the Johannesburg Sustainable Development Summit (2002). He is at present a member of the National Security Advisory Board and the Prime Minister's Council on Climate Change in India, and remains an active participant in the national and global dialogue on climate policy.

What is the principal challenge facing humanity in the twenty-first century? Is it the challenge of lifting billions out of poverty into a life of dignity? Or is it one of ensuring that we do not transgress the boundaries beyond which the risks of catastrophic environmental change are unacceptably large? In my view the word 'or' in the previous question is misleading. The two challenges are now so connected that coping with one requires that we cope also with the other. That is what sustainable development is all about – how poverty eradication and environmental protection can be mutually supportive.

The persistence of poverty[1] can be attributed to many factors, but, of these, resource poverty is the crucial one. A large proportion of the world's poor live in the rural areas of the developing world and face a growing scarcity of land and water. Many of them are in ecologically fragile regions such as arid and semi-arid zones, mountain areas, coastal areas exposed to violent weather, and so on. A critical dimension of resource poverty is the lack of access to safe and sustainable energy. In developing countries some 2.5 billion people are forced to rely on biomass – fuelwood, charcoal, and animal dung – to meet their energy needs for cooking. Indoor air pollution claims the lives of 1.5 million people each year, more than half of them below the age of five. 1.6 billion people – a quarter of humanity – live without electricity.

For all of these people in poverty, as well as for policymakers in the developing world, development that raises productivity, production and income is understandably the highest priority. Slowing down economic growth is not an option that they can consider. But I would argue that growth that is more mindful of the local environment is something that they can and should pursue, for it is the poor who are most exposed to environmental stress and resource poverty. Hence, when it comes to climate change, mitigating the risks and adapting to the changes that are unavoidable have to be components of any long-term strategy for poverty eradication.

The issue is not what we do first. Climate change is a threat that could worsen global inequality because it will affect low-latitude developing countries to a greater extent, and mostly in an adverse manner. Changes in water availability, the increase in vector-borne diseases such as malaria, and the greater risk of extreme climate events are some of the consequences that will affect the poor more than the wealthy. Therefore, the real challenge is to find solutions that address both problems simultaneously. This is the goal of sustainable development.

---

[1] The facts about poverty are well known: 2.6 billion people live on less than two US dollars per day, 800 million go to bed hungry every day, 26 000 children die every day because of poverty, a billion people entered the twenty-first century unable to read or write, 72 million children should be but are not in school, 1.1 billion people in developing countries have inadequate access to water, 2.6 billion lack basic sanitation, a billion urban dwellers live in slum conditions, and 1.4 million children die every year due to lack of access to safe water and sanitation (http://www.globalissues.org/article/26/poverty-facts-and-stats. Accessed 14 January 2009).

According to the Brundtland Commission, 'Sustainable development is a process of change in which the exploitation of resources, the direction of investments, the orientation of technological development, and institutional change are all in harmony and enhance both current and future potential to meet human needs and aspirations' (World Commission on Environment and Development, 1987, p. 46). The risks associated with climate change are clearly not consistent with this notion of sustainability. As the present author, who was involved in the writing of the Brundtland Commission's report, has stated elsewhere that 'Environmental resources like biodiversity or the delicately balanced chemistry of the atmosphere are resources which are critical to the maintenance of life on Earth. In such cases the objective of sustainability would require conservation in a stricter sense since compensation to preserve options may not be possible' (Desai, 2007, pp. 506–9).

A sustainable development strategy that addresses this risk must involve changes that mitigate the risk, and measures that help people to adapt to the climate change that is unavoidable even with mitigation efforts.

## Mitigation

The key to mitigation lies in rethinking energy policy. The carbon dioxide emitted by fossil fuel use is not the only greenhouse gas, but it is by far the most important, and the one most amenable to policy influences. In 2005 humans emitted some 27 billion tonnes of carbon dioxide into the atmosphere as a result of fossil fuel use – this is a little over 4 tonnes per capita. This aggregate hides huge differences – the per-capita figures are 20 tonnes for the USA, 12 for Russia, 8 for Europe, around 3.5 for China, and 1 tonne for India (Energy Information Agency, 2006).

The scientific consensus is that to contain climate change risks to a manageable level, by 2050 carbon dioxide emissions will have to be 50–75% lower than the business-as-usual level. The challenge of energy policy is to bring the global per-capita emission of carbon dioxide down to about 1 to 1.5 tonnes within this time frame. Another way of stating the challenge is that we need to increase our carbon productivity tenfold from the current level of around USD 740 of GDP per tonne of carbon dioxide emitted – an effort comparable in scale to the increase in manufacturing labour productivity over a century during the Industrial Revolution (McKinsey Global Institute, 2008, pp. 10–11).

Climate change is a global externality and requires a depth of cooperation between countries that goes far beyond anything we have experienced so far. The challenge is to agree on a fair sharing of environmental space between those who have occupied it first and those who are now in need of room to grow. A control on emissions will be required and, as Amartya Sen states in his interview at the Potsdam Nobel Laureate Symposium, the key questions are, who should do how much, and how should the costs be shared (Sen and Stern, 2007)? Should the long-term goal be to converge towards equal per-capita emissions at a level consistent with manageable climate change risk, say 1–1.5 tonnes by 2050? Or should those who have occupied the space with their past emissions do more to create space for the newcomers?

The risks of climate change depend on the cumulative emissions of greenhouse gases, and judgements about the fairness of alternative proposals on limits should take this into account. An illustrative calculation, presented in Table 1, shows that, even with limits greater than what are on offer at the moment, the developed world with less than one-sixth of the world population will occupy roughly one-half of the incremental space.

Energy consumption in the developing world is rising – as it should, given the present level of energy poverty. The big question for sustainability is whether the increase in energy demands that will necessarily accompany the move out of poverty can be met by low-carbon supply alternatives that are environmentally sustainable. In the developed world, adjustment must extend to already established energy consumption patterns. Thus, the real challenge is to manage demand. Is there a price that will be paid in terms of growth as we move to alternate energy paths? Or can the low-carbon alternatives provide an opportunity for new growth possibilities, particularly in regions deficient in fossil fuel resources, just as electricity did when it was first introduced?

Low-carbon growth may provide new business opportunities. But it also involves additional costs over and above the business-as-usual scenario. Some of the savings that arise from improved efficiency may have negative or zero costs (or collateral benefits, which amounts to the same thing). However, the required emissions reduction of 50–75% by 2050 will involve moving beyond low-carbon growth to measures that involve net additional costs. A recent McKinsey study has estimated that the abatement required to stay below 500 ppm of greenhouse gases will cost EUR 500–1100 billion in 2030, or about 0.6–1.4% of that year's projected global GDP (McKinsey Global Institute, 2008, pp. 15–16). However, 40+% of the abatement potential exists in the developing countries (excluding China), and will not be realized unless the transfers of finance and technology are substantially larger, more predictable, and more robust than at present. In addition, the developing world will need financial support for adaptation actions, which poses an even greater challenge because the rich countries do not see any direct return in terms of risk mitigation in this case. The application of the polluter-pays principle requires that the rich countries accept this obligation on the grounds that they are responsible for around 70% of accumulated carbon dioxide emissions from fossil fuel use since 1850.

**Table 1.** *Cumulative emissions of carbon dioxide from fossil fuel use (percentage share of world total).* (*Source:* Author's calculations based on data and BAU projections in Energy Information Administration, 2006).

|  | Developed countries | Developing countries | Developing Asia | Absolute amount (Gt CO$_2$) |
|---|---|---|---|---|
| 1980–2005 | 66.2% | 33.8% | 21.2% | 568 |
| 2005–2030 BAU | 53.8% | 46.2% | 33.3% | 895 |
| 2005–2030 with cuts* | 50.9% | 49.1% | 34.6% | 841 |

*Cuts: In 2030 Europe 30% below 1990 level; North America and OECD Asia at 1990 level; business as usual (BAU) in Russia and developing countries.

## Adaptation

Adaptation actions in the developing world have to address the links between poverty, ill health, population growth, and the deterioration of land, water and biotic resources at the local level. In the villages of the Asia, Africa and Latin America poverty eradication requires that the productivity of poor households is raised. This in turn requires a systematic effort to rehabilitate degraded land and water resources, and an integrated approach to land, water, and biotic management that respects climatic and other ecological constraints. The climatic changes which now are unavoidable will require that this will be even more necessary as a condition for poverty eradication.

In the rural areas of the developing world the impact of climate change will be felt directly through changes in precipitation, groundwater recharge, and river flows. Our knowledge about impacts in developing countries is still sketchy, and not all of the impacts will be negative. However a major change, even a favourable one, in something as basic as climate will require substantial societal, technological, and economic adaptation. The key instrument for such adaptation is water resource management. If we can get that right, many other things will also fall into place.

Balancing water use and availability in a watershed or river basin, setting priorities between competing demands, ensuring adequate drainage of used water, and maintaining water quality necessarily require that we get the land, forest, and settlement policies right. Rational land use, forest conservation in catchment areas, restoration of degraded lands, and land engineering for water retention and drainage are all aspects of water management. Public spending programmes for agriculture and rural development must be tailored to agro-climatic regions, and water resource planning must move away from civil engineering projects to become an element in integrated land and water management.

Climate change will change the physical geography of the planet, and this will lead to changes in its human geography too. One dimension of this is migration. The 60% of the world population that lives within 100 km of the coast will be affected by rising sea levels, worsening storm surges, saline intrusion, and so on. Many people will migrate, and much of this migration will be from one poverty-stricken area to another, as we already see among conflict refugees in Africa. But, as Amartya Sen points out, this will involve a slow process rather than sudden large-scale migration (Sen and Stern, 2007).

The population movements induced by climate change will come on top of a huge rural-urban shift. More and more of the population in the developing world will live in cities, which are already under pressure. Ensuring sustainable urbanization may be the most important challenge for coping with climate risks. The critical areas that need to be addressed are water, sanitation and energy use, particularly in transportation.

## Economic and technological solutions

Energy, water and human settlements are the critical sectoral areas both for poverty eradication and for mitigation and adaptation actions to cope with climate change risks. The policies and programmes in these areas have to operate in a market economy where the most important challenge is to get prices right so that they reflect full social costs from the beginning to the end of the production and consumption process, including, particularly, the costs of waste disposal. Unfortunately the three sectors of greatest concern are precisely the ones where markets are distorted by subsidies and often operate inequitably.

The most important policy challenge for mitigation is carbon pricing (see Edenhofer *et al.* and Mirrlees, this volume). There is, at present, no cost attached to carbon emissions in most countries, the few exceptions being those where some form of carbon taxation is in force. The market in carbon credits that has emerged with the establishment of emission caps fulfils a similar purpose. In a market economy the most effective instrument for promoting mitigation is to ensure through taxes or cap-and-trade systems that the global social cost of carbon is reflected in the calculations of companies, which decide on investments and develop new techniques, and of individuals, who consume goods and energy.

In the long run, the scale of adjustment required is such that we have to look to radically new technologies. Our past experience shows that a single, objective-oriented approach to technology development often leads to new problems. For instance, when CFCs were first introduced for refrigeration, aerosols, and foam rubber manufacture, they were considered safe chemicals because they are stable, non-corrosive, do not involve any explosion hazard, and are not directly toxic to human beings. It was only later that their impact on the ozone layer and the consequences of this were understood (see Molina, this volume). A more germane example is that of biofuels whose indiscriminate promotion has led to inappropriate land use and unintended increases in food prices (see Creutzig and Kammen, this volume). Hence, any mission-oriented approach to carbon-saving technologies must be accompanied by a system of technology assessment that takes ecological and economic dimensions into account and keeps the principle of equity in sight.

## The elements of a potential climate agreement

What are the elements of a potential climate accord that could address these problems in a manner that is, in Nicholas Stern's words, 'effective, efficient and equitable'?

First, we must agree on a long-term goal corresponding to an acceptable risk level for global warming. It has to be realistic enough to be attainable, yet ambitious enough to avert the more catastrophic consequences of temperature change. This will involve both an assessment of likely risks and value judgments about the level and distribution of the costs and benefits of mitigation measures. One point worth noting is the growing concern among scientists about potential tipping points that could cause serious change to the organisation and appearance of the Earth system, and produce consequent challenges for human society. Runaway climate change, which would make human life on Earth difficult if not impossible, is not a part of any projection; but we cannot currently rule out scientifically that it could be triggered.

Second, the most elementary notions of fairness require that the burden for immediate action must fall on those who are most culpable in terms of past emissions. The calculations presented above on how future cumulative emissions would be distributed suggest that the immediate commitments by the developed countries would need to be greater than what is being talked about at present. If the developed countries, USA and Russia included, fail to demonstrate a responsible sense of purpose, it will be difficult to persuade poorer countries, who have only just started on the path of energy consumption growth, to take on any serious commitment.

Third, the developing countries will also have to contribute to mitigation measures in the long-term. But their exemption from immediate commitments does not mean business as usual. Their energy consumption and emissions may grow. But they can and should be assisted in using all economically viable means to promote energy efficiency, to use lower-carbon energy alternatives, and to implement appropriate forms of demand management. It is in the global interest to provide concessional finance and technology transfer, first through means like the Kyoto

Protocol's Clean Development Mechanism (CDM, see Liverman, this volume), and second through the direct provision of soft loans and grants for mitigation efforts, including for deforestation avoidance and reforestation. One could even integrate the two strands by providing the soft grants for mitigation in the form of the purchase of carbon credits from developing countries that add to global mitigation because, unlike the CDM purchases, they are not used to offset developed country mitigation obligations.

Fourth, a certain degree of climate change is inevitable based on any realistic assumption of what the long-term agreed mitigation goal will be. The burden of adjustment to this change will be very unevenly distributed. Much of it will fall on countries that have limited financial and technical capacity to take on the additional effort required. These adaptation costs must be paid for in strict proportion to the responsibility for the problem (for example, as defined by cumulative emissions), and distributed according to need so that small island countries, for instance, receive much more in per-capita terms, because of their greater need, than large continental countries.

Finally, new technologies that save on carbon dioxide or other greenhouse gas emissions, or which sequester the emissions in some way, will be needed as we move to a point at which we do not add to the stock of greenhouse gases in the atmosphere. This will require cooperative arrangements beyond normal commercial exchanges for the development, dissemination, and sharing of these technologies.

All of these elements are envisaged in the agreements reached in the UNFCCC at Bali. They are being negotiated at present and are to be finalized by the end of 2009. The difficulty now is not the lack of a mandate but the willingness to recognize that time is running out and we do not have the option of concluding a weak agreement now in the hope that the next agreement a decade from now will be better.

## Time to change our thinking

We need to change how we think, and move beyond inherited concepts to develop a common language of discourse between economists, ecologists, engineers, and ethical philosophers. Like an ecologist, we must respect the integrity of natural systems; but, like an engineer, we must be willing to intervene in these systems to meet human needs. The solutions proposed have to work in a market economy, and this is where the economist's concerns about balancing costs and benefits and choosing optimally between alternatives comes in. Every solution that is proposed will involve some distribution of responsibility within and between generations, and within and between the political jurisdictions into which the human population and our planetary ecosystem are divided. This is where ethics comes in with its judgments of what is just and fair.

Effective global action on climate change will require such a synthesis. To an extent this has been achieved already as scientists, engineers and technologists look for creative solutions to climate change. The consensus-building process in the Intergovernmental Panel on Climate Change (IPCC) and the structured dialogue that it has promoted have clearly contributed to this. The economics of climate change are also receiving attention, while the recent seminal exercise led by Nicholas Stern has contributed hugely to the debate (Stern, 2007). But the degree of agreement that prevails in this area is well short of a consensus. The really difficult area is the ethical concern about burden-sharing, which has largely been left to the cut and thrust of diplomatic negotiations, where we have not moved beyond a few general principles such as 'common but differentiated responsibility'.

What we need is what Amartya Sen has called 'public reasoning'– a process of raising awareness not just about the problem but also about how it affects people differently, who has the capacity to cope and who needs help, the solutions that are available and those that still need to be found, and so on (Sen and Stern, 2007). But more than that, we need a sense of urgency. Ten years from now it may be too late to prevent catastrophic climate change. That will be a disaster both for sustainability and for development. The time to act is now.

## References

Desai, N. (2007). Sustainable development. In Basu, K., ed., *The Oxford Companion to Economics in India.* New Delhi.

Energy Information Administration (2006). *International Energy Annual 2004.* Washington.

McKinsey Global Institute (2008). *The Carbon Productivity Challenge: Curbing Climate Change and Sustaining Economic Growth.* Washington D. C. Available at http://www.mckinsey.com/mgi/reports/pdfs/Carbon_Productivity/MGI_carbon_productivity_full_report.pdf.

Sen, A. and Stern, N. (2007). *Amartya Sen – Nicholas Stern.* Interview. Potsdam Nobel Symposium, 10 October. Available at http://www.nobel-cause.de/webcasts.

Stern, N. (2007). *Stern Review: The Economics of Climate Change.* Cambridge.

World Commission on Environment and Development (1987). *Our Common Future.* Oxford.

# Chapter 15

## Development and sustainability: conflicts and congruence

---

## Kirit S. Parikh

Kirit S. Parikh studied economics and obtained his PhD in civil engineering at the Massachusetts Institute of Technology. He has held a professorship in economics since 1967 and was Director of the Indira Gandhi Institute of Development Research in Mumbai from 1986 to 2000. From 2004 to 2009 he was a member of the Planning Commission of the Government of India and also a member of the Energy Coordination Committee, the Committee on Rural Infrastructure, and the Committee on Infrastructure constituted by the Indian Prime Minister. In the past he was a member of Economic Advisory Councils serving five different Prime Ministers. Currently he is Chairman of Integrated Research and Action for Development, based in New Delhi. In 2003 he was made an honorary life member of the International Association of Agricultural Economists. Parikh has also edited the India Development Report.

*Note:* This chapter is a commentary on chapter 14.

Nitin Desai has described clearly the development challenge that the world faces. The extent of global poverty, ill health, illiteracy and ill-being is such that one cannot question the need for development. As Mahatma Gandhi said, even God would not dare to appear before a hungry person in any form other than food.

The threat to sustainability arises mainly from the unsustainable consumption patterns of the rich. In a paper prepared for the Secretariat of the UN Conference on Sustainable Development at Rio in 1992, Parikh *et al.* (1991) pointed out that the bulk of global resource use was by the people living in developed countries (Annex I countries[1]), who constituted 25% of the global population but consumed more than 70% of most resources (see Tables 1 and 2). Even their consumption of cereals accounted for nearly half of the total global consumption. One would have thought that the human stomach has a limited capacity and food consumption would saturate, but if we count as human consumption the grain consumed by the cow that becomes the hamburger, then food consumption keeps growing with income. Figure 1 shows that while direct consumption of food in terms of calories per person saturates, total use of cereals including for animal feed continues to increase in line with income.

The share of global resource use by developing countries has increased over the years compared to the data in Table 1 as poorer nations aspire to the consumption patterns of the rich. This is clearly seen in Table 3 which provides more recent data on consumption. The disparity ratio of per-capita cereal consumption has changed little while the ratios for milk and meat consumption have been reduced, largely due to economic growth in China and India. In spite of larger populations in developing countries, in 2007 the developed countries still consumed 39% of cereals, 50% of milk, 41% of meat, 40% of round wood, 74% of sawn wood and 71% of paper. The shares of fertilizer and cement use by developing countries have increased due to development of modern intensive agriculture and infrastructure.

Table 4 shows data for primary energy consumption and carbon dioxide emissions in 2005. It shows that disparity ratios of per-capita consumption have come down compared to Table 2 but the developed countries still consume 63% of total primary energy in the world and produce 59% of global carbon dioxide emissions.

Globalization and the communication and information revolution have made people all over the world aware of the lifestyle of the rich. The rapidly growing economies of an increasing number of countries are bringing such consumption within the reach of an ever increasing number of people. Preaching to them to forego goods they have long strived for (to not own cars, to live in small crowded

---

[1] Annex I countries are industrialized countries and economies in transition that have signed the United Nations Framework Convention on Climate Change (UNFCCC, 1992).

**Table 1.** *Consumption patterns for selected commodities in 1987: distribution among developed and developing countries. (Source:* Parikh *et al.,* 1991)

| Category | Products | World total (Mt or Mm³)* | Share of developed countries (%) | Per capita (kg or litre)** | | Disparity ratio of per-capita consumption | |
|---|---|---|---|---|---|---|---|
| | | | | Developed | Developing | Developed/ developing | USA/ India |
| a) Food | Cereals | 1801 | 48 | 717 | 247 | 3 | 6 |
| | Milk | 533 | 72 | 320 | 39 | 8 | 4 |
| | Meat | 114 | 64 | 61 | 11 | 6 | 52 |
| b) Forestry | Round wood | 2410 | 46 | 888 | 339 | 3 | 6 |
| | Sawn wood | 338 | 78 | 213 | 19 | 11 | 18 |
| | Paper, etc. | 224 | 81 | 148 | 11 | 14 | 115 |
| c) Industry | Fertilizers | 141 | 60 | 70 | 15 | 5 | 6 |
| | Cement | 1036 | 52 | 451 | 130 | 3 | 7 |

* Mt = million tonnes for food and industry, Mm³ = million cubic metres for forestry, ** kg = kilograms for food and industry; litres for forestry

**Table 2.** *Patterns of primary energy consumption and related carbon dioxide emissions in 1987. (Source:* Parikh *et al.,* 1991)

| Item | World total (Mt) | Share of developed countries (%) | Per capita (kg) | | Disparity ratio of per-capita consumption | |
|---|---|---|---|---|---|---|
| | | | Developed | Developing | Developed/ developing | USA/ India |
| Primary energy consumption (OE*): | | | | | | |
| Solid | 2309 | 66 | 1278 | 199 | 6 | 14 |
| Liquid | 2745 | 75 | 1720 | 175 | 10 | 61 |
| Gas | 1611 | 85 | 1147 | 61 | 19 | 227 |
| **Total** | **7009** | **75** | **4376** | **453** | **10** | **35** |
| Emissions ($CO_2$): | | | | | | |
| **Total emissions** | 20984 | 70 | 12.5 | 1.5 | 8 | 27 |
| Solid | 8848 | 64 | 4.8 | 0.7 | 6 | 14 |
| Liquid | 8085 | 70 | 4.8 | 0.7 | 8 | 54 |
| Gas | 3326 | 82 | 2.2 | 0.1 | 21 | 228 |

*Oil equivalent

**a)**

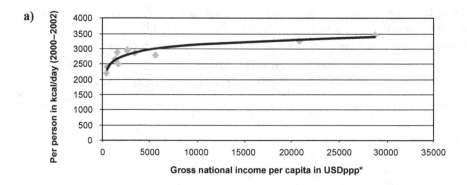

**b)**

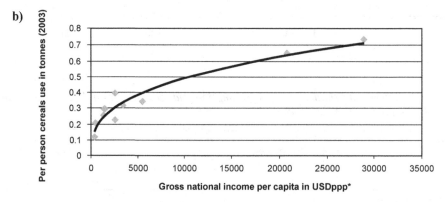

\* purchasing power parity

**Fig 1.** a) Daily per capita food calorie intake (2000–2002) and b) annual cereals use including for animal feed (2003) as a function of per-capita income. Grey diamonds: data points; black lines: fitted trends. (*Source:* FAO, 2005)

homes, to eat only vegetarian food, etc.), is unlikely to be effective. The greatest challenge facing a world of rapidly developing nations is to convince those people who have become prosperous to consume in a sustainable manner. The conflict here is essentially between development and sustainability.

Population growth on its own does stress the climate. However, the proximate cause is our unsustainable consumption patterns. A global population of six billion emitting greenhouse gases (GHGs) at the average level of US citizens would produce as many as 120 billion tonnes of carbon dioxide. On the other hand, 12 billion people emitting at the rate of the average Indian citizen would produce only 12 billion tonnes of carbon dioxide, well within the Earth's absorptive capacity (see

**Table 3.** *Consumption patterns for selected commodities in recent years: distribution among developed and developing countries.* (*Sources:* FAO, 2005; Population Reference Bureau, 2007; cement data: http://www.indexmundi.com/minerals/?product=cement&graph=production)

| Category | Products | World total (Mt or Mm3)* | Share (%) | | Per capita (kg or litre)** | | Disparity ratio of per-capita consumption | |
|---|---|---|---|---|---|---|---|---|
| | | | Developed | Developing | Developed | Developing | Developed/ developing | USA/ India |
| a) Food | Cereals (2007/08) | 2 126 | 39 | 61 | 678 | 240 | 3 | 6 |
| | Milk (2007) | 677 | 50 | 50 | 279 | 62 | 5 | 3 |
| | Meat (2008) | 278 | 41 | 59 | 93 | 30 | 3 | 21 |
| b) Forestry | Round wood (2003) | 3 346 | 40 | 60 | 1 112 | 393 | 3 | 5 |
| | Sawn wood (2003) | 401 | 74 | 26 | 249 | 20 | 12 | 38 |
| | Paper, etc. (2003) | 328 | 71 | 29 | 193 | 19 | 10 | 68 |
| c) Industry | Fertilizers (2002) | 142 | 35 | 65 | 41 | 18 | 2 | 4 |
| | Cement (2005) | 2 310 | 23 | 77 | 349 | 357 | 1 | 3 |

\* Mt: million tonnes for food and industry, $Mm^3$: million cubic metres for forestry
\*\* kg: kilograms for food and industry, litres for forestry

Table 4). This is not to argue that the world's population should live like Indian citizens do, but rather to emphasize the importance of consumption patterns. Once we recognize that the poor also aspire to the lifestyle of the rich, it is clear that population growth needs to be contained as much as possible as a larger population will ultimately put greater stress on the Earth's resources.

While most religions preach contentment and restraint, current levels of greed and consumption do not suggest that they have succeeded in modifying the behaviour of most people. There are unfortunately few who follow Mahatma Gandhi, who practised *aparigraha* (i.e., not taking anything more than what one needs). Even when at a river, Gandhi did not use a drop of water more than he needed.

Technological development can reduce the need for resources. However, an attitude of conservation and lifestyle changes can also be very important. Using mass transport wherever possible, walking or cycling for short distances, cutting

**Table 4.** *Patterns of primary energy consumption and related carbon dioxide emissions in 2005.* (*Source:* EIA, 2005)

| Item | World total (Mt) | Share (%) | | Per capita (kg) | | Disparity ratio of per-capita consumption | |
|---|---|---|---|---|---|---|---|
| | | Developed | Developing | Developed | Developing | Developed/ developing | USA/India |
| **Primary energy consumption (OE*)** | | | | | | | |
| Solid | 3087 | 46 | 54 | 935 | 335 | 3 | 10 |
| Liquid | 4269 | 64 | 36 | 1812 | 306 | 6 | 33 |
| Gas | 2706 | 74 | 26 | 1323 | 141 | 9 | 65 |
| Total | 11647 | 63 | 37 | 4860 | 860 | 6 | 24 |
| **Emissions ($CO_2$)** | | | | | | | |
| Total emission | 28051 | 59 | 41 | 10.9 | 2.3 | 5 | 20 |
| Solid | 11378 | 46 | 54 | 3.5 | 1.2 | 3 | 10 |
| Liquid | 10996 | 63 | 37 | 4.6 | 0.8 | 6 | 33 |
| Gas | 5666 | 74 | 26 | 2.8 | 0.3 | 9 | 66 |

*Oil equivalent

consumption of meat, reducing waste, and recycling can be very effective in reducing resource use.

Yet, we cannot preach *aparigraha* to the nearly 300 million people who live below the poverty line in India. In 2007 half of India's children were underweight (moderate to severe undernutrition) or stunted. About 30% of all adults had a BMI (Body Mass Index) under 18.5, which defines adult malnutrition (Planning Commission, 2008).

If the world is to be socially and politically sustainable, we must deal with poverty and deprivation. Sustainability requires economic development until a sufficient level of wealth is achieved. The poor often depend on natural resources for food, fodder and fuel. As populations grow use of these resources often exceeds their natural regenerative capacity. Natural resources thus become depleted and resource use becomes unsustainable. Development can help arrest such degradation by providing alternatives and by improving the productivity of such resources.

India needs to grow rapidly for a number of years if it is to eradicate poverty and offer its people a satisfactory standard of living. Only rapid and sustainable growth can generate the resources needed to provide the social and physical infrastructure for education, health services, clean water, sanitation, transport and energy. Only a rapidly developing economy can create adequate opportunities for gainful employment for all of India's people. However, India, like most tropical countries, is likely to face increasing constraints due to global climate change, restricting the attainment of its short- and long-term development goals.

The Government of India has restructured policies to achieve a new vision based on faster, more broad-based and inclusive growth. The key goal is to reduce poverty rapidly and focus on bridging the various divides that continue to affect our society. Inclusive growth is needed for social and political sustainability. India's need and right to develop cannot be denied. India recognizes that development requires an efficient energy sector. To ensure that development is sustainable India will eventually need to make a transition to a largely renewable energy system. Yet it must be accepted that India's emissions will grow and that the required share of the global environmental space must be provided.

An even greater challenge lies in resolving conflict around the use of common global resources. The industrialized countries have emitted two thirds of all cumulated GHG emissions. Table 5 shows the cumulative carbon dioxide emissions from 1950–2005 and 2000–2005. The share of emissions produced by Annex I countries was around 56% in 2000–2005. Over these six years Annex I countries have emitted 85 458 million tonnes of carbon dioxide ($MtCO_2$) compared to India's emissions of 6614 $MtCO_2$. India's emissions were 1020 $MtCO_2$ in 2000 and 1222 $MtCO_2$ in 2005, indicating a growth rate of 3.7% per year. Even with an emissions growth rate of 5% per year the sum total of India's emissions over 30 years from 2006–2035 would be less than what Annex I countries have emitted between 2000 and 2005.

The developed countries, which have occupied a disproportionate share of the environmental space, have a special responsibility not only to compensate the poor on whom they have inflicted heavy adaptation burdens but also to reduce their resource consumption as soon as possible. While adaptation can reduce the burden of climate change it cannot completely eliminate it, nor is it cost-free. A person living in a coastal area adapts when the sea level rises by moving. That saves his life but not his property. In fact, by migrating he may impose cost on others. By delaying action, the rich are occupying more of the global space at the expense of the poor (Parikh and Parikh, 1998). For example, the annual carbon dioxide emissions of the USA alone have increased between 1990 and 2005 by about as much as India's total annual emissions. The USA today emits five times as much as India (see Fig. 2).

The right to the global atmospheric carbon space does not belong to the initial occupiers. Unlike land, which can be fenced, global space cannot be fenced. There is no way to prevent developing countries from emitting GHGs except through a mutually acceptable global contract. This will require that industrialized countries reduce their emissions and make space for developing countries. All countries have a stake in achieving sustainability.

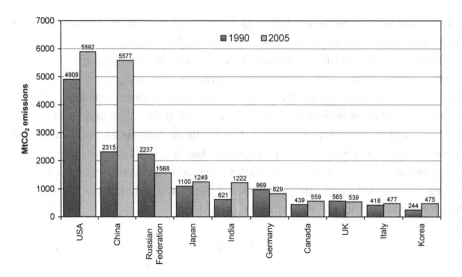

**Fig. 2.** Annual carbon dioxide emissions from fuel combustion of selected countries in 1990 and 2005. (*Source:* Parikh, 2007)

It is clear that India has not contributed to the threat of climate change and is not responsible for it. India's cumulative emissions over the period 1950–2005 constitute less than 2% of global emissions (see Table 5) and are well within any reasonable share of the global environment's absorptive capacity. Even the rich in India emit less carbon dioxide per capita than the average of developed countries. The per-capita emissions of the richest 10% of India's urban population, which constitutes only 3% of India's total population, was less than 4 tonnes in 2003 (see Table 6), compared with the US average of 19.9 tonnes and the European Union average of 8.5 tonnes. However we should mention that the rich in India would like to consume more but are unable to do so due to lack of infrastructure such as motorways.

India is extremely vulnerable to climate change. As a responsible nation India has taken the initiative to stimulate action on climate change. Prime Minister Manmohan Singh stated at the Heiligendamm G 8 + 5 Conference (Government of India, 2008) that

> 'We are determined that India's per-capita GHG emissions are not going to exceed those of developed countries even while pursuing policies of development and economic growth. [...] We must work together to find pragmatic, practical solutions, which are for the benefit of entire humankind'.

**Table 5.** *Cumulative carbon dioxide emissions from fuel combustion excluding emissions from land use changes.* (*Source:* WRI, 2009)

|  |  | Annex I countries | G 77 + China | China | India | Brazil |
|---|---|---|---|---|---|---|
| 1950–2005 | $MtCO_2$ | 875 158 | 518 989 | 130 067 | 22 581 | 69 723 |
|  | % of world total | 66.8 | 33.2 | 10.5 | 1.8 | 5.6 |
| 2000–2005 | $MtCO_2$ | 85 458 | 56 333 | 25 285 | 6 614 | 2 023 |
|  | % of world total | 56 | 37 | 17 | 4 | 1 |

The implications of this are worth noting. It implies a huge commitment. If global warming is to be limited to less than $2\,°C$, this requires the stabilization of GHG concentrations at 450 ppm (parts per million) carbon dioxide equivalents. This, in turn, will require an 80–90% reduction of emissions in industrialized countries by 2050. Thus, their per-capita emissions would need to be lowered to around 2.5 tonnes of carbon dioxide per year. Given the implementation of ambitious energy efficiency measures and promotion of renewables, India will reach this level of per-capita emission by 2030. Given current growth projection and presently available technologies India runs a risk of exceeding this level by 2050 and thus it will have to take steps to curtail its emissions. The ball, nevertheless, is indisputably in the court of the industrialized countries. The more they reduce their emissions, the lower a limit India will accept on its emissions. India should no longer be used as an excuse by industrialized countries for delaying mitigation action.

An effective agreement on mitigation at a global level is needed and we hope that Copenhagen will produce it. The agreement will have to be based on the principles of equity and differentiated responsibility.

It is sometimes argued that, even though per-capita emissions of India and other non-Annex I countries are low, India's industries compete on the world market and so we should have sectoral standards for emissions. There are many difficulties in implementing sectoral standards. First, we need to decide what should be compared: carbon dioxide per tonne of product, carbon dioxide per dollar value of output, or carbon dioxide per unit value added? Should we take sectoral averages or only consider new capacity? Should we account for the specific circumstances of a country such as ambient air temperature, which affects the fuel efficiency of a machine or a plant? If a country has a relatively large carbon dioxide emissions quota this is part of that country's comparative advantage, just like skilled labour, large capital stock or technological knowledge. Sectoral standards thus contradict the very basis of free trade.

The principle of equal per-capita emissions, at least in the long-run, has been widely accepted. The acceptance of this principle and the immediate allocation of

**Table 6.** *India's per-capita carbon dioxide emissions of 2003 by expenditure class.*(*Source:* calculated by the author using the Social Accounting Matrix of India for 2003–4 based on emissions from direct consumption of energy as well as indirect emissions embodied in consumption items)

| Expenditure class | Poorest | 2 | 3 | 4 | Richest |
|---|---|---|---|---|---|
| Rural populations | | | | | |
| (millions) | 75 | 150 | 293 | 150 | 75 |
| (% of total) | 7 | 15 | 29 | 15 | 7 |
| Carbon dioxide emissions | | | | | |
| (tonnes/person) | 0.1 | 0.2 | 0.3 | 0.6 | 1.2 |
| Urban populations | | | | | |
| (millions) | 29 | 57 | 114 | 57 | 29 |
| (% of total) | 3 | 6 | 11 | 6 | 3 |
| Carbon dioxide emissions | | | | | |
| (tonnes/person) | 0.3 | 0.5 | 0.8 | 1.5 | 4.0 |

tradable emission quotas on a per-capita basis would indeed be fruitful. Not only would it bring about a desired emission reduction, it would also stimulate technology development, reduce the costs of technology, increase incentives to rationalize GHG emission in all countries, and ensure equitability across nations.

Instead of allocating *annual* emission quotas it may be more rational to allocate global environmental space. For example, to ensure stabilization at 450 ppm carbon dioxide equivalents we should estimate the total GHG emissions from 1990–2050 or till 2100 that can be emitted in terms of 'tonne years' of emissions, taking into account how many years the emissions occupy the space. Quotas should be allocated on a per-capita basis in a tradable way. Alternatively, a rent could be charged from all users for every tonne year' of space occupied. This rent could then be distributed on a per-capita basis to all citizens of the world in inverse proportion to their per-capita income and per-capita emissions. This is like a carbon tax levied on a country's cumulative emissions from 1990 onwards.

In addition to mitigation, a further major challenge is posed by the burden of adaptation. Adaptation can help mitigate some adverse impact of climate change. However, adaptation in the form of migration out of submerged areas to urban areas can – as Nitin Desai points out – threaten sustainability. A rise in sea level, changes in the hydrological regime, salination ingress, coastal submergence and resulting migration further aggravate the problems already created by rapid urbanization.

Congestion already causes huge traffic jams in Indian cities leading to wasteful burning of fossil fuels and air pollution. A large proportion of India's population lives in slums without adequate sewerage facilities. Less than half the effluent from Indian cities is treated before it is discharged into lakes, rivers and oceans. Limited resources make it almost impossible to develop water, sanitation and transport infrastructure in pace with rapid urbanization. Mass migration induced by climate change would be catastrophic. We must find ways to deal with these problems. Mass transport systems must be built in large cities. Anticipating the need for them in smaller cities, long-term transport plans should be developed and rights of way for future mass transport corridors should be acquired now. Private builders and developers must be required to provide proper water and sewerage infrastructure. These, however, cannot be maintained without appropriate user charges.

While the ill-effects of urbanization on air and water quality are all too visible, one should not forget the impact it can have on rural areas. Rural-urban migration relieves the pressure on agricultural land. Farmers who stay behind can have more land to till. Pressure on rural commons for fuel may decrease and some regeneration can take place. On the other hand, more intensive cultivation can also have negative consequences for environmental sustainability. Sustainable urbanization will have to accompany sustainable agriculture.

Viewed from a long-term perspective, in order to sustain consumption at acceptable levels we must develop technologies using renewable resources. New technologies have to be sustainable and that requires multi-disciplinary approaches and involvement of engineers, scientists, ecologists and social scientists, as rightly emphasized by Nitin Desai. The challenge is to develop these technologies and adapt them before we cause irreversible damage to the Earth's biosphere. This will require that such technologies are shared among all as global public goods. 'Public reasoning' at a global level is called for, as suggested by Amartya Sen.

However, technologies, while critical, are by themselves not enough. Lifestyles will also have to be modified. Attitudes to consumption will have to change. As the Indian sages have advised, *'Ten Tyakten Bhunjithah'* ('you must give something up in order to enjoy it'). The rich in developed and developing countries alike will have to set examples of sustainable lifestyles for the poor to emulate. The sooner this happens the better is the chance of avoiding catastrophic climate change.

## References

EIA (2005). *International Energy Annual 2005* (June–October, 2007). Available at http://www.eia.doe.gov/iea.

FAO (2005). *Summary of World Food and Agricultural Statistics 2005.* Rome.

Government of India (2008). *National Action Plan on Climate Change.* Available at http://pmindia.nic.in/Pg01-52.pdf.

Parikh, J. (2007). Climate change and India – a perspective from the developing world. In *Climate Change Getting it Right,* Growth 59. Melbourne. http://ceda.com.au.

Parikh, J. and Parikh, K. (1998). Free ride through delay: risk and accountability for climate change. *Journal of Environment and Development Economics* **3**(3), 347–409.

Parikh, J., Parikh, K. S., Gokarn, S., Painuly, J. P. and Saha, B. (1991). *Consumption Patterns: The Driving Force of Environmental Stress.* Report prepared for UNCED. Mumbai.

Planning Commission (2008), Government of India. *The Eleventh Five Year Plan.* New Delhi.

Population Reference Bureau (2007). *World Population Data Sheet.* Available at http://www.prb.org/pdf07/07WPDS_Eng.pdf.

UNFCCC, United Nations Framework Convention on Climate Change (1992). http://www.unfccc.org.

WRI (2009). *Climate Analysis Indicators Tool (CAIT),* Version 6.0. Washington, D. C. http://cait.wri.org.

# Part III

Institutional and economic incentives

# Chapter 16

# Robust options for decarbonization

---

## Thomas Bruckner, Ottmar Edenhofer, Hermann Held, Markus Haller, Michael Lüken, Nico Bauer, and Nebojsa Nakicenovic

Thomas Bruckner, born in 1966, completed his doctoral studies in theoretical physics at the University of Würzburg, Germany. After four years at the Potsdam Institute for Climate Impact Research (PIK), he led the research group on Energy System Optimization and Climate Protection at the Institute for Energy Engineering at the Technical University of Berlin (2000–2008). Currently, he holds the Chair of Energy Management and Sustainability at Leipzig University, Germany. He is a fellow of the Heisenberg Programme of the German Research Foundation (DFG), a visiting scientist at the Potsdam Institute for Climate Impact Research, and a member of the Intergovernmental Panel on Climate Change. His current research focuses on integrated assessment modelling and the investigation of liberalized energy markets facing climate protection constraints.

*Note:* Photos and biographies of co-authors can be found in the appendix.

Since 2007, the perceptions of the international community all over the world about the dangers of climate change and about the need for vigorous response strategies have changed dramatically. This change was triggered by the release of the Fourth Assessment Report (AR4) *Climate Change 2007* by the Intergovernmental Panel on Climate Change (IPCC) and by the ongoing scientific progress in the field of global climate change. The scientific consensus reported in the AR4 received an unprecedented echo in the media and subsequently raised the public awareness concerning global climate change and its adverse impacts to an extent never seen before. As a result, the report encouraged numerous initiatives to combat global climate change – most notably the European Union's decision to reduce greenhouse gas (GHG) emissions by 20% by 2020 (compared to the amount of GHGs emitted in 1990). In addition, more than 100 countries followed the European example and adopted a global warming limit of 2°C or below (relative to preindustrial levels) as a long-term climate protection goal.

In order to assess the opportunity to stabilize carbon dioxide ($CO_2$) concentrations at a level that is compatible with the EU climate protection goal, the following issues need to be addressed. Which temperature changes are to be expected in the business-as-usual case, in other words, if no specific measures directed at mitigating climate change are implemented? Is there thus a real necessity to change course? If there is a real necessity, could cheap energy efficiency improvements solve the problem? If we need other, additional climate protection options, then which technologies are available and how great are the potential and available resources for the respective options? And finally, how should these options be combined in order to achieve least-cost climate protection?

## Projected energy demand and associated business-as-usual greenhouse gas emissions

An extensive review of recent long-term scenarios (Fisher *et al.*, 2007) revealed that enhanced economic growth is expected to lead to a significant increase in gross domestic product (GDP) during the twenty-first century (see Fig. 1a) – throughout the world but especially in the developing countries and emerging markets. The expected rise in prosperity will reveal itself in a significant increase in the demand for energy services. Motivated by the first oil crisis, humankind was able to reduce the primary energy input required to produce one GDP unit (the so-called primary energy intensity) und is expected to do so further in the future (see Fig. 1b). Unfortunately, the historical improvements in energy intensities were not sufficient to fully offset the GDP growth, resulting in increased energy consumption.

The respective increase in energy efficiency in the scenarios is more than compensated by the anticipated huge economic growth. In the business-as-usual case,

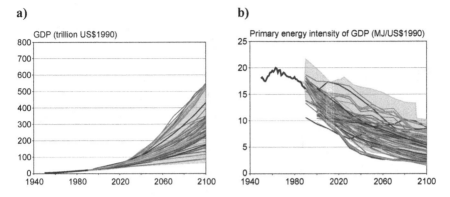

**Fig. 1.** a) Projected global economic growth and b) changes in primary energy intensity. (*Source:* adapted from Fisher *et al.,* 2007, pp. 180 and 184)

the demand for global primary energy is therefore projected to increase substantially during the twenty-first century (see Fig. 2 a).

Similarly to the development of the primary energy intensity, the carbon intensity (the amount of carbon dioxide emissions per unit of primary energy) is – with few exceptions – projected to decrease as well (see Fig. 2 b). This development reflects the global tendency to initially replace coal by oil and subsequently oil by gas, nuclear energy, and renewable energies.

Despite the substantial decarbonization projected to take place during the entire twenty-first century, even in the reference scenarios that do not include any explicit policies directed at mitigating climate change, the overwhelming majority of the emission projections exhibit considerably higher emissions in 2100 compared with those in 2000 (see Fig. 3 a). Due to the long life-time of carbon dioxide, this implies increasing carbon dioxide concentrations and in turn, increasing changes in global mean temperature throughout the twenty-first century. Figure 3 b shows the respective changes (together with the uncertainty range due to differences in the applied general circulation models, right-hand bars) for representative emission scenarios (so-called SRES scenarios, see Nakicenovic *et al.,* 2000) taken from the set of emissions scenarios shown in Figure 3 a.

## The threat of global climate change:
## avoiding the unmanageable

Compared with the preceding Third Assessment Report, the IPCC AR4 reflects a considerable improvement in our understanding of global warming. The report itself and the ongoing scientific progress achieved since then show an increasing recognition that the severity of the global climate change problem has been significantly underestimated in the past (Smith *et al.,* 2009; Meinshausen *et al.,* 2009).

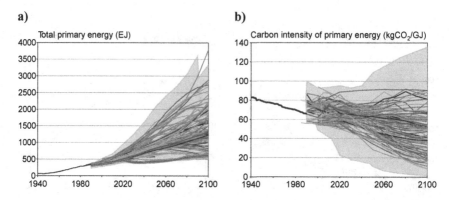

**Fig. 2.** a) Projected increase in primary energy supply and b) expected carbon intensity changes. (*Source:* adapted from Fisher *et al.,* 2007, pp. 183–4)

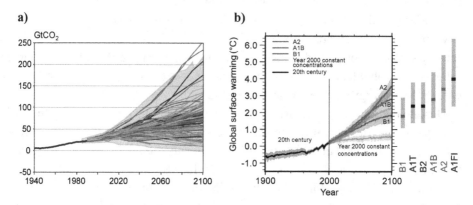

**Fig. 3.** a) Projected growth of carbon dioxide emissions and b) associated global mean temperature changes (relative to the temperature in 2000). (*Source:* adapted from Fisher *et al.,* 2007, p. 187; IPCC, 2007, p. 14)

According to its mandate, the IPCC is charged with summarizing the published scientific findings on global warming, its potential impacts, and opportunities to mitigate them. As a scientific council, the IPCC itself is not allowed to give specific policy recommendations concerning a suitable ceiling on global mean temperature rise to avoid dangerous interference with the climate system. Nevertheless, the information provided in AR4 (see Yohe *et al.,* 2007) supports the prominent climate protection goal that seeks to constrain global mean temperature change to less than 2 °C. This temperature threshold has been recommended by various advisory groups (e.g., the German Advisory Council on Global Change) in the past and became the official climate protection goal of the European Community in 2005.

Since then, more than 100 countries have adopted this global warming limit (Meinshausen *et al.*, 2009).

Assuming a best-guess climate sensitivity, staying below 2 °C implies that the $CO_2$-equivalent concentration would need to be stabilized at below 445 ppm (see Fig. 4a), compared to current concentrations of about 430 ppm $CO_2$-equivalent. That effectively means that we are already right at the limit of acceptable GHG concentrations in our atmosphere. Consequently, global emissions must decline significantly over the coming decades, with a global peak in emissions in the next five years. By 2050, emissions need to be reduced well below 50% (compared with the emissions in 2000). Halving emissions by 2050 would still bear the risk of exceeding 2 °C with a probability of up to 50%. Stronger emission reduction and more stringent stabilization goals are obviously necessary to decrease this probability.[1]

The boundaries of the corresponding emissions corridor shown in Figure 4b are based on the range of scenarios discussed in the literature that stabilize at 2 °C (with high probability), and are not necessarily admissible emissions paths themselves. Those paths that exhibit high values in the first half of the century have to decline rapidly thereafter and to become low-lying trajectories in the second half of the twenty-first century. A delay in implementing effective emission mitigation measures at an early stage might even require negative emissions in the long term, and would be extremely difficult to achieve. One possibility to achieve negative emissions is by using biomass energy in combination with carbon capture and storage technologies (BECCS) – an option that has recently attracted increasing scientific interest.

## Energy efficiency improvement: necessary, but not sufficient

Achieving the deep emission reductions discussed above requires a comprehensive global mitigation effort. Existing climate protection strategies in industrialized countries need to be further tightened. Simultaneously, ambitious mitigation measures need to be implemented in developing countries, where most of the increase in greenhouse gas emissions is expected in the coming decades (Fisher *et al.*, 2007, p. 199). Fortunately, numerous options are available that can facilitate the achievement of this goal:

- Improvement in energy efficiency
- Switching between fossil fuel types (e.g., replacement of coal by gas)
- Zero- or low-carbon energy conversion technologies (e.g., renewable energies)

---

[1] A recent discussion of this issue was provided by Meinshausen *et al.* (2009).

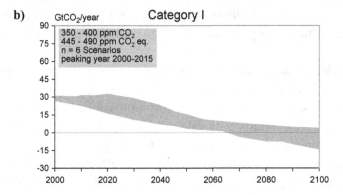

| | Global mean temperature increase above preindustrial at equilibrium, using "best estimate" climate sensitivity (°C) | Peaking year for $CO_2$ emissions | Change in global $CO_2$ emissions in 2050 (% of 2000 emissions) |
|---|---|---|---|
| $CO_2$-eq concentration (ppm) | | | |
| 445-490 | 2.0-2.4 | 2000 - 2015 | -85 to -50 |
| 490-535 | 2.4-2.8 | 2000 - 2020 | -60 to -30 |
| 535-590 | 2.8-3.2 | 2010 - 2030 | -30 to +5 |
| 590-710 | 3.2-4.0 | 2020 - 2060 | +10 to +60 |
| 710-855 | 4.0-4.9 | 2050 - 2080 | +25 to +85 |
| 855-1130 | 4.9-6.1 | 2060 - 2090 | +90 to +140 |

**Fig. 4.** a) Different carbon dioxide concentration ceilings (expressed in terms of $CO_2$-equivalent concentrations), their corresponding global mean temperature changes, latest point in time to shift to declining emissions and percentage emissions reduction in 2050 required to obey the stabilization goal. b) Set of emission pathways which meet the most stringent goal listed in the left-hand table. To meet the low stabilization level some scenarios deploy removal of carbon dioxide from the atmosphere (negative emissions). (*Source:* adapted from Fisher *et al.,* 2007, p. 199 and p. 229)

- Capture and storage of carbon from fossil fuels
- Reduction of non-$CO_2$ greenhouse gases (multi-gas strategy)
- Mitigation through improved land-use (e.g., reduced deforestation and afforestation)

Strategies to reduce multi-gas emissions can help achieve climate protection targets at substantially lower cost compared with emission mitigation efforts that address the release of carbon dioxide only. This is especially the case during the first half of the century, but in the long run it is essential to achieve deep reductions of carbon dioxide in any case, since carbon dioxide has a very long life-time (more than 20 % of emissions remain in the atmosphere over thousands of years, Archer *et al.,* 2009). In addition, land-use mitigation options could provide 15–40 % of the total cumulative abatement over the twenty-first century. Most such options are projected to be cost-effective strategies across the entire century (Fisher *et al.,* 2007, p. 172).

A tremendous decrease in energy intensity in the coming decades is essential if we are not to transgress the aforementioned 2 °C guardrail. Technological improvements and structural changes are expected to result in considerably lower greenhouse gas emissions than would otherwise be experienced. Assuming energy and carbon intensities frozen at current levels, for instance, would imply hypothetical average cumulative business-as-usual emissions that are roughly twice as high (see Fig. 5 a) as the baseline emissions projected for the suite of emissions trajectories depicted in Figure 3 a. The same message is visualized in Figure 5 b. Once again, assuming no improvement in the energy intensity (for instance, in the case of the SRES A2 scenario considered here), would result in considerably higher hypothetical emissions, even under business-as-usual conditions.

Many low-cost options to improve energy efficiency and to change the relative shares of fossil fuels in the provision of end energy are already contained in the baseline development. Therefore, there is restricted potential to achieve deep emission reductions by additional cost-effective energy efficiency improvement and fossil fuel switching measures.

An example showing a stabilization of the carbon dioxide concentration at 550 ppm is given in Figure 5 b where the (additional) contribution of demand reductions is small compared with the shares achieved by switching to low-carbon fuels (including shifts to nuclear energy and renewables) and carbon sequestration technologies (scrubbing). In order to achieve deep emission reductions (e.g., more than 50 % by 2050 compared to 2000), energy efficiency improvement and fossil fuel switching measures do not suffice. In addition, the application of low-carbon technologies becomes imperative.

### Innovative low-carbon technologies

Fortunately, numerous technologies exist which are capable of providing final energy while producing no or significantly less carbon dioxide compared with conventional fossil fuel burning (renewables, nuclear energy, and carbon capture and storage).

As Table 1 shows, there is abundant technological potential for renewable energies worldwide that would, in principle, suffice to meet even the highest projections of the total global primary energy demand in 2100 (see Fig. 2). The available wind potential (600 EJ/yr) alone would hypothetically be able to cover the entire primary energy demand of the world in 2005 (490 EJ). Even higher potentials are estimated for solar and geothermal energy (see Kohn, this volume).

Some important sources (especially wind and solar energy) exhibit an intermittent

a)                                          b)

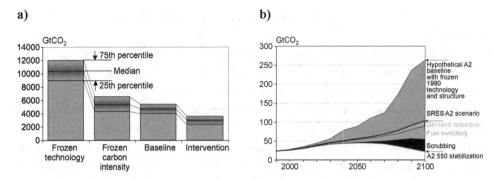

**Fig. 5.** a) Range of the global cumulative emissions emitted until 2100 for the different non-intervention scenarios shown in Fig. 3 a. 'Frozen technology' refers to the range hypothetical scenarios would exhibit without any improvement in energy and carbon intensities. The next bar shows the cumulative emissions assuming frozen carbon intensities and the third bar exhibits those emissions expected in the baseline (non-intervention) scenarios – where both energy and carbon intensity improvements are taking place. b) Contribution of different mitigation options to achieve a stabilization of carbon dioxide concentrations at 550 ppm (assuming a baseline development according to the IPCC SRES A2 baseline scenario without measures and policies directed at reducing GHG emissions). (*Source:* Fisher *et al.,* 2007, pp. 219–20)

availability dependent on daytime, season and weather conditions. In addition, renewable energy sources (with the exception of large-scale hydro-energy) are widely dispersed compared to fossil fuel deposits. Innovative concepts are available which can mitigate these drawbacks considerably by a combination of distributed usage (including appropriate communication strategies), storage, demand response, increased power system stability through the use of flexible alternating current transmission systems (FACTS) and interregional exchange (see Luther, this volume). Although renewables are in principle able to substitute fossil fuels completely, further research is needed to design integrated systems that exhibit low costs for the systems services envisaged here.

Nuclear energy is able to produce electricity with no (if only emissions at the power plant site are considered) or medium to low carbon emissions (if upstream emissions related to fuel supply and the construction of the power plants are taken into account). Under the present design of light-water reactors with a 'once-through' fuel cycle, however, the finite uranium resources (see Table 1) constrain the ability of nuclear energy to be the main lasting alternative to fossil fuel usage. Fast-spectrum reactors operated in a 'closed' fuel cycle by extracting the unused uranium and plutonium produced would solve this problem, albeit by accepting that reprocessing of the spent fuel increases the proliferation risks and security concerns. Beyond the long-term fuel resource constraints without recycling, there

**Table 1.** *Summary of global energy resources (including potential reserves) and their share of primary energy supply in 2005 (490 EJ). For renewable energies the technical potential is shown which takes into account conversion efficiencies as well as constraints on the available area. In contrast to the economic potential no explicit reference to cost is made.* (*Source:* Sims *et al.,* 2007, p. 264)

| Energy class | Specific energy source | Estimated available energy resource (EJ) | 2005 share of total supply (%) |
|---|---|---|---|
| Fossil energy | Coal (conventional) | > 100 000 | 25.0 |
| | Coal (unconventional) | 32 000 | |
| | Gas (conventional) | 13 500 | 21.0 |
| | Gas (unconventional) | 18 000 | |
| | Coalbed methane | > 8 000 | 0.3 |
| | Tight sands | 8 000 | 0.7 |
| | Hydrates | > 60 000 | |
| | Oil (conventional) | 10 000 | 33.0 |
| | Oil (unconventional) | 35 000 | 0.6 |
| Nuclear | Uranium | 7 400 | 5.3 |
| | Uranium recycle | 220 000 | |
| | Fusion | * $5 \times 10^9$ | |
| Renewable | Hydro (>10 MW) | 60/yr | 5.1 |
| | Hydro (< 10 MW) | 2/yr | 0.2 |
| | Wind | 600/yr | 0.2 |
| | Biomass (modern) | 250/yr | 1.8 |
| | Biomass (traditional) | | 7.6 |
| | Geothermal | 5000/yr | 0.4 |
| | Solar Photovoltaics | 1600/yr | < 0.1 |
| | Ocean (all sources) | ** 7/yr | 0.0 |

* estimated ** exploitable

are major barriers to an extended usage of nuclear energy. They comprise huge investment costs associated with investment uncertainties, unresolved waste management issues, security aspects in general, and – for some countries – the resulting adverse public opinion (Sims *et al.,* 2007, p. 254). As in the case of renewables, for advanced nuclear systems to make a higher contribution to the total share of energy would also require substantial cost reductions. Worldwide, only a few consortia are able to build nuclear power plants. With the current generation of power plants rapidly approaching the end of its lifetime, a significant share of the capacity of the nuclear industry is already needed even to secure a constant contribution made by nuclear energy to overall electricity production. On a global scale, sharing nuclear know-how is significantly constrained by commercial interests and security concerns. This could cause a significant bottleneck in attempts to solve the climate problem involving a pronounced contribution from nuclear energy.

Fossil fuel usage in combination with carbon capture and storage (CCS) technologies is a further option whereby a share of the future global energy supply could be produced with significantly lower carbon dioxide emissions. From a resource perspective, lower power plant efficiencies would result in an accelerated depletion of the fossil fuel resources. Due to the abundant availability of coal and potentially also hydrates (see Table 1), this, however, would not impose a major restriction on extensive application of coal-fired CCS technologies.

Although CCS can play a role in mitigating global climate change – at least as a transitional technology – its actual contribution may nevertheless be limited by the restricted availability of suitable geological disposal opportunities as well as by concerns about unintended leakage, risks associated with an accidental release of carbon dioxide, and environmental consequences. While deep ocean sequestration is another option, ocean eddy diffusion could potentially lead to a much larger region being affected with undesirable consequences than would be the case for sequestration in geological formations. Moreover, residence times of sequestered carbon dioxide are expected to be in the order of hundreds of years in the ocean, while potentially orders of magnitudes larger in formations. Finally, some of the authors (Edenhofer *et al.*, 2005; Held *et al.*, 2006) have suggested bond schemes to utilize the investigative power of the capital market to search for the most trustworthy combinations of CCS operators and geological formations. Such schemes are much harder to envisage for ocean sequestration. For all of these reasons, current schemes to operationalize CCS focus on geological formations rather than the deep ocean. CCS technologies imply higher costs compared to conventional fossil conversion, so that substantial cost reductions would be necessary to make this option an attractive one.

## Low-concentration stabilization scenarios
### *The role of oil/gas prices*

Currently the world experiences significant changes in the prices of raw materials and energy in particular. Though primary energy prices have returned to moderate levels, the future availability of fossil energy carriers is unclear. Scarcity of resources is reflected in high extraction costs, which in turn imply high energy prices. Increasing oil and gas prices influence technological change in the following ways. First, they foster additional investments in exploring and exploiting new and more costly oil fields including those holding non-conventional oil. Second, increasing oil prices make options like coal-to-liquid profitable if coal is relatively abundant and cheap. In a climate protection scenario, the extensive use of coal can only become an option if it is combined with CCS. In a scenario assuming relatively cheap coal and expensive oil and gas, the 'clean' coal option becomes more important compared to a scenario exhibiting low costs for all fossil fuels (see Fig. 6). Third, high oil prices may also improve overall energy efficiency, reducing the emissions up to the end of the century even in scenarios without any explicit mitigation policies or measures. It should be noted that long-term price trajectories of fossil fuels are quite uncertain. It is less uncertain that prices of oil and gas will increase faster than the price of coal because of the large coal reserves. However, large negative externalities associated with coal production and coal usage are likely to increase the cost of coal in the long run.

Figure 6 reveals the relative importance of different emission mitigation options in achieving a stabilization of the carbon dioxide concentration at 450 ppm as obtained with the model REMIND, developed at the Potsdam Institute for Climate Impact Research (see Bauer *et al.*, 2008; Leimbach *et al.*, 2009).[2]

The upper boundary of the corridor shows the business-as-usual emission trajectory which is dependent on the costs of fossil fuels. It is noteworthy that the increase of oil and gas prices does not alter the portfolio of mitigation measures substantially. Energy efficiency improvements (here including shifting between use of different fossil fuels, co-generation, and changing demand for final energy) play an important role in meeting this goal. A further considerable reduction of the emissions is realized through the application of CCS technologies, applied to both fossil fuels and biomass. Other renewables, especially solar photovoltaics and wind energy, as well as nuclear energy (light-water reactors), contribute significant shares. Although included in the general analysis, fast breeder reactors did not find application here because of their high capital costs compared to other mitigation options.

---

[2] REMIND comprises a top-down optimal growth model of the world economy combined with a bottom-up technology-rich description of the global energy supply system. In addition, the model contains a carbon cycle and climate system sub-module. Taken together, these modules are able to determine least-cost climate protection paths that are compatible with prescribed ceilings on global mean temperature change (e.g., the 2 °C EU climate protection guardrail). In contrast to traditional integrated assessment models, the model especially takes into account the possibility of induced technological change. In order to achieve this goal, learning curves are used in an endogenous way. This specific feature allows the determination of long-term cost-efficient strategies that minimize the integral climate protection cost over the entire time span considered (e.g., 150 years).

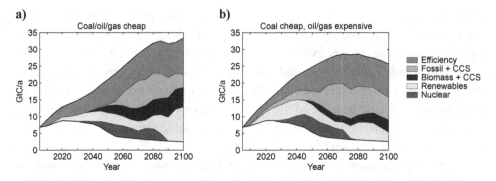

**Fig. 6.** The contribution of various mitigation options computed with the model REMIND for achieving the climate change stabilization target (450 ppm carbon dioxide). The upper boundary indicates the business-as-usual emissions, while the lower boundary represents the emissions in the mitigation scenario. The gap in between is filled by the contributions (so-called 'wedges') of emission mitigation options distinguished by the differently shaded areas. Panel a) shows the results for the case with cheap fossil fuels; panel b) shows the case with high costs for oil and gas. The pure rate of time preference for both cases is 3 % per year (see below).

## *The role of discounting*

The Stern Review (2006) has launched an exciting debate about the appropriate pure rate of time preference.[3] The report argued that the pure rate of time preference is an ethical value judgment about the weight and importance of future generations in current investment decisions. It points out that there is no ethical reason why future generations should be regarded as less important in current investment decisions than the current generation. However, the pure rates of time preference observed on capital markets are much higher than the rate derived from ethical considerations. The Stern Review states that a pure rate of time preference of 0.1 % is in accordance with intergenerational justice. Some authors choose a pure rate of time preference of 3 % in accordance with empirically observed behaviour on capital markets (see for example Toth, 1995). However the issue is much more complex, as Frederick *et al.* (2002) showed in an overview on the concept and measurement of discounting.

A lower pure time preference rate (1 % per year) favours – already in the business-as-usual (BAU) scenario – the application of emerging technologies for using renewable energies (especially wind and biomass energy sources in early decades of

---

[3] In economics, the pure rate of time preference is used to quantify how present consumer utility is valued compared to future consumer utility. Someone with a high time preference is focused substantially on his well-being in the present and the immediate future, while someone with low time preference places more emphasis on his well-being in the distant future. In this subsection only the issue of the pure rate of time preference is discussed and not the related issue of the inter-temporal elasticity of substitution, which is assumed to be equal to one.

the twenty-first century, see Fig. 7) while reducing, in part, the necessity to use CCS technologies.

Figure 8 shows the influence of excluding some of the different low-carbon technologies discussed above. As can be clearly seen, the exclusion of CCS technologies would result in a significant increase in the emission mitigation costs computed with the model REMIND. Compared to that, abstaining from applying additional renewables in order to combat global climate change would have a small influence, whereas the exclusion of nuclear energy would result in additional costs that are almost negligible compared to the overall mitigation burden.

## Creating a novel global energy system: the challenge ahead

As already pointed out above, achieving deep emission reductions requires a comprehensive global effort which includes both a complete change in the energy supply of industrialized countries and the establishment of low-carbon systems in developing countries and emerging markets – in short, nothing less than the creation of a completely novel global energy supply system. This would represent a true paradigm change compared with the current fossil-based energy systems and would take several decades to implement. In order to achieve this goal, the emissions mitigation measures must start immediately and rapidly engage the entire world. There is no time to waste. In a common effort, industrialized countries have to use their scientific capacity and creativity to develop and apply low-carbon technologies and to prove that a high standard of living can be sustained while producing considerably lower emissions in order to facilitate the early adoption of these technologies in the fast-growing emerging markets. The ultimate goal is a global carbon-free society.

Designing a cost-effective strategy to meet the climate protection targets discussed above (e.g., to limit global mean temperature increase to less than 2 °C relative to the preindustrial value) is a complex and dynamic problem. Although some conventional technologies (most notably, combined heat and power) might become economically viable once the costs of emission certificates increase, a major contribution towards achieving deep emissions reductions must be provided by the application of innovative low-carbon technologies. Unfortunately, some of these technologies are still prohibitively expensive. Anticipating learning capability and associated cost-reduction potential, however, is a key to resolving this problem.

While from an aggregated economic point of view, instantaneous massive investments into low-emission technologies seem to be optimal (Edenhofer *et al.,* 2006), more myopic agents (such as energy suppliers) may collectively act in such a way that the present-day energy system is conserved and consequently the global economy remains trapped in a suboptimal state.

a)

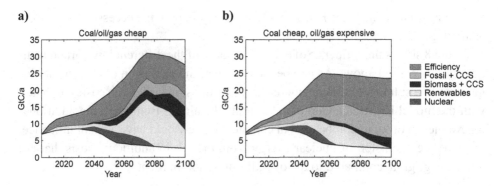

**Fig. 7.** Results of REMIND computations based on the same model assumptions as in Fig. 6 with the difference that a pure rate of time preference of 1% per year is applied.

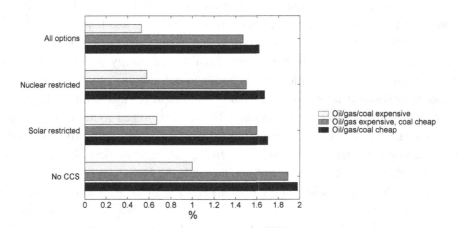

**Fig. 8.** (Monetary) consumption differences (i.e., relative reduction of per capita consumption in the stabilization case compared to the business-as-usual scenario). In the 'all options' case, all greenhouse gas mitigation opportunities discussed in Figures 6 and 7 (energy efficiency improvement combined with fuel shifting, renewables, nuclear energy and the application of CCS) are taken into consideration irrespective of their business-as-usual usage. In the other cases, some options are restricted to their respective usage in the business-as-usual scenario (REMIND model results, pure rate of time preference of 3 % per year).

Therefore, low-carbon technologies can only enter the market place if the cost of fossil fuel usage is increased significantly (e.g., through a worldwide carbon certificate market or carbon tax, see Edenhofer *et al.*, this volume). Without a reasonable price for carbon there are simply not enough incentives for firms and investors to foster a search process for the most cost-effective low-carbon technologies.

Fortunately, there are some recent promising initiatives in this direction: Chancellor Merkel has proposed a global carbon trading system, which would allow the reduction of emissions according to the 2 °C limit, at the same time implementing an allocation scheme that endows each citizen with the same emission rights. This proposal presupposes a global carbon market – otherwise the costs imposed on industrialized countries would not be acceptable. Negotiations have already started to harmonize and link the European Emission Trading Scheme with emission trading schemes emerging in California and elsewhere in the United States. The appropriate timing is essential because of the need for a continued signal to the carbon markets. Emissions trading, and related flexible mechanisms, are likely to remain a core element of any post-2012 regime.

Admittedly, emissions trading is only one necessary condition for achieving low stabilization targets. In fact, the Stern Review found that only 40 % of the low-carbon future can be financed through the carbon market (Stern, 2006). What is needed is a comprehensive suite of policies to shift the International Energy Agency's estimated figure of USD 20 trillion of energy investments by 2030 into low-carbon technologies and to assure these investments in the first place. On the national level, policy frameworks such as quota schemes or feed-in tariffs – or even a reasonably designed technology policy supporting demonstration projects for CCS but also for solar thermal power plants and other innovative technologies – are recommended. These would in particular allow the cost reductions inherent in technologies with high learning potential to be realized. On the international level, new innovative technology co-operation mechanisms will be required to both deploy existing technologies in emerging economies and develop and share new low-carbon technologies.

From a long-term perspective, a comprehensive global emission mitigation effort requires enhanced innovation to create novel low-carbon technologies, incentives to support their initial diffusion and the internalization of external costs (e.g. through emissions trading). Such a response to the dangers of global climate change would induce a transition towards a truly sustainable global energy system as a glorious 'side effect'.

# References

Archer, D., Eby, M., Brovkin, V. *et al.* (2009). Atmospheric lifetime of fossil-fuel carbon dioxide. *Annual Reviews of Earth and Planetary Sciences,* **37,** 117–34.

Bauer, N., Edenhofer, O. and Leimbach, M. (2009). Low-stabilisation scenarios and technologies for carbon capture and sequestration. *Energy Procedia,* **1**(1), 4031–8.

Edenhofer, O., Held, H. and Bauer, N. (2005). A regulatory framework for carbon capturing and sequestration within the post-Kyoto process. In Rubin, E. S., Keith, D. W. and Gilboy, C. F., eds., *Greenhouse Gas Control Technologies: Peer Reviewed*

*Papers and Overviews.* Vol. I of *Proceedings of the 7th International Conference on Greenhouse Gas Control Technologies (5–9 September 2004, Vancouver, Canada).* Amsterdam, 989–97.

Edenhofer, O., Carraro, C., Köhler, J. and Grubb, M., guest eds., (2006). Endogenous technological change and the economics of atmospheric stabilization. *Energy Journal,* **27**(Special Issue 1).

Fisher, B. S., Nakicenovic, N., Alfsen, K. *et al.* (2007). Issues related to mitigation in the long term context. In: Metz, B., Davidson, O. R., Bosch, P. R. *et al.,* eds., *Climate Change 2007: Mitigation of Climate Change. Contribution of Working Group III to the Fourth Assessment Report of the Intergovernmental Panel on Climate Change.* Cambridge, 169–250.

Frederick, S., Loewenstein, G. and O'Donoghue, T. (2002). Time discounting and time preference: a critical review. *Journal of Economic Literature,* **40**(2), 351–401.

Held, H., Edenhofer, O. and Bauer, N. (2006). How to deal with risks of carbon sequestration within an international emission trading scheme. In *Proceedings of the 8th International Conference on Greenhouse Gas Control Technologies (19–22 June 2006, Trondheim, Norway),* issued on CD-Rom (ISBN 0-08-046407-6), Amsterdam.

IPCC – Intergovernmental Panel on Climate Change (2007). Summary for policymakers. In Solomon, S., Qin, D., Manning, M. *et al.,* eds., *Climate Change 2007: The Physical Science Basis. Contribution of Working Group I to the Fourth Assessment Report of the Intergovernmental Panel on Climate Change.* Cambridge, 1–18.

Leimbach, M., Bauer, N., Baumstark, L. and Edenhofer, O. (2009). Mitigation costs in a globalized world – climate policy analysis with REMIND-R. *Environmental Modeling and Assessment* (in press).

Meinshausen, M., Meinshausen, N., Hare, W. *et al.* (2009). Greenhouse-gas emission targets for limiting global warming to 2 °C. *Nature,* **458**(7242), 1158–62.

Nakicenovic, N., Alcamo, J., Davis, G. *et al.* (2000). *Special Report on Emissions Scenarios.* A special report of Working Group III of the Intergovernmental Panel on Climate Change (IPCC). Cambridge.

Sims, R. E. H., Schock, R. N., Adegbululgbe, A. *et al.* (2007). Energy supply. In Metz, B., Davidson, O. R., Bosch, P. R. *et al.,* eds., *Climate Change 2007: Mitigation of Climate Change. Contribution of Working Group III to the Fourth Assessment Report of the Intergovernmental Panel on Climate Change.* Cambridge, 251–322.

Smith, J. B., Schneider, S. H., Oppenheimer, M. *et al.* (2009). Assessing dangerous climate change through an update of the Intergovernmental Panel on Climate Change (IPCC) "reasons for concern". *Proceedings of the National Academy of Sciences of the United States of America,* **106**(11), 4133–7.

Stern, N. (2007). *The Economics of Climate Change: The Stern Review.* Cambridge.

Toth, F. (1995). Discounting in integrated assessments of climate change. *Energy Policy* **23**(4–5), 403–9.

Yohe, G. W., Lasco, R. D., Ahmad, Q. K. *et al.* (2007). Perspectives on climate change and sustainability. In Parry, M. L., Davidson, O. F., Canziani, J. P. *et al.,* eds. (2007). *Climate Change 2007: Mitigation of Climate Change. Contribution of Working Group II to the Fourth Assessment Report of the Intergovernmental Panel on Climate Change. Cambridge,* 811–41.

# Chapter 17

# Price and quantity regulation for reducing greenhouse gas emissions

## Ottmar Edenhofer, Robert Pietzcker, Matthias Kalkuhl, and Elmar Kriegler

Ottmar Edenhofer is an economist who studies climate change policy, environmental and energy policy, and energy economics. He is currently Professor of the Economics of Climate Change at the Technical University of Berlin, and Deputy Director and Chief Economist at the Potsdam Institute for Climate Impact Research. Since 2008 he has also served as Co-Chair of Working Group III of the Intergovernmental Panel on Climate Change (IPCC), which was awarded the 2007 Nobel Peace Prize.

*Note:* Photos and biographies of co-authors can be found in the appendix.

## The challenge of climate change[1]

Climate change is a market externality.[2] Market actors emit greenhouse gases (GHGs), leading to costs in terms of climate change damages that are not paid by the emitters themselves, but by others. The result of this market failure is that more than the optimal level of GHGs is emitted. If the external costs were included in the costs of emitting GHGs ('internalizing the costs'), it would become unprofitable to continue emitting GHGs at the current rate. Internalization of the costs is thus essential for effective long-term reductions in GHG emissions.

Two major types of market instruments have been proposed to internalize the cost of pollution: Pigovian pollution taxes[3] (a price signal), and tradable pollution permits (a quantity signal). The idea of Pigovian taxes is to make the polluter pay the external costs of pollution, thus bringing together the social and private costs of polluting, and therefore adjusting pollution to the efficient level. The key difficulty with Pigovian taxes is calculating which level of tax will counterbalance the pollution externality (i.e., calculating the marginal damages[4] of pollution). In contrast, tradable pollution permits give rise to a price on pollution that reflects the relative scarcity of pollution permits; for example, the quantity of the permits will determine its price. The key difficulty here is in setting the quantity of permits, and thus the overall pollution, to the efficient level. It is a long-standing debate in environmental economics which of the two instruments is superior in varying circumstances (Hepburn, 2006).

Based on the concern that there are tipping points in the Earth's climate system, the triggering of which could dramatically increase climate change damages (and the uncertainty about them), policymakers need to decide to avoid dangerous interference with the climate system (as expressed in Article 2 of the Framework Convention on Climate Change). This decision would most likely involve setting a climate protection target, for example in terms of a maximum temperature rise.

---

[1] This text focuses on the design of climate policy instruments. It does not derive a global cumulative carbon budget that would allow us to achieve either an optimal temperature goal or an optimal cost-benefit ratio. To do this, questions of ethics, equity and environmental effectiveness would have to be discussed. Within this text we instead assume that these questions have already been resolved by a careful application of welfare economics and ethics. Therefore, we limit our analysis to the design of policy instruments necessary to address the market failures associated with man-made climate change. The results presented here remain valid in a cost-benefit-analysis (CBA) in which the damages are taken into account explicitly. Such a CBA would be one method to derive the optimal carbon budget (for further discussion of this point see Edenhofer and Kalkuhl, 2009).

[2] A market externality is the impact (positive or negative) of a market transaction on a third party that is not directly involved in the transaction. In terms of climate change, this means that the price paid for energy does not reflect the climate change due to energy production and the resulting damages to all people suffering from climate change.

[3] A Pigovian tax is designed to raise a market activity's price to its true costs, including external costs.

[4] In economics, the term 'marginal' is used to describe the change of an aggregated value associated with the last unit produced or emitted. The marginal cost is the change in total cost that arises when the quantity produced changes by one unit, thus it is the cost of producing one more unit of a good. The marginal damage of carbon dioxide would be the additional damage caused by emitting one additional tonne of carbon dioxide.

Such a target can be converted to a total maximum carbon budget that may be used without incurring an unacceptably high probability of violating the climate protection target (Meinshausen *et al.*, 2009). Once the carbon budget is set, the question remains how to cost-effectively allocate its usage over time.

Moreover, climate protection requires the transformation of the existing energy and transport infrastructure into an energy-efficient, low-carbon infrastructure. This transformation is an ongoing project involving huge long-term investments, for example in low-carbon power plants and the energy-efficient refurbishment of existing buildings. These investments will only occur if stable long-term expectations about the carbon price persist. Research has shown that early investments into efficient energy use and clean technologies can greatly reduce the economic cost of climate protection (Grubb *et al.*, 1995, Edenhofer *et al.*, 2009b). Therefore, creating stable, long-term expectations about future carbon prices – implemented through either a quantity or price regulation – and designing credible long-term road maps for climate protection are central tasks for policymakers.

### Introduction to the debate on price versus quantity instruments

To contribute to the debate about climate change policy instruments, we developed a conceptual computer-based economic model. Before using the model for a detailed analysis of the economic properties of tax and quantity instruments in the subsequent sections, we begin by stating three arguments that cannot be treated in our single-region model because they relate to international concerns:

1. **International harmonization of carbon prices:** Since climate change can only be tackled globally, a meaningful effort will have to rely on the implementation of carbon pricing mechanisms in most regions of the world. It is a clear advantage of emission trading schemes (ETSs) that mechanisms creating (i) an integrated international cap-and-trade system, and (ii) incentives for reducing emissions in regions without an emissions cap (as attempted by the clean development mechanism, CDM) are conceivable. This would lead to the emergence of a globally harmonized carbon price.[5]

2. **International burden sharing:** Another advantage of implementing carbon markets rather than carbon taxes is that international burden sharing of the costs of climate change and emissions abatement can be more easily achieved

---

[5] Taxes are a policy instrument that most nations and political parties are very sensitive about. The ongoing difficulties encountered in the process of harmonizing taxes among EU countries demonstrate how complicated international tax harmonization would be. Emissions trading systems do not yet carry a similar ideological burden. Therefore, it seems plausible that introducing and linking ETSs will be more feasible. Furthermore, most nations already levy energy taxes, some of them justified by climate change. It is not clear if a harmonized carbon tax would replace or complement existing taxes.

by adjusting regional caps and allowing for interregional trade in permits. Admittedly, the tax revenues could also be recycled to yield the same outcome as ETS burden-sharing schemes. However, the institutional prerequisites might be more demanding for an international tax scheme in which an international body has to be endowed with the power to transfer the tax income from one nation state to another, a mechanism that has proven difficult in the past.

3. **Setting the baseline:**[6] Closely linked to the question of burden sharing is the question of baseline setting. While all evidence speaks in favour of auctioning permits at a national level, how should permits be distributed between the states participating in an ETS? The possibility of changing this distribution by setting different baselines allows for international burden sharing, but at the same time it creates a very difficult negotiation topic: As it is necessary to set an individual baseline for each country, each country will try to influence the negotiations to increase its own baseline. A tax, by contrast, does not necessarily create this problem. Setting an equal tax without tax exemptions can therefore be appealing due to its simplicity and perceived equal treatment of all parties. Whether this difference is seen as an advantage or disadvantage compared with an ETS depends on the assumptions about the political process leading to an international agreement, and the negotiation position of the different nations involved.

## Frameworks to explore price and quantity policies
### *Cost-benefit analysis versus carbon budget constraint*

The difference between price and quantity instruments has been mostly discussed within a cost-benefit analysis framework. Under such a framework both the economic costs and benefits of a given strategy are evaluated. The difficulty of such an analysis is that it raises many questions about the value of goods that cannot be bought or sold, such as 'what is the value of clean air?'

Weitzman (1974) has shown within a static framework that price instruments are superior to quantity instruments if marginal abatement costs increase faster than marginal damages. The extension of Weitzman's famous framework to a stock-pollutant problem such as climate change, in which not the annual emissions themselves but the cumulative stock of all previous emissions produces climate change damages, was undertaken by Newell and Pizer (2003). Under their – quite specific

---

[6] A baseline is the amount of emissions against which efforts of countries to decrease GHG emissions are measured. A country with a fast-growing population might have a growing baseline to reflect the fact that it will find reducing total emissions more difficult than a country with decreasing population.

and partly questionable – assumptions,[7] taxes will usually be preferred in the first periods when marginal damages do not change much as GHG concentrations are still low and severe climate damages are still far away and are therefore reduced through discounting; in later periods marginal damages of emissions rise due to higher GHG concentration and discounting will have less effect. Then, the marginal damages increase faster than marginal mitigation costs and a quantity instrument like an ETS performs better.

In contrast to Weitzman's cost-benefit framework, we do not perform a full cost-benefit analysis. Instead, we assume a given and fixed carbon budget and discuss instruments to achieve this target with minimum costs. Such a framework circumvents the need to estimate an appropriate damage function required for cost-benefit analysis, which would be very difficult because the exact future damages resulting from an incremental amount of emissions are extremely sensitive to future emission paths, climate sensitivity and available technologies (Stern, 2008). Furthermore, other side effects of high carbon dioxide concentrations in the atmosphere, such as ocean acidification, would have to be considered. To complicate the problem, valuation of damages is not possible without normative assumptions about the needs and preferences of future generations. Finally, the Earth system as a whole has a value of its own that exceeds its economically quantifiable value. Hence, we will compare taxes and ETSs in the context of achieving a given cumulated carbon budget ('all nations together may not emit more than a certain amount of carbon dioxide – for example, 1000 gigatonnes carbon dioxide equivalents – over the next few hundred years') at maximum welfare.[8]

### Social planner model versus game theory

The debate about prices versus quantities has mostly been discussed within the framework of a social planner. Such a model assumes a benevolent planner with full foresight who takes all decisions. While the social planner framework defines

---

[7] They allow negative net emissions, assume exponential decrease of abatement costs (the costs associated with reducing emissions), decay of carbon dioxide with a half-life of 84 years (newer scientific research claims a half-life of temperature change of >1000 years, see Matthews and Caldeira, 2008), and set damages from global warming to 1.85% of GDP at 3 °C temperature (a survey among environmental economists estimated the loss at 6.5% GDP at 3 °C temperature increase, see Roughgarden and Schneider, 1999).

[8] Welfare is here calculated as the time-discounted sum of the logarithm of consumption over the next hundred years. While this indicator does not encompass all that is included in the common usage of the term 'welfare', it is one of the main measurements used in economics due to the methodological difficulties of including more complex concepts like 'sustainability' or 'happiness'. Different efforts have been made to create a more holistic indicator for welfare like the Index of Sustainable Welfare (ISEW), the Genuine Progress Indicator (GNI), the Gross National Happiness Product (GNHP) or the Happy Planet Index (HPI). However, these alternative concepts all suffer from limitations (Lawn, 2005) and have not succeeded in replacing purely monetary measures like GDP or consumption.

a benchmark of 'first best'[9] solutions, it does not allow the assessment of policy instruments when multiple externalities – such as market imperfections, technological spillovers or incomplete futures markets – require correction.

In contrast, a game theoretic model with different actors who all maximize their own welfare allows the inclusion of market failures and is therefore better suited to the analysis of policy instruments targeting multiple externalities.

### General features of our model

To address the above-mentioned concerns, we developed a model with the following main features.[10] First, it is an endogenous growth model; saving rates and the resulting economic growth are internally calculated by the model according to certain production equations, and not directly prescribed by the programmer. Second, the model allows the analysis of further market externalities besides climate change, such as monopolistic market power or property risk. Third, it is a general equilibrium model that comprises multiple economic sectors that interact with each other. Fourth, the model reproduces the existing asymmetry between government regulation and reactions of the economic sector by explicitly representing the government as the leader of a Stackelberg game.[11] Finally, it is a qualitative model that is not calibrated to data from a specific country.

Starting from a given inter-temporal carbon budget there are two different policy design options to achieve an economically efficient emissions reduction. *Price instruments* (taxes) reduce demand for economic factors and thus decrease emissions. In contrast, *quantity instruments* (ETS) limit emissions directly by restricting the available amount of permits and thus cumulative emissions. After first analyzing a deterministic setting in which all parameters are fixed and known by all actors, we will discuss what happens when uncertainty comes into play, for example about resource extraction costs or the learning potential of renewable energy.

For the sake of simplicity, we do not distinguish between various types of fossil resources. Therefore, emissions are proportional to resource consumption, and the problem of climate protection is reduced to the problem of fossil resource conservation.

---

[9] 'First best' meaning the optimal solution in a world in which all markets function properly.

[10] For a detailed description of the model, see Edenhofer *et al.* (2009a).

[11] A Stackelberg game assumes a hierarchical asymmetry: one player (Stackelberg leader) makes his decision before the other players (Stackelberg followers) by considering information about the expected reaction of the followers to his move. Here, the government (leader) assumes profit-maximizing behaviour of the economic sectors (followers), who react to the tax path announced by the government.

### *Observations in a deterministic setting*

The results of our model may at first seem surprising, but they are in fact in line with economic intuition; both types of market instruments – optimally implemented price and quantity instruments – can have the same economic efficiency. If the government possesses all necessary information for estimating economic development and no further market failures occur, an optimal emission tax as well as a cumulative permit trading scheme both achieve climate protection at minimal cost. Both instruments result in the same carbon price, which increases until backstop technologies[12] are competitive and replace their carbon-based alternatives (see Fig. 1). As expected, the price grows with the net interest rate corrected by an extraction cost term (Hotelling, 1931).

Different institutional requirements arise from choosing either a tax or an ETS. The tax requires that the government is able to impose the optimal time path of the tax (see Fig. 1), which is often hampered by political conflicts. Otherwise the private sector cannot reach its inter-temporal market equilibrium. In the case of tradable emission permits, the government has to be able to enforce the cap. Furthermore, to reach the optimal price path for the permits, the futures markets for the fixed stock of permits must be complete; it must be possible to trade permits for each time step in the future.

### *Distribution of rents*

The carbon budget creates a scarcity rent for the permit owners. Scarcity rents are profits to the owner of a scarce good that arise from the fact that the price of the good increases when supply of the good decreases. In this case, the government decreases supply by limiting the total amount of emissions.

In this perspective, creating rents is at the heart of environmental policy. The translation of resource scarcity into rents is the reason why purely economic agents care about the environment. It is a common understanding within welfare economics that rents can be removed from private agents without distorting the efficiency of resource allocation. One advantage of an emission tax is that it transfers the rent to the government. These revenues can then be redistributed or used to reduce existing tax distortions.

In contrast, if permits are freely allocated according to previous emissions, the ETS leaves this rent to permit owners, thereby decreasing social welfare. This effect was observed during the first period of the European ETS when power companies

---

[12] Backstop technologies are energy technologies that do not produce any carbon dioxide and are assumed to have infinite potential. In our simplified model, renewable energies are modelled as a backstop technology.

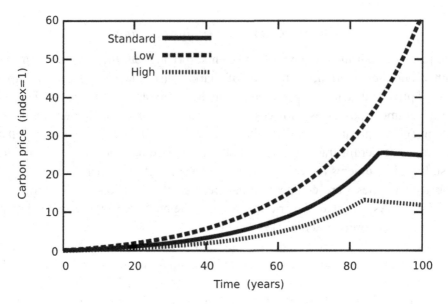

**Fig. 1.** Optimal carbon price in order to achieve the carbon budget (values are indexed with regard to the first year of simulation). The curve shows a kink once the backstop technology has replaced its carbon-based alternatives. Dotted lines show the sensitivity of the optimal resource tax with respect to different parameterizations of economic factors; here the cost-decreasing learning effects within renewable energy production, which are assumed to be low, high or standard. (*Source:* based on calculations in Edenhofer *et al.,* 2009a)

made billions in windfall profits[13] by incorporating market prices for emission permits into their electricity prices without actually having to pay for these permits (Sijm *et al.,* 2006). However, if the permits are fully auctioned, the rent is again transferred to the government, so the outcome is totally symmetric to using a resource tax.

### *Input and output regulation*

Taxation or quantity regulation can be imposed on goods with different levels of refinement along the production process (for example, on the amount of fossil fuel resources, of secondary energy or of final output). To achieve efficient emission reductions, an instrument must be directly related to the economic factor causing the emission. An energy tax (output instrument) that does not discriminate between different sources of the taxed energy is generally not efficient. Although an energy

---

[13] Windfall profits are unexpected profits through unforeseen changes in the market; e.g., through changed government regulation.

tax reduces emissions due to a decrease in energy consumption, it has almost no influence on factor allocation or resource substitution within the energy production process. In contrast, a resource tax (input instrument) leads to optimal factor real-location as energy is partly replaced by capital or labour.[14]

Thus, internalizing an externality is most efficient when the polluting factor with most substitution possibilities is regulated, rather than some aggregated good for which no environmentally friendly substitute exists. If only the aggregated final product is regulated, (for example, by a value-added tax), consumers have no sub-stitution possibilities; they can only reduce their demand. If energy in general is taxed, production firms can decrease secondary energy use by either decreasing output or switching to less energy-intensive production processes, so they have at least some substitution possibilities. If GHG emissions are directly taxed, many more substitution possibilities are tapped; power producers can increase power plant efficiency or use less emission-intensive options like natural gas or renewable energy, and production firms can decide to use less energy-intensive production processes or buy energy from power producers using renewable sources.

### *Sectoral coverage*

It is worth mentioning that a regulatory instrument has to cover all relevant sectors; i.e., all resource flows through the economy (Hargrave, 2000). This can be done by an upstream system where the resource extracting sector is regulated, or by a down-stream system where the producer of the final product has to report the total carbon content along the production chain of a product, and either pay taxes or buy permits for this amount of carbon (see Fig. 2). In an idealized world of complete sectoral coverage and zero monitoring and transaction costs these approaches are equivalent. If transaction costs exist, it seems plausible that regulating few actors (resource mining companies) through upstream regulation will prove easier than regulating many actors (production companies or even households) through downstream reg-ulation. In real life, transaction costs are widely persistent and substantial, which is reflected in the difficulties of the different carbon footprint projects that try to de-termine how much carbon was emitted all along the production chain to produce a final good.

In real-world policy implementations, it is commonly observed that individual sectors are exempt from tax or quantity regulations (Rupp and Bailey, 2003; Bach, 2005). This decreases the coverage of production sectors by the regulation, thereby reducing substitution possibilities and strongly increasing total cost. Hence,

---

[14] Investing in energy efficiency would be an example of replacing energy with capital, while the replacement of automated production by manual labour would represent a shift towards labour.

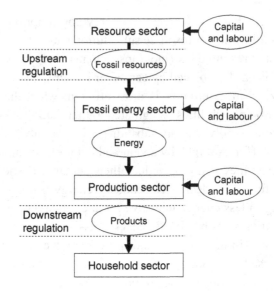

**Fig. 2.** Exemplary production chain. (*Source:* adapted from Edenhofer *et al.*, 2009 a)

exempting sectors from the regulation will lead to much higher costs for society compared to a regulation covering all sectors.

## Supply-side dynamics and the green paradox

In his 2008 paper on global warming, Hans-Werner Sinn develops the 'green paradox'. With regard to the strategic behaviour of resource owners he concludes that rising resource taxes accelerate extraction and therefore worsen global warming. His analysis relies fundamentally on the assumption that resource owners take only the resource budget given by nature into account. Thus, resource owners will extract the entire resource stock, and resource taxes will only change the timing but not the total amount of extraction. Within Sinn's framework, an asymmetry of price and quantity instruments arises, since an ETS in which the number of permits is lower than the potential resources that could be extracted automatically restricts the total amount of resources that will be extracted. In contrast, only a few price instruments will be able slow down resource extraction. Possible market-based policy instruments (in contrast to command-and-control instruments, such as a moratorium on coal power plants), suffer from credibility problems or high transaction costs, or imply huge, politically unfeasible transfer payments to resource owners (Edenhofer and Kalkuhl, 2009).

In our model, however, both the resource tax and the ETS will impose the carbon budget onto the resource owners' extraction problem.[15] Our resource tax is high enough and rises in such a way that it removes the rent from resource owners. As the demand-price relation for the resource is known by the regulator, the tax is fixed to the right level so that the pure extraction costs plus the tax yield a resource price at which demand is reduced to the amount allowed by the carbon budget. Thus, resource owners cannot sell more resources than the carbon budget allows without incurring losses. Another important difference between our model and Sinn's is that the mitigation target is not derived endogenously from cost-benefit analysis, but externally as a resource budget. Thus the concept of 'internalization' gains a new meaning: price as well as quantity instruments transform the resource scarcity rent into a climate rent that protects the atmosphere as a global common.

It follows from our analysis that a successful climate protection policy instrument manages to (i) devaluate the resource owners' scarcity rent, and (ii) establish an optimal resource price by a public authority that governs the global common on behalf of humankind. The quantity instrument directly transforms the resource rent into a climate rent by announcing a fixed permit budget. Thus, resource owners realize that the scarce permit stock has already devalued their – now abundant – resource stock and that there is almost no room left for rent-making.

An optimal price instrument also implicitly fixes a carbon budget. However, it does not directly communicate the politically-set carbon budget; resource owners only perceive the tax rate and might ignore the fact that the government imposes the tax in such a way that it fixes the carbon budget. Thus the tax obscures the devaluation of the resource rent. If resource owners do not perceive the intended quantity effect of the tax, they cannot determine the resource extraction path correctly. The resulting extraction path then is non-optimal, which could possibly result in too much resources being extracted.

The ETS and the tax are thus only equivalent if the resource owners anticipate the correct time path of the tax and believe that the public authority is committed to safeguarding the carbon budget.

## Introducing uncertainty

In real life, we do not know too much about the future – the development of oil prices or the future enforcement of energy efficiency standards are examples where our knowledge is limited and uncertainty comes into play. We therefore analyzed exemplarily the effect of wrong estimation of important parameters on our model results. To demonstrate the sensitivity of results to model parameters, Figure 1 shows

---

[15] A more formal discussion about the explicit assumptions and technical implementations of specific policy instruments can be found in Edenhofer and Kalkuhl, (2009).

the changes in the optimal carbon price path when the cost-decreasing learning effect within renewable energy production is varied.

## *Optimal resource taxing*

If the regulator implements a price instrument, the calculation of the optimal resource tax requires exact estimation of supply, demand, technology and substitution options – at least for the next century. These informational requirements are highly demanding and probably beyond the computational capacity of a real-world government or research institution. If the government errs in predicting crucial parameters that are related to resource consumption, it misses either the protection target (accompanied by overconsumption) or, through too restrictive climate protection, the optimal consumption path (see Fig. 3).

## *Optimal issuing of permits*

In contrast to direct resource pricing, a quantity restriction directly controls the amount of emissions, and hence prevents violation of the climate target. However, the regulator has to decide about the timing of permit issuing, and thus faces the same uncertainties about future demand as in the tax model. Wrong estimation of economic parameters leads to suboptimal timing and causes welfare losses.

If the regulator allows banking and borrowing of permits, he shifts the uncertainties about future demand to the private sector; private agents risk their profits if they cannot predict these parameters correctly (Krysiak, 2008). Permits can be used at any time in the future. It is up to the private firms to decide *when* to use their permits according to their estimation about future permit prices.

## *Futures markets and institutional equivalents*

For a permit market to function successfully, it is necessary that future prices are already known or that traders believe that they can predict them (Dasgupta and Heal, 1979, p. 108). Futures markets can be distorted by insecure property rights, imperfect information, limited access to markets in the future, or uncertainty about regulator's future policies. For example, the collapse of permit prices within the first trading period of the EU ETS was caused by an over-allocation of permits and the absence of banking, which would have allowed the transfer of permits to the future (see also Brunner *et al.,* 2009).

As a successful ETS will cover all relevant economic sectors and activities, the permit market will be highly fragmented and private agents will have difficulty coordinating their plans. Furthermore, assessment of futures markets requires

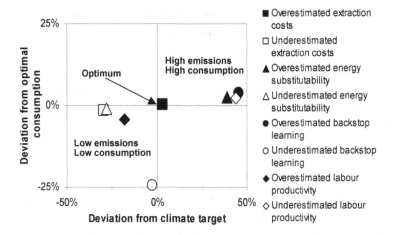

**Fig. 3.** Deviation from climate target and optimal consumption if the government does not estimate economic parameters correctly in order to calculate the optimal resource tax. Wrong estimation of parameters risks violating both the climate protection target as well as the cost-effective consumption target. (*Source:* Edenhofer *et al.,* 2009 a)

research that is always costly to undertake. Hence, in a completely deregulated permit market, only economically powerful enterprises could afford private market research and information collection. However, markets are not efficient if not all relevant information is freely available for all market participants. Therefore, an institution is required to provide information about future carbon markets, such as the costs and risks of long-term abatement options. With the Intergovernmental Panel on Climate Change (IPCC) there already exists an institution that has a very strong reputation for compiling relevant data on technologies and their costs. The reports of the IPCC could be enhanced in such a way that its content can be better captured by investors, firms and banks for financing the long-term transition to a low-carbon economy.

One possible institution that could improve the planning security of enterprises would be a carbon bank endowed with a carbon budget. Such a bank would manage permits by maximizing net present value of its permit stock. It could define trading ratios to influence the time-path of mitigation if market discount rates differ from socially optimal discount rates.[16] As an independent institution like a central bank, the carbon bank reduces regulatory uncertainty about future policies that

---

[16] The discount rate describes how future assets (bonds, capital stocks, investments, etc.) are devalued just because their pay-off lies in the future. It equals the interest rate on capital markets and depends on the economic growth rate and normative aspects about distribution of wealth over time and the valuation of future consumption compared to current consumption. A high discount rate implies a high devaluation of future consumption; a discount rate of zero values present and future consumption equally.

might be exposed to political pressure (elections or public finance). Nevertheless, it should react with flexibility to new insights into the climate system.

A market combined with research and banking institutions might respond in a more effective way to parameter changes than a government with only limited capability for fine-tuning due to the nature of the political decision-making process. Experience shows, however, that markets are not always efficient and also often suffer from failure.

### *A symmetric safety valve*

Another possibility for reducing short-term volatility of permit prices and thereby investor risk would be to establish a symmetric safety valve as proposed by Roberts and Spence (1976), and Burtraw *et al.* (2009). Such a safety valve would take the form of a regular ETS with two constraints:

- If the permit price drops below a certain value, say EUR 15 per ton of carbon dioxide, the issuing government buys permits until the permit price rises above the price floor.
- If the permit price rises above a certain value, say EUR 300 per ton of carbon dioxide, the government sells further permits until the price drops below this price ceiling.

The price floor would reduce the risk of investment in clean technologies as investors will always receive a minimum return for their investment. The price ceiling would weaken one of the main advantages of a cap, namely that the environmental goal is reached at all times. Yet, it could soften the economic impacts of unexpected events by loosening the cap. It could thus increase the credibility and stability of the ETS; if temporal relief systems for critical times are defined in advance, the political promise of sticking to the system even through a crisis becomes more plausible.

Such a symmetric safety valve would reduce short-term market fluctuations, but not in itself lead to optimal inter-temporal permit allocation. To reach this goal, the safety valve has to be combined with the above-mentioned measures to promote functioning futures markets.

### Regulation of additional market failures

In this section we discuss other forms of market failure and the policies required to correct them. A main characteristic of taxes is their capability to directly influence price. Hence, a tax is more flexible than an ETS and can often correct additional market failures that are caused by sub-optimal pricing of a single factor. It turns out, however, that similar welfare improvements can be achieved with an ETS if it is complemented by additional policy instruments.

## *Monopolistic market power*

Monopolistic market power in the resource sector increases the resource price above the optimal level, thus leading to a more conservative resource extraction path (see Fig. 4). Although this might contribute to climate protection, it is not an economically efficient approach as the monopolist provides resources on a suboptimal level in order to generate substantial rents. Furthermore, it does not guarantee compliance with the carbon budget in the long run, as the resource owners will extract their whole resource stock if it is profitable to do so.[17] Hence, market power in the resource sector cannot replace climate policy. On the contrary, in the case of climate protection it enhances welfare if governments not only reduce emissions but also regulate a monopolistic resource owner.

The advantage of a resource tax lies in its ability to address two market failures at the same time: the climate protection target (which is not anticipated by resource extractors) and monopolistic market power. A quantity policy cannot directly correct the effects of monopolistic market power. However, if the permit market is competitive and the total amount of emission permits is less than the total amount of resources, competition between resource owners will be increased as they will not be able to sell all of their resources. Therefore, a reduction of monopolistic power can be expected.

## *Expropriation risk – when ownership of resources is insecure*

If resource owners expect that their property rights are insecure,[18] they will change their extraction timing. As considered by Sinn (2008), risk of expropriation results in resource owners discounting their revenues at a higher rate (they add a risk premium onto the discount rate), leading to accelerated extraction (see Fig. 4). This behaviour is plausible. For example, if I am not sure that I will still be the owner of a certain oil field in 20 years, I will prefer to extract and sell the oil now at a slightly cheaper price and invest the money elsewhere rather than risk losing the oil. One option to remove the effect of expropriation risk and flatten the extraction path is to subsidize the resource price after an initial period of taxation. This makes

---

[17] The complete extraction of the resource stock in the absence of climate policy depends on some basic assumptions about the substitutability of fossil resources and the dynamics of extraction costs, as well as on the timeframe considered.

[18] One example of insecure property rights might be authoritarian regimes of oil-exporting countries that are under a certain threat of losing control over their oil resources.

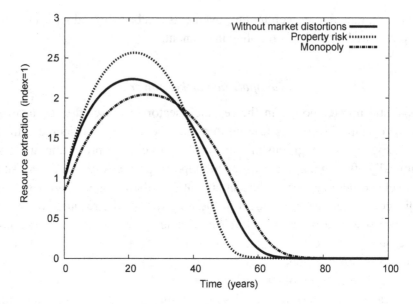

**Fig. 4.** Impact of several market imperfections on resource extraction. If property rights for resources are insecure, resource owners will extract the resources faster than the economic optimum, while a monopolistic resource owner will slow down resource extraction (values are indexed with regard to the first year of simulation). (*Source:* based on calculations in Edenhofer *et al.,* 2009 a)

future extraction more attractive than immediate extraction. However, such a subsidy shifts income from households to resource owners.

Another option is to institute an optimal price or quantity instrument that effectively expropriates the resource owners, thereby removing both the resource rent and the uncertainty of resource property rights. The problem of insecure property rights then only persists for permit owners who face regulatory uncertainties about future trading ratios or permit caps. Although the carbon budget is always adhered to, the timing in this case is suboptimal because higher effective discount rates are used to compensate for uncertainty.

## *Policies to push technological change*

Is carbon pricing the only important action that a government should take in order to avoid dangerous climate change? Conventional economic wisdom would say yes, as *The Economist* (2008) did when it criticized subsidies for clean technologies. Admittedly, a high carbon price is an incentive for investing in clean technologies. However, carbon prices alone fail to push clean technologies towards an optimal level because usually there are additional market failures with respect to innovation-driven technologies (Edenhofer *et al.,* 2006).

Typical market failures result from the nature of knowledge; while research has to be funded by someone, the gains from the resulting knowledge will not be fully captured by the funding firm. Intellectual property rights such as patents exist, but beyond direct marketing, knowledge is spread through formal and informal channels, and advancements in production processes are copied by other firms. Therefore, society as a whole benefits from research much more than the company funding the research and development (R&D). As a consequence, individual companies will invest less than the economic optimum in R&D (Jones and Williams, 2000). Also, other spillover effects exist, such as 'learning by doing'; for many goods, the production cost decreases by a certain amount each time that total cumulated production capacity of this good is doubled. Accordingly, all companies of a certain industry can profit from the total experience gained in that industry. This effect is readily observable for photovoltaic modules, where the cost per watt has fallen from about USD 50 to less than USD 3 over the last 33 years (Junginger *et al.*, 2008). To overcome these externalities and reach the economic optimum, economists recommend subsidies for investments that are related to spillover effects (Romer, 1986) or public R&D expenditures (Jones, 1995; Popp, 2004; Edenhofer *et al.*, 2005).

Our model supports the thesis that it is important to apply further instruments in addition to the tax or ETS. In particular, these comprise public R&D expenditures, both for energy efficiency and renewable energy technology, and investment subsidies to internalize spillovers of 'learning by doing' effects within the renewable energy sector. Although underinvestment in clean technology markets can be addressed by specific technology subsidies, one might ask if an additional increase of the carbon tax could induce sufficient higher investment. However, we calculated that without explicit technology subsidies, the effect of a further increase of the resource tax is not significantly different from the effect of basic quantity regulation (see Fig. 5a). As renewable energy production remains far below its optimal level, long-term consumption is reduced remarkably in comparison to a world in which an explicit technology subsidy is implemented (see Fig. 5b).

## Summary and conclusions

It is widely accepted that a price on carbon dioxide is required for successful climate protection. This can be achieved either through price mechanisms such as taxes on emissions or through quantity mechanisms such as emissions trading schemes. In this text we discussed and compared the effects of and the issues surrounding the implementation of different price and quantity regulations under a carbon budget constraint. The following conclusions apply to the design of all instruments:

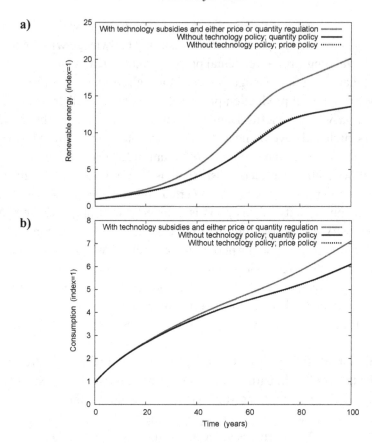

**Fig. 5.** a) Renewable energy production and b) total consumption of goods with and without technology policy; i.e., explicit technology subsidies (values are indexed with regard to the first year of simulation). (*Source:* based on calculations in Edenhofer *et al.*, 2009 a)

- It is important to **stabilize investor expectations** about the stringency of future carbon constraints by providing a credible long-term signal of future carbon prices. Future **carbon prices** need to be (i) **sufficiently high** and (ii) **consistent,** so that long-term investments are adjusted accordingly. This is the prerequisite for making the energy, production and transport infrastructure less carbon-intensive.
- For a price instrument, this requires a credible long-term commitment to a **rising carbon tax trajectory.** For a quantity instrument, the requirements are a **fixed total cap** and either **well-functioning futures markets** or an **institution** that **allocates** the total permits in all future times.
- Governments should capture the **scarcity rent of carbon.** Revenues from taxes or permit auctioning should be used to (i) offset distortionary taxes, (ii) subsidize abatement technologies to offset other market externalities from technological

spillover, or (iii) counteract the regressive effect of the carbon constraint (distributional equity). While carbon taxes will directly deliver annual revenues to governments, a cap-and-trade system will require auctioning of permits to raise a comparable revenue stream. Therefore, **auctioning is strongly preferred** to handing out permits for free.

- Regulation of fossil fuel **input** (e.g., a resource tax) is better than **output regulation** of secondary energy (e.g., an electricity tax) because the regulation directly addresses the pollution externality, exploiting all substitution and efficiency options along the production chain.
- The point of **regulation** should be **upstream** (at the level of fossil fuel producers and importers) rather than **downstream,** to allow broad coverage of sectors with low transaction costs.
- It is important to take into account **additional market externalities** besides climate change that affect the efficiency of taxes and quantity instruments. This includes, among other factors, the risk of **expropriation** and **monopolistic** energy markets.
- Technology spillover effects require additional policy instruments such as **subsidies for clean technologies.**
- The climate protection target will only be achieved if the **scarcity rent of resource owners is devaluated.** A cap on cumulative emissions directly communicates this devaluation, while a tax only achieves the devaluation if the regulator convinces resource owners that he will adjust the tax in such a way that he safeguards the total carbon budget. Otherwise, the resource owners might not extract resources along the optimal path, which could possibly result in excessive extraction of resources.
- If a **carbon bank** is entitled to issue allowances according to a publicly known cumulative carbon budget, the budget is made explicit and transparent, and can be anticipated easily by resource owners. If this is the case, the resource owners cannot increase their rents by deviating from the social optimal extraction path. It should be noted that even if the long-term credibility of a carbon bank can be taken for granted, the short-term volatility of prices remains a daunting issue. Thus, the main challenges for an ETS are reducing the **volatility of spot permit prices** and creating **stable expectations about future permit prices.**

The preference for a tax or a quantity instrument in a **realistic setting with uncertainty** hinges on the assessment of whether governments or markets are better suited to bear risks and make predictions about the future:

- A **price instrument** places the risk of misjudging the right tax rate on the **government**. Possible consequences of predicting the wrong mitigation costs are either

economic losses (if taxes are too high), or environmental losses due to non-compliance with the carbon budget (if taxes are too low).
- In contrast, a **quantity instrument** always achieves the environmental goal by observing the carbon budget. It moves the risk to the **economic agents,** with profit losses as a consequence of wrong predictions of future permit prices.

Although markets are often seen as more capable of collecting information than a centralized authority, this will entirely depend on the implementation of an efficient carbon market, including mature futures markets or other institutions for stabilizing future price expectations, such as insurance schemes, hedging strategies, or an international carbon bank. The choice of a quantity instrument can also provoke new market failures, as a new permit market is created that may be subject to speculations and myopic investment decisions.

Finally, since climate change is a global problem, the effort to reduce greenhouse gas emissions must be global. The long-term goal therefore should be the international harmonization of carbon prices. This will probably be more difficult to achieve with a system of national carbon taxes than with a global system or regionally interlinked systems of emissions trading.

# References

Bach, S. (2005). *Be- und Entlastungswirkungen der Ökologischen Steuerreform nach Produktionsbereichen.* Band I des Endberichts für das Projekt: Quantifizierung der Effekte der Ökologischen Steuerreform auf Umwelt, Beschäftigung und Innovation. Forschungsprojekt im Auftrag des Umweltbundesamts, FuE-Vorhaben Förderkennzeichen 204 41 194. Berlin.

Brunner, S., Flachsland, C., Luderer, G. and Edenhofer, O. (2009). *Domestic Emissions Trading Systems.* Background paper for the project 'China economics of climate change – toward a low carbon economy'. To be published by EarthScan.

Burtraw, D., Palmer, K. L. and Kahn, D. B. (2009). *A Symmetric Safety Valve.* Resources for the Future discussion paper No. 09–06. Available at http://ssrn.com/ab stract=1352002.

Dasgupta, P. S. and Heal, G. M. (1979). *Economic Theory and Exhaustible Resources.* Cambridge.

*Economist, The* (2008). Green, easy and wrong: why a verdant New Deal would be a bad deal. Issue 6 November. Available at http://www.economist.com/opinion/display story.cfm?story_id=12562343.

Edenhofer, O., Bauer, N. and Kriegler, E. (2005). The impact of technological change on climate protection and welfare: insights from the model MIND. *Ecological Economics,* **54**(2–3), 277–92.

Edenhofer, O., Carraro, C., Koehler, J. and Grubb, M., eds. (2006). *Endogenous Technological Change and the Economics of Atmospheric Stabilisation.* Cleveland.

Edenhofer, O., Pietzcker, R., Kalkuhl, M. and Kriegler, E. (2009a). *Taxation Instruments for Reducing Greenhouse Gas Emissions, and Comparison with Quantity Instruments.*

Background paper for the project: 'China economics of climate change – toward a low carbon economy'. To be published by EarthScan.

Edenhofer, O., Carraro, C., Hourcarde, J.-C. and Neuhoff, K. (2009b). *The Economics of Decarbonization.* Report of the RECIPE project. Potsdam.

Edenhofer, O. and Kalkuhl, M. (2009). *Prices vs. Quantities and the Extraction of Exhaustible Resources.* In preparation.

Grubb, M., Chapuis, T. and Ha Duong, M. (1995). The economics of changing course, implications of adaptability and inertia for optimal climate policy. *Energy Policy,* **23**(4–5), 417–31.

Hargrave, T. (2000). *An Upstream/Downstream Approach to Greenhouse Gas Emissions Trading.* Washington D.C.

Hepburn, C. (2006). Regulation by prices, quantities or both: a review of instrument choice. *Oxford Review of Economic Policy,* **22**(2), 226–47.

Hotelling, H. (1931). The economics of exhaustible resources. *Journal of Political Economy,* **39**(2), 137–75.

Jones, C. I. (1995). R&D-based models of economic growth. *Journal of Political Economy,* **103**(4), 759–84.

Jones, C. I. and Williams, J. C. (2000). Too much of a good thing? The economics of investment in R&D. *Journal of Economic Growth,* **5**(1), 65–85.

Junginger, M., Lako, P., Lensink, S., van Sark, W. and Weiss, M. (2008). *Technological Learning in the Energy Sector.* Petten.

Krysiak, F. C. (2008). Prices vs. quantities: the effects on technology choice. *Journal of Public Economics,* **92**(5–6), 1275–87.

Lawn, P. A. (2005). An assessment of the valuation methods used to calculate the Index of Sustainable Economic Welfare (ISEW), Genuine Progress Indicator (GPI), and Sustainable Net Benefit Index (SNBI). *Environment, Development and Sustainability,* 7(2), 185–208.

Matthews, H. D. and Caldeira, K. (2008). Stabilizing climate requires near-zero emissions. *Geophysical Research Letters,* **35**(4), L04705.

Meinshausen, M., Meinshausen, N., Hare, W., Raper, S. C. B., Frieler, K., Knutti, R., Frame, D. J. and Allen, M. R. (2009). Greenhouse-gas emission targets for limiting global warming to 2 °C. *Nature,* 458, 1158–1162.

Newell, R. G. and Pizer, W. A. (2003). Regulating stock externalities under uncertainty. *Journal of Environmental Economics and Management,* **45**(2), 416–32.

Popp, D. (2004). ENTICE: endogenous technological change in the DICE model of global warming. *Journal of Environmental Economics and Management,* **48**(1), 742–68.

Roberts, M. J. and Spence, M. (1976). Effluent charges and licenses under uncertainty. *Journal of Public Economics,* **5**(3–4), 193–208.

Romer, P. M. (1986). Increasing returns and long-run growth. *The Journal of Political Economy,* **94**(5), 1002–37.

Roughgarden, T. and Schneider, S. H. (1999). Climate change policy: quantifying uncertainties for damages and optimal carbon taxes. *Energy Policy,* **27**(7), 415–29.

Rupp, S. and Bailey, I. (2003). *New Environmental Policy Instruments and German Industry.* Working paper. Plymouth.

Sijm, J. P., Neuhoff, K. and Chen, Y. (2006). $CO_2$ cost pass-through and windfall profits in the power sector. *Climate Policy,* **6**(1), 49–72.

Sinn, H. W. (2008). Public policies against global warming: a supply side approach. *International Tax and Public Finance,* **15**(4), 360–94.

Stern, N. (2008). The economics of climate change. *American Economic Review, Papers & Proceedings,* **98**(2), 1–37.

Weitzman, M. L. (1974). Prices vs. quantities. *Review of Economic Studies,* **41**(4), 477–91.

# Chapter 18

# Controlling climate change economically

---

## James Mirrlees

James Mirrlees was born in Minnigaff, Scotland, in 1936. He studied mathematics at Edinburgh and Cambridge Universities, and received a PhD in Economics at Cambridge University in 1963. He has been Professor in Oxford and Cambridge, where he is Emeritus Professor of Political Economy, and is now also Distinguished Professor-at-large in the Chinese University of Hong Kong. In 1996 he was awarded the Sveriges Riksbank Nobel Memorial Prize in Economic Sciences 'for his contributions to the economic theory of incentives under asymmetric information'. Mirrlees had focused on the design of income taxation but soon realized that his method could be applied to many other similar problems where decision-makers are faced with asymmetrical or incomplete economic information. Up to the present day, optimal public economic policy remains his main research interest, as well as development economics.

*Note:* This chapter is a commentary on chapter 17.

When people do things that emit greenhouse gases into the atmosphere, they cause damage – damage to other people; in the present, in the future, and all over the world. To an economist, it is clear that emitters should pay a price equal to the value of the damage caused. That is what we call the carbon price (though non-carbon greenhouse gas emitters need to be charged too). With carbon prices in place, now and in the future, people will burn fossil fuels only if the value of doing so is greater than the cost of the fuel and the combustion device, and the damage caused by the increases in temperature it brings about. Supplying energy by less damaging processes will pay off. The invention and development of new energy technologies will be more profitable. However, leaving aside general principles it is not easy to estimate the damage caused by global warming, and there is considerable disagreement about what should count as damage, and about the economic cost of that damage.

As many non-economists have recognized, our relatively carbon-free atmosphere is an exhaustible resource. This is the approach taken by Edenhofer and his co-authors who, instead of computing the social cost of carbon, consider the simplified problem of optimally allocating a maximum admissible quantity of emissions – a carbon budget. According to Hotelling's principle, the price of an exhaustible resource should, uncertainties apart, rise at a rate equal to the rate of interest. Consequently, Edenhofer and his co-authors report their finding that the carbon price should rise over time, at least for an initial period until backstop technologies become more competitive. But if global warming is already causing damage, and going to cause more in the immediate future, Hotelling's principle has to be modified. I am sure that is the case. It is quite possible that the carbon price should not be rising; it should perhaps already be as high as it is ever going to be. In particular, rough calculations suggest that it should already be much higher than the prices that have so far emerged in cap-and-trade markets.

Carbon prices can be determined in markets, in order to equate the emissions people wish to make to the quantity of emissions that is compatible with mitigation goals. Many have argued that it is easier to estimate the desired quantity of carbon emissions than to estimate the value of the damage caused by emissions; others that it is easier to estimate the carbon price. Neither position is tenable. If we seek an optimal solution, prices and quantities have to be estimated together, as I shall explain. It is possible to develop a plan for present and future emission levels that would have a good chance of keeping damage from warming within tolerable bounds. Such a plan can be resolved into target quantities for individual countries or industries, as in the case of the Kyoto Protocol. While proceeding in this way has the advantage that it is readily understood and relatively easy to discuss, it is, however, unlikely to gain the universal coverage necessary to achieve the desired effect.

Even taking the future course of emissions as a given, it is a daunting task to estimate the damage. Not only do we need to estimate the extent to which a unit of emissions will reduce national incomes (the total of the individual incomes that we actually care about), we need to do that for long periods of time, far into the future. We also need to assess forms of damage that are not easily quantifiable in monetary terms, particularly loss of life and environmental destruction. Certainly, future economic damage should be discounted. Equating the discount rate to the average rate of return to capital investment may be helpful in estimating it; but that rate of return cannot be estimated independently of considerations of future climate change damages. In any case, the discount rate should not be applied to the value of life and environment lost. That matters, because it means that the equivalent value of such destruction is important now, even if it happens a century hence.

There is also a strong case for increasing damage estimates to allow for uncertainty about the future size of national incomes, which will partially offset the discount rate. Anticipated lower national incomes in the future would require us to adjust the discount rate downwards. That is another reason why damage must be estimated a long way ahead. I agree with Nicholas Stern and Martin Weitzman that the effective discount rate for standard economic damage should be low, perhaps two or three percent, though I think we all have different reasons.

The level of damage at any future time depends on how much global warming actually takes place. Our estimate of the carbon price therefore also depends on the extent of global warming, which will be influenced by the controls and carbon markets and taxes that are put in place. Certainly very different levels of damage are possible, up to widespread loss of life. The question is whether marginal (i.e., additional) damage increases with the level of damage.

It might be thought that, following the initial decades in which global warming begins to have a significant impact (i.e., the present day), the marginal damage from further unit increases in emissions will not vary much with the level of future damage. This view is supported by estimates quoted in the Stern Report, which imply that the marginal damage will not increase from 2 °C to 4 °C of warming. These figures may well reflect the simple ways of estimating future damage that have so far been employed in attempts to quantify the effects of warming. However, when one considers urban reconstruction, population displacement, and likely death tolls from high and rising water levels, all costs and damages that should not be discounted, it is hard not to believe that marginal damage will increase with the level of future carbon concentration, and will be very high if concentrations reach a level where the temperature increase is 4 °C or more.

If it is true that marginal damage from global warming will increase with the extent of warming, serious consequences follow for countries that wish to effectively address the climate change problem. If it becomes clear that many countries are not going to radically reduce their emissions, the marginal social cost of warming will be all the greater. It might then be the duty of compliant countries to reduce their emissions to zero as quickly as possible. The value for them of a global deal is great, as are the advantages of subsidising carbon savings in other countries.

The claim that marginal damage increases with carbon concentration provides an answer to the following question: how can we be sure that a carbon price calculated by estimating future marginal damage will be high enough to assure effective mitigation that keeps the carbon concentration from pushing temperatures above a tolerable level, say 3 °C? The price estimate based on future concentrations above that level would surely be more than high enough to prevent it. Effective mitigation measures that draw down carbon concentrations would surely be undertaken since the cost of radical cuts in greenhouse gas emissions is comparatively low (it is reliably estimated to be only a modest fraction of national incomes, less than 5 %). This argument also shows that we cannot estimate the appropriate carbon price simply by discounting future marginal damage, since that depends on what future carbon prices will be. We need a full dynamic model, to be solved simultaneously for optimal emissions and optimal carbon price. A cruder version of this model is to estimate the desirable carbon price on the basis of realistic forecasts of temperatures and water levels, given policy commitments as shown by actual carbon prices, nation by nation, and adjust it year by year. I expect that the desirable carbon price would fall over time, as countries improved their mitigation policies.

Is it easier, then, to estimate the quantity of emissions that we should have? It is easier to propose a particular quantity plan than a price-and-tax plan, and that is what the global-warming community has done, and what most governments have accepted, in a rather quixotic way. Laying out a carefully estimated scenario is indeed a great achievement, and makes some kind of international agreement possible. But it is something very different from the calculation of an optimal plan, which (some) economists might prefer. It leaves open the question of how much pollution each nation or person may cause. Now that is certainly hard to estimate, in the sense of giving an optimal prescription to each nation. The nearer decisions come down to the individual level, the more we must move to price rather than quantity, because different individuals should have different 'carbon footprints' – they have different tastes, live in places that create different needs, and have different possibilities for emission reductions.

The way that carbon markets work now creates many anomalies. The main one is that the level of carbon prices seems much too low, when compared with the sorts of figures that are estimated on the economic basis I outlined. The carbon price needs to be estimated on the basis of future damage, as a check on the level and effectiveness of the quantity constraints that carbon markets are supposed to embody. One reason for the low price is the provision for buying emission permits from emission-reduction projects in other countries. Of course that has the beneficial effect of reducing net emissions in other countries. But it is offset by increasing net emissions above what was supposed to be the agreed level for the country, or group of countries, operating the carbon market. There is no net advantage for global warming from that. This flaw in the current implementation of carbon markets could have been remedied if the quantity of permits made available each year had been reduced by the amount expected to be purchased from emission-reducers. As things are, these external purchases brought the carbon price down, reducing the price to users of the kinds of consumption and investment that have a relatively large impact on global warming, and reducing the incentive to introduce and develop green technologies.

I am not arguing that there should be no subsidy for emission-reducing projects, such as forests. On the contrary, it is most desirable to encourage projects that absorb greenhouse gases. We should talk of a carbon price, not a carbon tax, because it should operate as a subsidy to activities that absorb greenhouse gases from the atmosphere, as well as a tax on the emission of greenhouse gases. The low carbon prices in the cap-and-trade markets are evidence that, for the time being, expansion of forests (to take a major example) and some other flora is the most efficient way of reducing global warming. Since the carbon price ought to be, based on any estimate, considerably higher than the current price, there should be a very rapid expansion of carbon-absorbing flora. But, for obvious reasons, it cannot last for long, because there is only limited space before the value of producing food makes further expansion undesirable.

The more general problem with carbon markets is inclusiveness. Not everything can be covered by a carbon market, though it seems governments are very far from requiring permits wherever they should. Farmers should have to purchase annual permits for their farm animals, just as they should be able to sell annual permits on the basis of their woodland. It is a complicated issue, however: the quantity of permits should be related to the type of animal, its age, and other characteristics. This will no doubt seem quite impossible in the European Union. Can we envisage it in India and Africa? Yet, if not, how are we going to get emissions down to 80% of 1990 levels by 2050, with cuts increasing in intervening years; or even the somewhat more modest ambition of the G8 countries? In many cases, it is simply easier to apply the carbon price as a tax than to require purchase of permits. The tax rate could be based automatically on the price in the carbon market. One argument for setting quantities of carbon permits (within each country) and allowing them to be traded in a market is that allowable emissions are then produced by those for whom it is most valuable. If the market does not cover the full range of emission activities, this will not happen to the extent it should.

# Chapter 19

# What is the top priority on climate change?

---

## Paul Klemperer

Paul Klemperer, born in 1956, studied engineering at Cambridge University, and was awarded an MBA and a PhD in economics at Stanford University in Palo Alto, California. He is currently Edgeworth Professor of Economics at Oxford University. His main research interests are in industrial economics theory and policy, competition policy, and microeconomic theory, especially auction theory. He has been a Member of the UK Competition Commission, and has advised the US Federal Trade Commission, the European Commission, and many other government agencies and private companies.

Action on climate change is urgently needed. Substantial uncertainty about the importance of the problem remains,[1] but this uncertainty means we should worry more, not less, because while things may not be as bad as the most likely scenarios suggest, outcomes could also be a lot worse.[2] What, therefore, should be the West's top priority for climate change policy?

## The critical issue

The critical issue is that no strategy will work unless it is consistent with developing countries' continued economic growth.[3] So we are unlikely to be able to reduce the use of 'dirty' energy sufficiently unless we can find a cheap, clean, substitute.[4] And that requires innovation.

Developing countries are not going to give up the immediate aspirations of their (often growing) populations in exchange for environmental benefits that arise largely in the future. These nations simply do not have the luxury of worrying about preserving the environment for their great-grandchildren. China, for example, stresses even in the Foreword to its National Climate Change Programme that 'economic and social development and poverty eradication are [its] first and overriding priorities'.[5] Whether or not this is morally right (though it may be justified for a developing country) is irrelevant. It is a political imperative for the leadership of a country in which, according to the latest figures, about 200 million people live below the World Bank's 'dollar-a-day' poverty line, and in which 100 million are illiterate.[6]

---

[1] See Klemperer (2007), and also, e.g., Table 1 of Nicholas Stern's paper in this volume.

[2] See Klemperer (2008a), and also Topic 6.2 in the IPCC's 4th Synthesis Report, 2007.

[3] This point applies across the polluting sectors, including, for example, deforestation (see, for example, Angelsen and Kaimowitz (1999), Lambin *et al.* (2001)), but I am focusing here on energy, where policy may be most essential. See also Sunita Narain's essay in this volume for a discussion of developing countries' climate change priorities.

[4] I am *not* arguing that systems for pricing carbon, such as carbon taxes or a 'cap and trade' permit system, are not helpful. But they are not sufficient.

[5] However, the Chinese government is merely quoting the statement of the United Nations Framework Convention on Climate Change (UNFCCC, 1992) that 'economic and social development and poverty eradication are the first and overriding priorities of the developing country Parties' (article 4, paragraph 7).

[6] This is not, of course, to suggest the Chinese are less 'moral' than the West – on the contrary, their value system may place more weight on, and their culture offers more support to, intergenerational justice. And, of course, many people from all parts of the world are concerned about the effects of climate change on current as well as future generations, and regard environmental protection as a necessity, not a luxury. (In China, Pan Yue, deputy director of China's state Environmental Protection Administration, who was named the New Statesman's 'Person of the Year 2007', is just one notable example.)
I focus especially on China among the rapidly-developing countries, because of its size, and because of my focus on energy use; other countries are obviously especially important in the context of deforestation. Keidel (2007) estimated that 300 million Chinese lived below the 'dollar-a-day' poverty line (which is calculated on the basis of 'purchasing power parity' exchange rates that compare prices across different countries), but Chen and Ravallion's (2008) more recent work suggests an estimate of 200 million for 2005, and that this number is rapidly declining. The World Bank (2008) gives an illiteracy figure of 100 million for 2000; the UN (2007) give an estimate of 130 million for 2003.

Thus, although China has probably now overtaken the US to become the world's number one polluting nation,[7] its officials emphasize that it has no obligation to cut emissions under the Kyoto Protocol. Moreover, it seems unlikely to do so voluntarily, at least on the scale required – consider, for example, China's recently-announced plan to build 97 new airports in the next 12 years (while the UK has agonized about whether to build a single extra runway at Heathrow!).

## The challenge

Much recent research suggests that we need to stabilize greenhouse-gas concentrations below 400 parts per million (ppm) carbon dioxide equivalent if we wish 'to preserve a planet similar to that on which civilization developed and to which life on Earth is adapted'.[8] Indeed, the IPCC says that stabilization at around 380 ppm carbon dioxide equivalent would yield a more than 20% probability that global warming will exceed 2 °C, the level that is commonly referred to as the threshold for 'dangerous' warming (and the EU has adopted the target of keeping the temperature increase below this level).[9]

Perhaps these estimates are pessimistic. But even stabilizing greenhouse gas concentrations at 500 ppm carbon dioxide equivalent (whereby temperature increases above 2 °C would be very likely[10]) requires a roughly 50% reduction in greenhouse-gas emissions by 2050. Allowing for population growth, this requires a two-thirds fall in per-capita emissions to about 2–2.5 tonnes carbon dioxide equivalent by that date.[11]

---

[7] See Auffhammer and Carson (2007). Note also that China's energy intensity is 1.5 times the global average (World Bank (2008) figure for 2005).

[8] Hansen *et al.* (2008) write that achieving this objective requires stabilization at 350 ppm $CO_2$; including all greenhouse gases, this would correspond to a little under 400 ppm $CO_2$ equivalent if today's relative atmospheric concentrations of the different greenhouse gases were maintained. In 2005 levels were 380 ppm $CO_2$ and 430 ppm $CO_2$ equivalent (Stern 2007, Section 1.2). The campaign to set a target of 350 ppm $CO_2$ endorsed by, among others, Al Gore at the United Nations Climate Change Conference in Poznań, 2008 (see http://www.un.org/climate-change/blog/2008/121208.shtml) is based upon Hansen *et al.* (2008), and refers to levels of $CO_2$ alone, but most current debate refers to $CO_2$ equivalent levels.

[9] See IPCC (2007), Table 3.9, Working Group III Report 'Mitigation of Climate Change'. A probability of 20% is obtained at 378 ppm $CO_2$ equivalent even using calculations that 'do not take into account the full range of biogeophysical feedbacks that may occur'. The claim that temperature increases should not exceed 2 °C above pre-industrial levels is now routine in official documents (see, e.g., EU, 2005), as well as the media.

[10] Stabilization at 500 ppm $CO_2$ equivalent yields a temperature increase above 2 °C in the vast majority of scenarios according to the IPCC (2007, Figure 3.38, p. 228), and with a probability of 48–96% depending on the model used, according to Stern (2007, Box 8.1). Possible emissions pathways, and the associated risks, are discussed extensively in the eight articles forming Section VI of Schellnhuber *et al.* (2006).

[11] Stabilization at 500 ppm $CO_2$ equivalent is thought to require global emissions to be reduced to about 20 gigatonnes by 2050, at which date the global population is projected to be about nine billion (Stern, 2008). The scale of the challenge is illustrated by the fact that the latest edition of Shell's (2008) highly-respected Energy Scenarios implies that 650 ppm $CO_2$ equivalent is an optimistic outcome, and that 1000 ppm or more is also plausible – see Prinn *et al.* (2008) who analyse a range of reputable emissions scenarios and find that all lie between 550 and 1780 ppm in 2100. The UK has committed to cutting its $CO_2$ equivalent emissions by 80% by 2050 and to reducing them to between 2.1 and 2.6 tonnes per person (Committee on Climate Change Report, 2008).

The United States, Canada and Australia now each emit well over 20 tonnes of carbon dioxide equivalent per head annually, while the EU and Japan each emit a little above 10 tonnes per head. However, there are signs that these regions may reduce their emissions, because their already-rich populations can afford to worry about their children and grandchildren.

India's per-capita emissions are still below two tonnes of carbon dioxide equivalent, and most of sub-Saharan Africa is well below one tonne of carbon dioxide equivalent. However, China and several other rapidly-developing countries already emit more than six tonnes per head.[12] So the key challenge is: how do we persuade countries like China to more than halve their emissions when they are so focused on economic growth?

### The (limited) efficacy of trade policy

The West does have *some* leverage: the French President Nicolas Sarkozy was right to suggest, for example, that the EU should threaten to tax imports from countries that have neither a carbon tax nor a cap-and-trade permit system.[13] If the threat were carried out, and the EU taxed imports' embodied carbon emissions at a rate equal to the price of an EU Allowance, this would be equivalent to introducing these countries' export sectors into the EU's permit system, and would reduce emissions in exactly the same way.

Taxing 'dirty' imports would have other advantages too: it would reduce emitters' incentives to flee the EU for more lax jurisdictions;[14] it would solve any problem of EU firms being disadvantaged relative to non-EU competitors; and it would therefore also greatly weaken the case for giving free permits to firms,[15] thereby enhancing the EU's ability to raise revenues for other climate change mitigation activities. Many economists argue that import taxes undermine free trade. They are

---

[12] China's 2006 emissions are estimated to be 6.0 tonnes $CO_2$ equivalent per head. Other large rapidly-developing countries with high emissions include Turkey (5.7 tonnes per head), Mexico (6.4 tonnes per head), South Africa (10.6 tonnes per head), the Russian Federation (15.4 tonnes per head), Brazil (5.4 tonnes per head counting conventionally, plus 7.25 tonnes per head extra due to land use change, i.e., deforestation), and Indonesia (2.7 tonnes per head conventionally, plus 11.5 tonnes per head due to land use change). All the national per-capita emissions figures in this section are Ecofys (2008) estimates for 2006.

[13] See, for example, Barchfield (2008).

[14] This incentive can be exaggerated. It operates mostly in the long run, and is mitigated by the expectation of future carbon regulation in developing countries.

[15] Witness the comment of Sigmar Gabriel, German Environment Minister, in justifying Germany's recent backtracking on the principle of full auctioning of permits: 'As long as European companies are governed by stricter climate protection regulations than their competitors in countries like China, we have to seek to establish special rules' (Bryant *et al.,* 2008). Unless there is substantial foreign competition, giving permits to companies for free represents an unnecessary and improper handout of windfall profits, since consumer prices rise to reflect permits' value, independent of how they are allocated – see Binmore and Klemperer (2002, section 2), Fries (2008), Klemperer (2004; 2008b).

wrong in theory because the absence of any charge for carbon emissions is effectively a subsidy, for which the import taxes simply compensate. And they are also wrong in practice, because we should care more about carbon emissions than about the health of the WTO.

Of course, the practical problems of implementation would be substantial. So we would very much hope never to have to carry out Sarkozy's threat. However, if the EU promised not to tax imports from countries that introduced their own carbon taxes or permit systems for their exports, many countries would likely introduce these measures; the exporting country, rather than the EU, would then collect the revenues from the taxes or permit sales. Moreover, having introduced tax or permit systems for exports (and benefited from the revenues), developing countries might later extend them to other sectors of their economies.

However, even China's substantial export sector represents only around one third of its GDP, although a large proportion of these exports do go to developed-world countries that might plausibly impose an import tax.[16] But while the West can also make other threats, such as to exclude uncooperative countries from international organizations and sporting events, or to encourage consumer boycotts, etc.,[17] the bottom line is that it has only limited influence over the developing world.

China, in particular, seems unlikely to incur significant abatement costs unless it is compensated; this is probably the binding constraint on any global deal (India matters hugely too, of course, but its per-capita emissions are so much lower that it will probably participate in any agreement that China will accept[18]).

## The need for more research and development (R&D)

So what conclusions can we draw?

First, whether we like it or not, China (and India and others) will continue to develop nuclear energy. Therefore, unless the West continues to develop it too, the safety, storage and handling issues will be resolved in developing countries, in many of which there is both less democratic accountability than in Europe and the US, and also more pressure to take shortcuts than in richer countries.[19]

Second, China (and India and others) will continue to exploit its enormous coal reserves. Therefore, we urgently need research and development on low-cost Carbon

---

[16] The developed world (largely EU, USA, Japan, Canada, and Australia) accounts for about five-eighths of China's exports. About 40 % of this total goes to the EU, and a similar volume to the USA. See IMF (2008) data for 2007 for the export figures in this paragraph, which are calculated using nominal exchange rates (purchasing power parity rates are substantially different); using nominal values, China's exports 'are on average no more or less carbon-intensive than domestic consumption and investment' (Weber *et al.*, 2008).

[17] See Aldy, Orszag and Stiglitz (2001).

[18] As noted above, India is still below the commonly-suggested target of 2–2.5 tonnes of $CO_2$ equivalent emissions per head.

[19] Thomas Bruckner *et al.* discuss issues about nuclear energy, and also coal use and CCS, in this volume.

Capture and Storage (CCS) technologies to remove coal plants' emissions. The UK government is right to subsidize a demonstration CCS plant.[20] It should probably subsidize several. It is also right to focus on developing technology that can be retrofitted to traditional plants. China, after all, is building one such plant every five days.

Crucially, however, it will always be cheaper to burn coal (and oil and gas) without CCS than with it. We can encourage developing countries to use CCS through a revised Clean Development Mechanism[21], or – even better – by including these countries in an emissions trading scheme that allocates them enough permits so that they make money by participating. However, Western electorates will only be willing to transfer limited resources to the developing world. There may also be problems monitoring whether CCS technology is being used as claimed, or whether leakage occurs at the storage sites. So CCS alone will not suffice.[22] Only clean energy sources that are cheaper than those currently available are likely to prevent further emissions growth in the developing world.[23]

If large-scale nuclear power is politically unacceptable, substantial investment in clean energy R&D is the only alternative. But the private sector will not do this unaided. Businesses know that when an innovation is sufficiently important, the innovator gets little of the benefit; for example, the developers of drugs for AIDS, and of vaccines for Anthrax and bird flu, were threatened with compulsory licenses in many countries (including in the United States) until they 'voluntarily' licensed their innovations cheaply. The difficulties of getting effective patent protection in the first place (which means any innovator fears being copied, and then forced to compete with imitators), the riskiness of much energy R&D, and the large scale of some of the necessary investments (for example, research into fusion) are further reasons why business is reluctant to undertake the necessary R&D without subsidies.[24]

So it is catastrophic that – as the Stern Report emphasized[25] – public expenditure on energy R&D has been declining in most countries over the last 30 years, and it is shameful that most of Europe spends a much smaller fraction of its GDP on public energy R&D than even the USA and Japan. The UK is one of the worst offenders.

---

[20] See http://www.berr.gov.uk/whatwedo/energy/sources/sustainable/ccs/ccs-demo/page40961.html.

[21] Diana Liverman discusses various proposals to reform the Clean Development Mechanism in this volume.

[22] A dramatically cheaper 'geo-engineering' solution that sucks $CO_2$ directly from the sea or the atmosphere (in effect making all existing energy sources clean) might suffice. Here too, public money for R&D is essential for the reasons discussed below.

[23] For example, further development of solar energy may be a particularly promising avenue for the substitution of dirty energy – see the discussion by Walter Kohn in this volume.

[24] Even if these problems did not apply, private enterprise would accomplish less innovation than would be socially optimal, because – as argued above – it is implausible that the international community will make a credible commitment to set a price for greenhouse-gas emissions that equals their full social costs.

[25] See figure 16.3 of the Stern Review (2007), which draws on data from the IEA.

## Publicly-funded R&D

Calling for more publicly-funded R&D raises two questions: how should the funds be raised?; and how should they be targeted? Countries should agree that each will support more R&D if others do likewise, thus increasing all countries' incentives to do so. Furthermore, if the EU's cap-and-trade emissions permits were all auctioned, rather than largely given away free,[26] the expected revenues would be at least 30 billion euros per year (based on current carbon prices[27]), and could be greater still if the scheme were expanded to include more sources of emissions.[28] A large fraction of the auction proceeds could and should be pledged to R&D funding.[29] Similar approaches should be taken outside the EU.

Economics has less to say about how best to spend the money.[30] It seems that, even with a clear and apparently relatively easily achievable goal, innovative processes can be highly unpredictable.[31] That suggests distributing the money to a variety of different actors and approaches. Existing funding at both national and EU levels should be increased, especially for basic science (and science teaching).[32] There is probably a greater role for publicly-funded prizes for specific achievements than is now common – witness the success of the XPrizes.[33] The vagueness of these remarks demonstrates an urgent need for research into the economics of innovation!

[26] The permits will mostly be given out to companies free until 2012. (See note 15 for the (lack of) justification for this.) As of December 2008, the EU plans to auction 100% of permits for electricity generation in 2013, apart from some 'derogations'; it plans to auction 20% of industrial permits by then, rising to 70% by 2020, for industries not considered at risk of 'carbon leakage'; see EU (2008).

[27] The number of emissions allowances (EUAs) to be allocated annually (2083 million tonnes $CO_2$ in the period 2008–2012; see European Commission (2007) and Committee on Climate Change report (2008, p. 151)) multiplied by their market price (EUR 15.30 at mid-December 2008, see http://www.pointcarbon.com) yields about EUR 30 billion. Note, however, that this carbon price is low, relative both to the recent past and to some expectations. (The Committee on Climate Change report (2008, p. 169) uses a carbon price of EUR 51/t$CO_2$ in 2020 based on 'the assumption of an EU 30% GHG target and central fossil fuel prices [which] corresponds to the post-global-deal world [it is] expecting and planning for'.)

[28] The EU's Emissions Trading Scheme will cover aviation for the first time from 2012 (European Commission, 2008).

[29] As of December 2008, the EU plans to hypothecate the proceeds from the sales of 200 million emissions allowances in the post-2012 period to the development of CCS and renewable energy sources; see EU (2008). Hypothecation violates economic orthodoxy, of course, but it seems consistent with practical politics in this context.

[30] See, however, the useful discussion in Arrow *et al.* (2008).

[31] For example, Bresnahan (2008) documents that even though e-commerce was an obvious application of the PC, many of the obvious players – including Citibank who invested USD 300 million, and an IBM-Sears-Roebuck-CBS joint-venture – made very large R&D investments in unsuccessful attempts to develop it; e-commerce only eventually arrived after academics-turned-entrepreneurs developed the web browser.

[32] The danger is that special interests will misdirect funding to particular firms, industries, etc. One way to reduce the likelihood of this is to allocate funding through institutions such as the National Science Foundation in the USA, and the Royal Society and the Engineering and Physical Sciences Research Council in the UK.

[33] The prototype is the Ansari X Prize, which offered USD 10 million to the team who could most convincingly pioneer space tourism. This reportedly galvanized substantial private sector investment, which resulted in overcoming the technological challenges (Kalil (2006), p. 5–7, see also Masters and Delbecq (2008)). Further prizes have been announced in genomics, environmentally friendly vehicles, and moon transportation (see http://www.xprize.org). Publicly-funded prizes can also take the form of government purchase guarantees.

## Conclusion

More R&D of clean energy is probably the highest priority of all. There are other priorities too, of course. In particular, curbing deforestation is a cheap and cost-effective solution, and has the collateral benefit of preserving biodiversity. But finding a clean energy source that is cheaper than those currently available is the only politically plausible way of curbing growth in developing nations' emissions.

*Acknowledgements*: I have no special expertise on the scientific aspects of climate change. I am very grateful to all those who have tried to educate me, but they are not responsible for remaining errors. I am particularly grateful to many friends and colleagues, including Jeremy Bulow, Ken Coutts, Dave Frame, Steven Fries, Chris Goodall, Tim Harford, Cameron Hepburn, Cath Howdle, Max Tse, Peter Zapfel, and especially Elizabeth Baldwin, for numerous helpful comments and suggestions on the current paper; Elizabeth Baldwin also provided expert research assistance for it.

## References

Aldy, J. E., Orszag, P. R. and Stiglitz, J. E. (2001). *Climate Change: An Agenda for Global Collective Action*. Paper prepared for the conference on The Timing of Climate Change Policies. Washington, D. C.

Angelsen, A. and Kaimowitz, D. (1999). Rethinking the causes of deforestation: lessons from economic models. *World Bank Research Observer,* **14**(1), 73–98.

Arrow, K. J., Cohen, L. R., David, P. *et al.* (2008). *A Statement on the Appropriate Role for Research and Development in Climate Policy*. Reg-Markets Center Working Paper No. 08–12. Available at http://www.reg-markets.org.

Auffhammer, M. and Carson, R. (2008). Forecasting the path of china's $CO_2$ emissions using province-level information. *Journal of Environmental Economics and Management,* **55**(3), 229–247. Available at http://ideas.repec.org/p/cdl/agrebk/971.html.

Barchfield, J. (2008). France's Sarkozy calls for carbon tax. *International Business Times,* 15 January.

Binmore, K. and Klemperer, P. (2002). The biggest auction ever: the sale of the British 3G Telecom licenses. *Economic Journal,* **112**(478), C74–96.

Bresnahan, T. (2008). *Entrepreneurs, Large Firms and Innovation*. Clarendon Lectures, 3rd–5th November 2008, Oxford.

Bryant, C., Harvey, F. and Barber, T. (2008). Climate change fears after German opt-out. *Financial Times,* 22nd September.

Chen, S. and Ravallion, M. (2008). *China is poorer than we thought, but no less successful in the fight against poverty*. World Bank Policy Research Working Paper 4621.

Committee on Climate Change (2008). *Building a low-carbon economy – the UK's contribution to tackling climate change*. The First Report of the Committee on Climate Change. Norwich. Available at http://www.tsoshop.co.uk.

Ecofys (2008). *Factors Underpinning Future Action – Country Fact Sheets 2008 Update*. Report prepared for the UK Department of Energy and Climate. Available at http://www.ecofys.com/com/publications/reports_books.asp.

European Commission (2008). *Directive 2008/101/EC of the European Parliament and of the Council of 19 November 2008 Amending Directive 2003/87/EC so as to Include Aviation Activities in the Scheme for Greenhouse Gas Emission Allowance Trading within the Community.* Available at http://eur-lex.europa.eu/LexUriServ/LexUriServ.do?uri=CELEX:32008L0101:EN:NOT.

European Commission (2007). *Emissions Trading: Commission Adopts Amendment Decision on the Slovak National Allocation Plan for 2008 to 2012.* Document No. IP/07/1869. Brussels.

EU (2005). *Climate Change: Medium and longer term emission reduction strategies, including targets with Council conclusions.* The Council of the European Union, Document No. 7242/05. Brussels.

EU (2008). *Energy and Climate Change – Elements of the Final Compromise.* EU Document No. 17215/08. Available at http://www.consilium.europa.eu/uedocs/cms_data/docs/pressdata/en/ec/104672.pdf.

Fries S. (2008). Allocating emission allowances: towards a sustainable approach. *Oxonomics,* **3**(1), 36–9.

Hansen, J., Sato, M., Kharecha, P. *et al.* (2008). Target atmospheric $CO_2$: where should humanity aim? *The Open Atmospheric Science Journal,* **2**, 217–31.

IMF – International Monetary Fund (2008). *Direction of Trade Statistics (DOTS).* Updated 10 December. Manchester. http://www.esds.ac.uk/International.

IPCC (2007). *Climate Change 2007: Synthesis Report. Contribution of Working Groups I, II and III to the Fourth Assessment Report of the Intergovernmental Panel on Climate Change.* Geneva. Available at http://www.ipcc.ch/publications_and_data/publications_ipcc_fourth_assessment_report_synthesis_report.htm.

Kalil, T. (2006). *Prizes for Technological Innovation.* Brookings Institution Working Paper, Hamilton Project Discussion Paper. Washington, D. C.

Keidel, A. (2007). The limits of a smaller, poorer china. *Financial Times,* 13 November.

Klemperer, P. (2004). *Auctions: Theory and Practice.* Princeton, USA. (Chinese translation: (2006), Renmin University Press, Beijing).

Klemperer, P. (2007). Why economists don't know all the answers about climate change. Editorial. *Financial Times,* 11 May, p. 15. Available at http://www.gqq10.dial.pipex.com/PressArticles/Doc3.pdf.

Klemperer, P. (2008a). If climate sceptics are right, it is time to worry. Editorial. *Financial Times,* 29 February, p. 11. Available at http://www.gqq10.dial.pipex.com/PressArticles/Doc33.pdf.

Klemperer, P. (2008b). Permit auction will not affect prices. *Financial Times,* 9 June, p. 10. Available at http://www.gqq10.dial.pipex.com/PressArticles/EmissionPermitAuctions.pdf.

Lambin, E. F., Turner, B. L., Geist, H. J. *et al.* (2001). The causes of land-use and land-cover change: moving beyond the myths. *Global Environmental Change,* **11**(4), 261–269. Available at http://www.sciencedirect.com/science/journal/09593780.

Masters, W. A. and Delbecq, B. (2008). *Accelerating Innovation with Prize Rewards: History and Typology of Technology Prizes and a New Contest Design for Innovation in African Agriculture.* IFPRI (International Food Policy Research Institute) Discussion Paper 835.

Prinn, R. G., Paltsev, S., Sokolov, A. P. *et al.* (2008). *The Influence on Climate Change of Differing Scenarios for Future Development Analyzed Using the MIT Integrated Global System Model.* MIT Joint Program on the Science and Policy of Global Change, Report No. 163.

Schellnhuber, H. J., Cramer, W., Nakicenovic, N., Wigley, T. and Yohe, G., eds. (2006). *Avoiding Dangerous Climate Change.* Cambridge.

Shell (2008). *Shell Energy Scenarios to 2050.* The Hague. Available at http://www-static.
     shell.com/static/aboutshell/downloads/our_strategy/shell_global_scenarios/SES%20
     booklet%2025%20of%20July%202008.pdf.
Stern, N. (2007). The Economics of Climate Change: the Stern Review. Cambridge.
UN – United Nations (2007). *United Nations Common Database, UNESCO estimates.*
     Updated 22 November. Manchester. http://www.esds.ac.uk/International.
UNFCCC (1992). *United Nations Framework Convention on Climate Change.* Available at
     http://unfccc.int/essential_background/convention/background/items/1362.php.
Weber, C. L., Peters, G. P., Guan, D. and Hubacek, K. (2008). The contribution of Chinese
     exports to climate change. *Energy Policy,* **36**(9), 3572–7.
World Bank (2008). *World Bank World Development Indicators.* Updated in December.
     Manchester. Available at http://www.esds.ac.uk/International.

# Chapter 20

# Research and technology for sustainability – a global cause

Annette Schavan

Annette Schavan was born in 1955. She studied Catholic theology, philosophy and education. After holding leading positions in several Christian organizations, she served as Minister of Education, Youth and Sport of the state of Baden-Württemberg from 1995 to 2005. Schavan has been a member of the German parliament and Federal Minister of Education and Research since 2005. In February 2009, she received an honorary professorship of the Catholic Theology Department of the Freie Universität Berlin. In 1998 she was elected Vice Chair of the German Christian Democratic Union party (CDU).

The Earth has been entrusted in our care, and we are all responsible for it. This responsibility challenges the way we think and act on two different levels: first, we need to think beyond the local environment in which we live; second, we need to think beyond our own lifetimes here on Earth. In other words, acting responsibly at a global level also means thinking of those who have not yet been born, who will follow us in future generations. And it means thinking of those who suffer the worst consequences of our actions, even though they may live in other parts of the world.

The UN Millennium Development Goals underline the fact that all people share the need for healthy food, clean water and safety. Climate change will threaten each of these essential conditions of life, and will challenge our ability to adapt. We therefore need to find ways in which we can fulfil our responsibilities towards all life on Earth more effectively.

What is now quite obviously an essential and urgently necessary step for combating climate change has been looming on the horizon for quite some time: we need greater resource and energy efficiency and independence from fossil fuels, but also effective and fair ways to pursue welfare and prosperity. The most recent IPCC Report sent an unequivocal message: climate change is accelerating, and is almost certainly largely man-made. Although some uncertainties remain, nobody can seriously deny that the rate and intensity of change in key environmental parameters poses an unprecedented risk to the long-term stability of social, economic and environmental systems worldwide.

The international debate about climate change has finally acknowledged the urgent need for action. With her comments at the G8 Summit in Heiligendamm, Chancellor Angela Merkel put Germany's position in a nutshell: 'Accelerated climate change is a serious threat. … Therefore, we need determined action from the international community. … We need to work together to promote innovation and technological developments for climate protection.' The international community must treat the subject of climate change with priority. It is a problem that affects the wealthy, developed world, as well as emerging and developing nations.

Science has played a significant role in making us realize that urgent action is needed. Thanks to improved scientific understanding of global climate change, we have finally increased the pace of our response. Around the world, we are not only seriously discussing how to deal with climate change and its consequences for politics, the economy and society; we are also about to reach a global consensus that joint emission reduction targets are absolutely necessary. What we now need are ambitious climate protection goals in Germany, in the European Union and beyond. We also need more extensive research to strengthen the scientific foundations for our decisions and actions.

This essay aims to shed light on what climate change means for technological innovation policies in an increasingly globalized economy and society. What impact will climate change have on our efforts to achieve sustainable development? And how does this translate into scientific and technological progress?

## Climate change as a challenge for technological innovation policy

There is an urgent need today to find joint solutions to the emerging effects of global climate change. The need for a global solution to this global problem is one of the most important lessons that we have learned from the findings of climate change research. As an issue it has fully penetrated the international political agenda during the past decade.

The challenges we face as a result of climate change are highly complex. Extensive research is being carried out to find knowledge-based approaches that answer some of the main societal questions. Can climate change still be mitigated to such an extent that adverse outcomes are averted? How can societies adapt to the changes that are inevitable? Who will gain from climate change and who will lose, and how can we provide fair compensation? Are there ways to manage our common resources to the benefit of all, and to achieve long-term sustainability for human life on Earth?

Ever since the German government began supporting measures to reduce greenhouse gas emissions, it has also committed itself to playing a leading role in implementation of reduction strategies. In order to accelerate this process, the German Ministry of Education and Research commissioned a large number of experts from science, industry and politics to draw up a comprehensive 'High-Tech Strategy on Climate Protection' (BMBF, 2008). It was presented in October 2007 at a climate research summit in Berlin.

This strategy has involved pooling strengths and resources, and identifying areas where we believe renewed action, new strategies, and targeted support are needed to achieve technological advances. The core aim of the Strategy is to achieve sustainable energy supply and utilization alongside sustainable use of natural resources. We need to focus consistently on this aim to ensure that research makes a lasting contribution to attaining the climate goals that we have set.

We also have to realize that dealing with climate change requires more than technological progress. We also need to improve public understanding of the problem. A further important goal is to develop and promote fundamental changes in society, the economy, institutional structures, and lifestyle and consumption patterns of individuals. We need a change of consciousness in our society: individual citizens must accept their share of responsibility and recognize that their decisions also influence global processes and the environment.

Scientific research plays a key role in climate protection by providing a broad knowledge base for political decisions and for strategy and investment planning. To support this process we need novel forms of communication and collaboration between climate researchers and decision-makers (see Kadner, this volume). Germany's High-Tech Strategy on Climate Protection focuses on this information process. That is why a 'Climate Service Center' was established in January 2009 in Germany, allowing climate-related knowledge to be pooled, evaluated and disseminated.

Three considerations are particularly important for Germany's national and international strategies on climate change.

## 1. Climate protection as a global driver of innovation and economic growth

We need strategic alliances and partnerships between science, industry and politics. These innovation alliances should pursue joint strategies that enhance the existing potentials of each partner. These strategies will increase awareness in society, industry, and politics that climate protection does not merely require restrictions, but may offer new opportunities and prospects.

'Green markets' and environmental goods already account for 5 % of industrial production and 1.8 million jobs in Germany. To further promote this development, we need the business community to join the public sector in significantly increasing investment in research and technology. The current financial and economic crisis underlines the necessity of redirecting our investments and establishing new fields of innovation and business.

Research and investment in alternative energy sources and in mitigation of and adaptation to climate change will be among the main priorities in Germany in the next few years. Germany has a strong international reputation for scientific research and is a world leader in sectors relevant to climate protection, resource efficiency and new energy systems. For this reason, we in Germany believe that investing in climate protection is more than just a moral obligation. We believe that it will also pay dividends. Germany is already a leading exporter of environmental technologies (Fig. 1).

Germany's experience shows that climate protection measures can contribute to economic growth, prosperity, and the creation of new jobs.

Innovation policy plays a key role in this process. With the High-Tech Strategy on Climate Protection, Germany is helping to mobilize private research efforts and capital with the aim of accelerating critical innovation processes that enhance climate protection. To this end, we have initiated the following cross-industry innovation alliances:

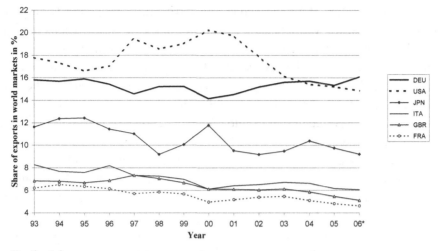

*) estimated

**Fig. 1.** World market share of major suppliers of environmental technologies from 1993 to 2006. (*Source:* adapted from Legler *et al.*, 2007, p. 110)

**Solar cells.** This alliance aims to develop new and significantly improved solar cells based on organic materials (see Heeger, this volume). The medium-term plan is to develop mobile solar energy converters that are cheap to produce, have an efficiency rate of 10%, (i.e., transforming 10% of the incident solar energy into exploitable electrical energy), and a lifespan of more than 20 years.

**Energy storage.** A second innovation alliance focuses on the development of highly efficient energy storage solutions. This is an important technological link for efficient use of renewable energy sources. Currently, lithium-ion batteries represent the most promising energy storage technology. However, the storage capacity and reliability of these batteries need to be improved, allowing for more flexible applications in mobile devices, for stationary energy storage in the energy sector, and for use in vehicles.

**Auto industry.** Because the car industry plays such a major role in Germany's economy, it comes as no surprise that we are supporting the development of innovative technologies such as automotive electronics systems that reduce carbon dioxide emissions and fuel consumption. Computer-based communication and data exchange technologies for cars also need to be developed. These will enable drivers to communicate with other road users and traffic infrastructure systems – for example, in the form of traffic congestion warnings, minimum distance regulations, and traffic control – thereby greatly reducing traffic-based emissions.

**Carbon nanotubes.** Carbon nanotubes (CNT) represent one of the most promising innovations in the field of materials research. They have higher electric conductivity than copper, lower thermic conductivity than diamonds, and an elasticity ten times greater than steel. If we succeed in transferring these properties to new, macroscopic materials, we would be able to improve numerous applications in energy and environmental technologies, light-weight construction, and energy storage.

**Financial investments.** For every euro that the German government invests in these alliances, the private sector has agreed to add a further five euros to help new innovative technologies become marketable. This should greatly speed up the integration of these new technologies into the market, and will in turn accelerate effective climate protection.

We have also begun speaking to partners from the financial sector about their role in climate protection. We believe that this step is long overdue. It will encourage investments in resource and energy efficiency and in renewable energy. So far, investment in this area has been limited, and substantial deficits in research and information still exist. In cooperation with major German financial services providers, we have established the 'Finance Forum: Climate Change'. Our goal is to enable financial markets to make an effective contribution to climate protection and adaptation.

**Establishment of public-private networks.** Last but not least, we are developing new instruments to support and finance the development of regional clusters in cutting-edge fields of technology such as energy efficiency and sustainable energy generation. The aim is to establish several highly integrated public-private networks that include commercial companies, research organizations and political institutions. They will work together to identify the potential commercial opportunities of new ideas and to translate research findings into marketable products and services.

### 2. Strategic partnerships with future generations

A second important area concerns the relationship between generations: We must make sure that each new generation is aware of – and passionate about – the issues and responsibilities that relate to climate protection. In other words, we also need to form strategic partnerships with future generations. Young people tend to be open-minded about issues relating to climate protection and are usually willing to face their individual responsibility in the global context. That is why we must give young people the tools and skills they need.

The support of young researchers is therefore an integral part of the High-Tech Strategy, and, indeed, is key to the success of our overall climate strategy. We must fill the next generation with enthusiasm for science and technology and offer young people interesting career opportunities in these fields.

Among young researchers, international exchange of experience and knowledge has become routine. After all, innovation comes about not only thanks to the wisdom of the old and experienced, but also thanks to the inquisitiveness of the young.

Modern means of communication greatly facilitate the exchange of ideas. However, we must not forget that personal contact with inspiring personalities will probably remain the most important source of enthusiasm for science and research. That is why we need more networks and platforms that facilitate encounters between leading researchers and young people. At the same time, we need to create settings in which children and young people can develop a fascination for research and technology.

Research funding and education are two sides of the same coin; it is all about securing our future. We need to structure our education system in such a way that it challenges and supports young people according to their individual talents and abilities. A society that loses interest in its talented young people has no future. That is why we need to ensure that young people do not see climate change only as a threat. It should also challenge them to think and act in new, innovative and unconventional ways. High-quality education programmes with plenty of transfer opportunities will be a key factor in attracting more young people to science.

### 3. International cooperation – the key to sustainability

To ensure effective climate protection, adaptation and resource management, the science and research communities need to act globally. In the future, the institutional and regulatory framework will no longer be created just at a national level. International cooperation – also beyond established partnerships – is becoming ever more important. That is why we need European and international innovation alliances. Germany's High-Tech Strategy on Climate Protection aims to create targeted links between European and international partners.

The European Institute of Innovation and Technology (EIT) is just one of many beacons of European innovation policy. Looking beyond Europe, we have a strong interest in involving developing and emerging countries, and engaging them in an intensive, open dialogue about the opportunities and risks of science and technology. The time has also come for us to enter into innovation alliances with the countries whose development will play an enormous role in our future decisions about global emissions. The partners in these alliances should all stand on an equal footing, acknowledging the role of industrialized nations in the past while also recognizing the present and future need for climate protection measures in all countries.

As one of the world's largest economies, Germany has a responsibility to engage in a mutual learning process. This is the only way we can improve our understanding of what it will take to achieve sustainable development. We can pool our strengths by increasing international research cooperation and developing joint research agendas. The science and research communities are giving us access to a large base of knowledge and experience, which is also aiding the development of effective global solutions.

## Outlook

Modern democracies are knowledge-based societies. Policy-makers derive their legitimacy not just from the democratic consent of citizens, but also by basing their political decisions on the most up-to-date knowledge available. Regardless of individual interests, the science and research communities have an obligation to deal with issues that are of fundamental relevance to common welfare and to the future of our society.

Given the complexity of many of the societal, economic and ecological challenges we face, we need solution-oriented research. We need reliable and honest advice on scientific matters as well as speedy access to new findings and technological advances, which will encourage the development and production of innovative technological applications.

The tasks and solutions confronting us today in the field of climate protection are so complex that they hardly ever fall neatly within one single discipline. They require an interdisciplinary, international and intercultural dialogue (see also Gell-Mann, this volume). That was one of the reasons why the *Leopoldina,* Germany's Academy of Natural Scientists, was renamed The National Academy of Sciences in mid-2008. It will represent German science at an international level. In addition to promoting the sciences, the Leopoldina sees its main mission in the interdisciplinary study and dissemination of scientific findings. It will offer a setting for encounters, discussions and exchanges in which the boundaries between disciplines and countries can be transcended more effectively.

To ensure that our research agendas are successful, we must increase our investments in science and research. The European Union has set itself a target for 2010: three percent of gross domestic product is to be invested in research and development (R&D). This will require both the public and the private sector to considerably increase their R&D spending. The money we invest today in research and development will form the basis for the prosperity of future generations. In light of the current financial and economic crisis this principle is more valid than ever.

Germany has a great tradition of scientific innovation, and we want to apply this experience more effectively as an instrument for achieving global sustainability. We cannot consider our efforts successful until we have reconciled the demands of preserving the Earth's resources not only with prosperity and welfare, but also with the development of a free, dynamic and informed society. We want a society that is capable of thinking in terms of integrated and interrelated systems, and is capable of achieving sustainability, both for its own benefit and that of future generations.

# References

BMBF – Federal Ministry of Education and Research (2008). *The High-Tech Strategy on Climate Protection.* Berlin. Available at: http://www.bmbf.de/pub/the_high-tech_ strategy_for_climate_protection.pdf.

Legler, H., Krawczyk, O., Leidmann, M. *et al.* (2007). *Zur technologischen Leistungsfähigkeit der deutschen Umweltschutzwirtschaft im internationalen Vergleich.* Studien zum deutschen Innovationssystem, 20. Hannover.

# Chapter 21

# Energy research and technology for a transition toward a more sustainable future

## Nebojsa Nakicenovic

Nebojsa Nakicenovic, born in 1949 in former Yugoslavia, studied economics and computer science at Princeton University, and completed his doctorate at the University of Vienna, Austria. In 1973 he joined the International Institute for Applied Systems Analysis (IIASA) near Vienna to conduct studies on long-term, global energy prospects and on the restructuring of the global automotive industry. Later he led programmes on the dynamics of technological and social change and environmentally compatible energy strategies. Nakicenovic is currently Deputy Director of IIASA, Professor of Energy Economics and Head of the Energy Economics Group at the Vienna University of Technology, and Director of the Global Energy Assessment. His research focuses on the diffusion of new technologies and their interactions with the environment.

*Note:* This chapter is a commentary on chapter 20.

The last fifty years of unprecedented development in the world have improved the human condition enormously but at the same time have resulted in widening gaps between rich and poor and in adverse environmental impacts on all scales, from indoor air pollution to climate change and biodiversity loss. Current patterns of development are thus clearly unsustainable. We need a fundamental paradigm change to produce a shift toward more sustainable development paths. This includes affordable access to adequate energy services. In her contribution to the present book, Annette Schavan also calls for fundamental innovations to help achieve structural changes in society, the economy, institutional structures, and in lifestyle and consumption patterns.

The recent financial crisis and the ensuing ever-deeper economic depression are no doubt going to bring additional hardship, especially to those without access to basic human needs. A predominant social issue that is increasingly becoming a major preoccupation for world leaders is how to address social inequality and poverty, especially in the developing world (Karekezi and Sihag, 2004). The longer the economic crisis deepens, the more threatened those living in poverty will be.

In response to the call to fight social inequality and poverty, world leaders endorsed the United Nations Millennium Development Goals (MDGs) as agreed upon by the Millennium General Assembly of the UN in 2000 and further advanced at the World Summit on Sustainable Development in 2002. The MDGs include eight specific goals but the primary objective is to halve extreme poverty by 2015 (Elliot, 2005). It is becoming increasingly evident that at current trends this goal will not be achieved even decades later in the poorest countries and regions of the world. Thus, it is urgent that significant effort is devoted toward the achievement of the MDGs.

Currently, it is estimated that about 2.6 billion people live on less than USD 2 a day and up to 3 billion on less than USD 2.5 per day (Chen and Ravallion, 2008; World Bank, 2008) – 75% of whom reside in rural areas (IADB, 2005). Furthermore, it is estimated that 1.4 billion people live in extreme poverty (World Bank, 2008). This estimate is an upward revision from the previous one of 1 billion people living in extreme poverty. This trend underscores the importance of increasing efforts to meet the MDGs.

Affordable access to modern energy services has a significant role to play in meeting development goals as it is a fundamental prerequisite for reaching virtually all MDGs. However, modern energy services in the majority of developing countries are characterized by inequality of access, notably between the poor and the affluent, but also between rural and urban areas. At the national level, this is demonstrated by the low levels of modern energy in the primary energy supply mix, and by low electrification and low electricity consumption levels.

About 2 billion people or approximately a third of the world's population are without access to modern energy and about 1.6 billion are without access to electricity – the very symbol of affluence and modernity – while about 2.4 billion still cook with traditional forms of biomass (Nakicenovic *et al.*, 1998; Saghir, 2005; UN-Energy, 2005). Limited access to cleaner energy services supplied by modern energy carriers is an important contributor to rising levels of poverty in some sub-Saharan African countries (UNDP, 2007; Takada and Fracchia, 2007).

It is estimated that the cost of connecting a household without prior access to electricity is in the order of USD 1000 (Goldemberg, J., personal communication), resulting in total capital needs of about USD 500 billion, assuming an average of four persons per household and two billion people without access. Distributed over twenty years, this translates into annual investment requirements of some USD 25 billion. This represents a huge investment that is lacking, yet it does not appear excessive in comparison to the gigantic scale of the government guarantees and debt cancellation in the financial sector since the economic crisis emerged. To be effective, this kind of investment would have to be enhanced initially by a certain level of free energy for the poorest, say 700–1000 kWh per year or about 2–3 kWh per day (WGBU, 2009).

Thus, there is a clear need to embark on a new development path toward sustainable and affordable access to adequate energy services. Fortunately, many policies and measures directed toward increasing access to modern energy services have multiple benefits for other development goals, from the reduction of indoor air pollution and its assaults on human health to reductions of greenhouse gas emissions.

Some may argue that this transformation toward more sustainable development paths and energy patterns in the world will be difficult to achieve because falling consumer demand leads to a vicious circle that results in ever-decreasing employment reducing further the demand for traditional goods and services.

At the same time, this crisis of the 'old' is an opportunity for the 'new' to emerge. It is an opportunity that needs to be seized and should not go to waste. Joseph Schumpeter referred to paradigm-changing transformations of this kind as 'gales of creative destruction' (Schumpeter, 1942). As old techno-economic and institutional development paths encounter their limits, the chances for fundamentally new development paths to emerge and eventually diffuse become more likely.

Decarbonization of the global economy toward a carbon-free future is such a paradigm-changing transformation. It appears to be a must, given the ever-more-threatening manifestations of global climate change. In her contribution to the present book, Annette Schavan quotes the unequivocal message of the IPCC Fourth Assessment Report (IPCC, 2007) that climate change is accelerating and is almost certainly largely caused by humans. Anthropogenic greenhouse gas emissions over the last two centuries, since the beginning of the industrial revolution, have increased atmospheric concentrations of carbon dioxide from some 280 ppm to over 380 ppm today. The IPCC estimates that the global average surface temperature has increased by some 0.8 °C during the last century. Annette Schavan also observes that the negative effects of climate change can already be felt, and quotes the Federal Chancellor, Angela Merkel, saying that determined action from the international community is required to promote innovation and technological developments to support climate protection.

The necessary change toward wider access to modern energy services together with climate protection and decarbonization of the global economy is effectively blocked today by the addictive dependence on fossil energy sources. This explains the need for the Schumpeterian 'gales of creative destruction'. Today, 80 % of global energy comes from fossil sources, and this situation needs to be reversed so that 80 % of energy would be carbon-free or carbon-neutral well before the end of the century. The old energy systems need to be replaced by innovative, environmentally and climate-friendly alternatives. In parallel, the reliance on inadequate, traditional energy by the poor, which constitutes some 10–20 % of primary energy today (see Fig. 1), also needs to be replaced by modern renewable and other clean energy sources as well as efficient end-use devices from modern stoves to advanced lighting, communication and information technologies. For that to occur, we need vigorous private and public research and development efforts and partnerships in order to create the necessary scientific foundations for the paradigm-changing transformations. In this context, Annette Schavan argues that we need science and research to gain a better understanding of the complexity of the processes and interactions within the climate and the Earth system. She further argues that the important aim is to create fundamental innovations to help achieve structural changes in society, the economy, institutional structures, and in lifestyle and consumption patterns. We need to establish a foundation for the deployment and adoption of new systems and services that lead toward complete decarbonization of the global economy, and that involve all the world's population, from those without access to energy today to those living in affluence at high levels of consumption.

In other words, research and development (R&D) that lead to the diffusion of new and advanced technologies and practices represent a possible solution to the double challenge of providing development opportunities to those who are excluded and allowing for further development benefitting the more affluent. As Annette Schavan points out, this needs to occur without risking irreversible changes in ecological, biophysical and biochemical systems. As regards energy, this implies a shift from traditional energy sources to clean fossils and modern renewable energy in the case of those currently excluded from access, and a shift from fossil energy sources to carbon-free and carbon-neutral energy services in the more developed parts of the world. In all cases this means a vigorous improvement of energy efficiency, from

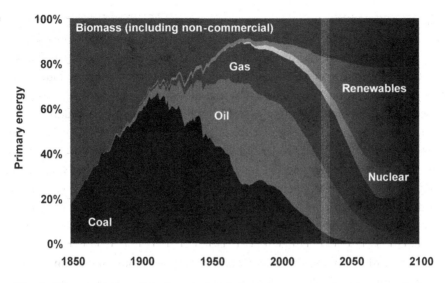

**Fig. 1.** History and possible future of global primary energy showing the relative shares of the most important energy sources. The future developments are consistent with stabilization of global temperature increase at about 2 °C above preindustrial levels. (*Sources:* Riahi and Nakicenovic, 2007; IIASA, 2007)

supply to end use, expanding shares of renewables, more natural gas and less coal, vigorous deployment of carbon capture and storage, and in some cases – where it is socially acceptable and economically viable – also nuclear energy.

Figure 1 shows the historical evolution of global primary energy and one possible future development path toward decarbonization. It is an illustration of the needed transformational change of the global energy system. New energy technologies and practices but also changes in lifestyles and behaviour are prerequisites in order to shift the energy system from its current dependence on fossil energy toward complete decarbonization well before the end of the century.[1]

This particular scenario describes a future world that stabilizes concentrations of greenhouse gases just above the current levels and thereby limits the temperature increase to about 2 °C. Even a global temperature increase of 2 °C would lead to significant disruptions of natural ecosystems, threatening water availability and communities in coastal areas (IPCC, 2007). The poor and those who are excluded would bear the brunt of such changes. Nevertheless, a 2 °C world would probably avoid the most severe adverse – and perhaps also irreversible – consequences associated with higher magnitudes of global warming. Therefore, this particular scenario can be characterized as a transition toward sustainability that enables the

[1] This will require vigorous introduction of carbon-free sources of energy and carbon capture and storage (CCS) from fossil energy, and perhaps also biomass, in order to reduce carbon emissions to zero or even turn them negative toward the end of the century.

fulfilment of the MDGs through provisioning energy services in most of the world while simultaneously avoiding more drastic climatic changes (UN-Energy, 2005).

The current investments in the global energy system are estimated at some USD 500 billion per year (Nakicenovic and Kimura, 2005). This includes investments in energy production, conversion and distribution but excludes most of the end use such as vehicles, heating systems or industrial facilities. Adding end-use investments would bring the estimate to some USD 750 billion per year. The sustainable scenario depicted in Figure 1 would require at least twice this investment effort during the coming decades to the tune of about USD 1 trillion per year or about USD 20 trillion until 2030. In comparison, the investments needed to provide access to modern forms of energy to the two billion people currently living without it are relatively small, at about USD 25 billion per year or about USD 500 billion until 2030.

The nature of technological change and the associated deep uncertainties require that innovations are adopted as early as possible in order to lead to lower costs and wider diffusion in the following decades. The longer we wait before introducing these advanced technologies, the higher the required emissions reduction will be. At the same time, we may miss the window of opportunity for achieving substantial cost reductions through technological learning as a function of cumulative experience and investments. This requires research, development and deployment (RD&D) as well as investments in order to achieve accelerated diffusion and adoption of advanced energy technologies.

Current global energy research and development (R&D) trends are unfortunately going in the opposite direction. Public annual expenditure in this area in OECD countries has declined to some USD 8 billion today from about USD 12 billion two decades ago, while private ones are estimated to have declined proportionally and are now about four times the public efforts (IEA and OECD, 2008). This means that today we are investing less than USD 10 per person in the world per year in energy-related R&D activities. Many studies indicate that this sum needs to increase by at least a factor of two to three in order to enable the transition toward new and advanced technologies in the energy systems (Bierbaum *et al.*, 2007). However, it needs to be noted that Finland, Japan and Switzerland constitute important exceptions with substantially higher public and private spending on energy R&D efforts. In her essay, Annette Schavan suggests 3% of gross domestic product (GDP) as a goal for future R&D efforts. Tripling global energy R&D and assuming the current 4% share of global energy in total R&D efforts translates into some 1.5% of global GDP.

As mentioned, the required investments in energy systems, an estimated USD 20 trillion needed between now and 2030, are at least a factor of a hundred greater than the needed R&D efforts. This translates into about twice the current level of investment, with most of the requirements being in developing parts of the world. To achieve a transition toward more sustainable development paths substantially larger investment in energy infrastructures and energy R&D is needed. All told, R&D efforts need to be tripled and energy investments at least doubled in order to assure the timely replacement of energy technologies and infrastructures.

The salient finding of a number of recent integrated assessment studies is that the additional costs needed to achieve a more sustainable future and climate stabilization are relatively small in comparison to these overall investment needs. Often they are 'negative', namely lower than those projected by traditional scenarios of future developments, sometimes called business-as-usual (BAU) scenarios. However, attaining a more sustainable type of future requires higher 'up-front' investments until about 2030. The great benefit of these additional investments in a future characterized by carbon-leaner energy systems and a more sustainable development path is that in the long run (by 2050 and beyond) the investments would be substantially lower compared to the BAU alternatives. The reason for this is that the cumulative nature of technological change translates the early investment in decarbonization and a sustainable energy future into lower costs of the energy systems in the long run, along with the co-benefit of climate stabilization. This all points to the need for radical change in energy policies to assure sufficient investment in our common future. Accelerated technological change in energy production and end use needs to be promoted. In other words, the global financial and economic crisis offers a unique opportunity to invest in new technologies and practices that would generate employment and affluence in most parts of the world. Seizing this chance today would pave the way for the eradication of poverty as well as a more sustainable future with lower rates of climate change. The crisis of the 'old' is a historic chance to sow the seeds of the 'new'.

## References

Bierbaum, R., Holdren, J. P., MacCracken, M. *et al.* (2007). *Confronting Climate Change: Avoiding the Unmanageable and Managing the Unavoidable.* Scientific expert group report on climate change and sustainable development for Sigma Xi and the United Nations Foundation. Washington D. C.

Chen, S. and Ravallion, M. (2008). *The Developing World Is Poorer than We Thought, but no less Successful in the Fight against Poverty.* World Bank policy research working paper no. 4703. Washington, D. C.

Elliot, D. (2005). *Employment, Income and the MDGs – Critical Linkages and Guiding Actions.* Briefing paper prepared on behalf of the Fauno Consortium. Bern. Available at http://www.springfieldcentre.com/publications/sp0501.pdf.

IADB – Inter-American Development Bank (2005). *The Millennium Development Goals in Latin America and the Caribbean: Progress, Priorities and IDB Support for their Implementation.* Washington. Available at http://www.iadb.org/sds/mdg/file/Cover,%20Foreword%20and%20Introduction.pdf.

IEA and OECD – International Energy Agency and Organization for Economic Co-opera-
    tion and Development (2008). *Energy Technology Perspectives, Scenarios and
    Strategies to 2050.* Paris.
IIASA – International Institute for Applied Systems Analysis (2007). *GGI Scenario
    Database.* Laxenburg. http://www.iiasa.ac.at/web-apps/ggi/GgiDb/dsd?Action=html
    page&page=series.
Karekezi, S. and Sihag, A., eds. (2004). *"Energy Access" Working Group Global Network
    on Energy for Sustainable Development: Final Synthesis/Compilation Report.*
    Roskilde.
Nakicenovic, N., Grübler, A. and McDonald, A., eds. (1998). *Global Energy Perspectives.*
    Cambridge.
Nakicenovic, N., Ajanovic, A. and Kimura, O. (2005). *Global Scenarios for the Energy
    Infrastructure Development.* Interim report IR-05-028 on work of the International
    Institute for Applied Systems Analysis. Laxenburg.
Parry, M. L., Davidson, O. F., Canziani, J. P. *et al.,* eds. (2007). *Climate Change 2007:
    Mitigation of Climate Change. Contribution of Working Group II to the Fourth
    Assessment Report of the Intergovernmental Panel on Climate Change.* Cambridge.
Riahi, K. and Nakicenovic, N., eds. (2007). Integrated assessment of uncertainties in
    greenhouse gas emissions and their mitigation: introduction and overview. *Techno-
    logical Forecasting and Social Change.* Special Issue, **74**(7), 873–86.
Saghir, J. (2005). *Energy and Poverty: Myths, Links, and Policy Issues.* Energy Working
    Notes 4. Washington, D. C. Available at http://siteresources.worldbank.org/INTEN
    ERGY/Resources/EnergyWorkingNotes_4.pdf.
Schumpeter, J. A. (1942). *Capitalism, Socialism and Democracy.* New York.
Takada, M. and Fracchia, S. (2007). *A Review of Energy in National MDG Reports.*
    Report for the United Nations Development Programme. New York. Available at
    http://www.energyandenvironment.undp.org/undp/indexAction.cfm?module=Librar
    y&action=GetFile&DocumentAttachmentID=2088.
UN-Energy – United Nations Energy (2005). *Energy Challenges for Achieving Millen-
    nium Development Goals.* New York. Available at http://esa.un.org/un-energy.
UNDP – United Nations Development Programme (2007). *Mainstreaming Access to
    Energy Services: Experiences from Three African Regional Economic Communities.*
    Dakar.
WBGU – Wissenschaftlicher Beirat der Bundesregierung Globale Umweltveränderungen:
    Schubert, R., Schellnhuber, H. J., Buchmann, N. *et al.,* eds. (2009). *Welt im Wandel:
    Zukunftsfähige Bioenergie und nachhaltige Landnutzung.* Berlin. Available at http://
    www.wbgu.de/wbgu_download.html.
World Bank (2008). *Poverty Data: a Supplement to World Development Indicators 2008.*
    Washington, D. C. Available at http://siteresources.worldbank.org/DATASTATIS
    TICS/Resources/WDI08supplement1216.pdf.

# Part IV

Technological innovation and energy security

# Chapter 22

# A world powered predominantly
# by solar and wind energy

---

## Walter Kohn

Walter Kohn, born in Vienna in 1923, majored in mathematics and physics at the University of Toronto and obtained his PhD at Harvard University. In 1957 he became a US citizen. Professor Kohn was awarded the Nobel Prize in Chemistry in 1998 'for his development of the density-functional theory'. His work revolutionized scientists' approach to the electronic structure of atoms, molecules and solid materials. As Emeritus and Research Professor at the University of California in Santa Barbara, Kohn is today collaborating with younger colleagues on research in this field. He was the executive producer of the documentary film The Power of the Sun, which dealt broadly with solar energy, was first shown in 2005, and was later shown internationally in 10 languages. The film presents the history, science and applications of solar energy, both in the developed and less developed world.

*Note:* An addendum to this chapter is available at http://www.nobel-cause.de/book/chapter22_addendum.pdf.

It is widely agreed that during this century humankind is facing two critical energy-related challenges:

**1. Decline in oil and natural gas production.** Total oil and natural gas production, currently providing about 60% of global energy needs (see Fig. 1), is expected to peak in 10 to 30 years, with oil likely to peak first (IEA, 2004, p. 129).[1] Oil production in current oil fields is estimated to drop by about one half within a mere 20 or 30 years after passing its peak (see Fig. 2). Natural gas is expected to follow a similar pattern with a delay of two to three decades.

**2. Increase in greenhouse gases.** By the end of this century, accumulation of anthropogenic carbon dioxide ($CO_2$) and other greenhouse gases (GHGs) in the Earth's atmosphere is expected to lead to a major increase of mean global surface temperature in a range from approximately 2 °C to approximately 7 °C above pre-industrial levels (see Rahmstorf *et al.*, this volume; IPCC, 2007), accompanied by significant acidification of ocean waters (WBGU, 2006; Hofmann and Schellnhuber, 2009) and a very substantial rise in the global ocean level (IPCC, 2007; Rahmstorf, 2007).

Both the exhaustion of oil and gas as well as global warming are due, in about equal measure, to two causes. First, the world's population is increasing rapidly, mostly in the less developed world (LDW) and in India, from 6.7 billion in 2009 to an estimated levelling off at 9 to 10 billion in about 2050 (see Fig. 3).

Second, per-capita consumption of fossil fuels has grown strongly since the Industrial Revolution in the developed world and is currently increasing rapidly in China, India (IEA, 2007) and the LDW. The governing simple mathematics for the global consumption of any commodity over a given period is:

(consumption) = (population) x (per-capita consumption)

At present, total consumption of energy is continuing to grow rapidly in China, India and in the LDW, but is fairly stable in the developed world.

The data shown in Figures 2 and 3 imply that, due to the continuing growth of world population, per-capita oil production will peak around 2015 (see Fig. 4), while the peak of total oil production, which (because much later) is harder to estimate, will occur some 15 or more years later.

The following are some broad principles for dealing with the global challenges of energy supply and climate change:

---

[1] The effects of the dramatic global economic downturn beginning in the summer of 2008 may not yet be fully reflected in data for the period after the middle of 2008.

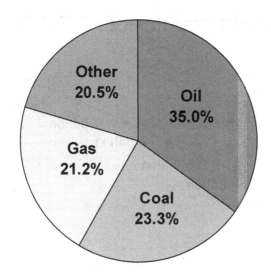

**Fig. 1.** Contribution of fuel types to global energy consumption in 2001. (*Source:* after Dell and Rand, 2004, p. 15)

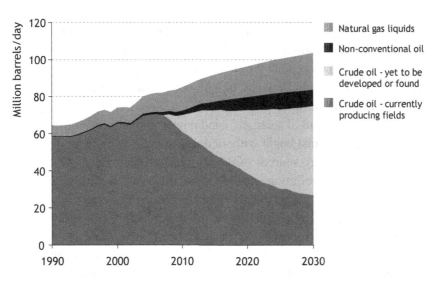

**Fig. 2.** World oil production by source. (*Source:* adapted from IEA, 2008)

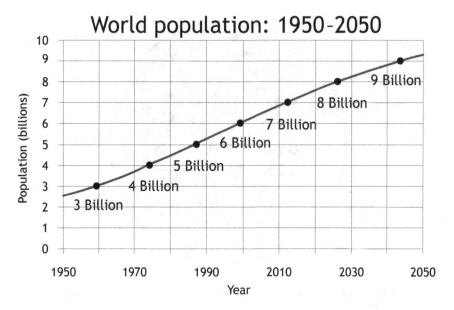

**Fig. 3.** World population growth. (*Source:* U.S. Census Bureau, 2009)

- They must be addressed without delay and with strong global cooperation.
- The growth of world population must be halted by about the middle of this century, or earlier, at no more than 9 billion, and gradually reversed.
- Energy conservation and efficiency in both consumption[2] and all forms of production must be substantially enhanced.
- Large-scale development of solar and wind power, and other established sustainable energy sources, must begin without delay.
- Four other major energy sources raise enormous problems that must be recognised. *Coal* generates pollution and is, without the costly capture and sequestration of $CO_2$, the greatest single cause of global warming. *Nuclear fission reactors,* with their as yet ineffective surveillance, are unacceptably easy stepping stones to nuclear weapons, as recent history has shown. Global-scale *bio-energy* production, which is $CO_2$-neutral in the steady state, strongly competes with food production for land and water. *Nuclear fusion,* while well established in the laboratory and in the hydrogen bomb, is still far from proven as a practical energy source.

---

[2] Examples include rapid replacement of SUVs and similar vehicles by much lighter, more fuel-efficient cars, and of incandescent by compact fluorescent lights; greatly expanded public transportation, especially in the USA and Canada; proper insulation of buildings; green architecture such as energy-neutral housing, and green city planning.

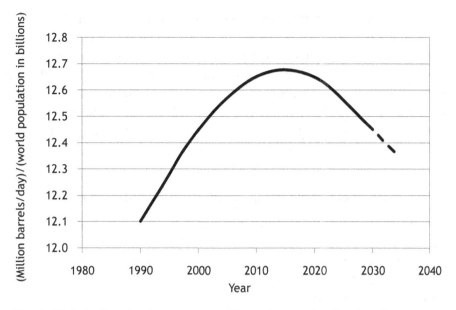

**Fig. 4.** Global oil production per person. Illustrative graph using data shown in Fig. 2 and Fig. 3. (*Source:* W. Kohn)

## The revolution in energy production

In 2001, oil, coal, gas, biomass, nuclear, and other energy sources respectively accounted for approximately 35%, 23%, 21%, 11%, 7%, and 3% of global energy consumption (Dell and Rand, 2004; see Fig. 1). By the time oil and gas supplies are effectively exhausted in about the middle of this century, there probably will not yet be a safe and cost-effective technology available for carbon capture and storage, at least not on the required scale for burning of coal. This naturally shifts attention to solar and wind energy.

**Solar energy** is by far the most abundant source of energy (Sawin and Moomaw, 2008), but it is still substantially more costly than oil- or natural-gas-derived energy (U.S. Department of Energy, 2005). It currently accounts for less than 1% of total world energy production (Worldwatch Institute, 2007). However, its share is likely to increase greatly as projected growth rates of production are tremendous: a 30% increase per year is a reasonable estimate (EPIA, 2008; BP, 2009).

**Wind energy** is currently much cheaper and more widely produced than solar energy. Its annual percentage increase is similar to that of solar energy (Global Wind Energy Council, 2008). The total average available wind energy up to the practical maximum height of about 80 metres above the ground (the typical hub height of large wind turbines) is much less than available solar energy. Nevertheless, according to current estimates, wind energy by itself could also supply the world's total energy needs several times over (Archer and Jacobson, 2005).

Given the limited amount of remaining fossil fuels and their dangerous impact on our climate, it is obvious that the infrastructure for acceptable alternative energies must be created rapidly, beginning immediately. Otherwise the world faces a frustrating choice between, on the one hand, a global economic meltdown and a violent scramble for dwindling oil and gas deposits, and, on the other, the dangerous use of coal and/or nuclear energy on a vast scale.

## A world powered predominantly by solar and wind energy

Of course, the real-world problems of coping with continuing population growth, disappearing oil and natural gas resources, and continuing global warming, are enormously complex and intricately connected. Global warming is fairly uniform across the globe. However, energy demands and availability vary enormously. Available solar and wind energy depends strongly on geography and local climate, and varies strongly with season, time of day, and weather. This creates additional, subsidiary challenges of cost-efficient energy *storage* and *transportation*.

I shall, of course, not even attempt to deal with all these issues within a few pages, but instead shall discuss a greatly simplified model for a world powered predominantly by solar and wind energy, which I shall call 'sol-wind energy', combining solar and wind energy into a single entity. The expression 'sol-wind energy' reflects the fact that these two energy sources are complementary, plentiful, clean, GHG-neutral, and are likely to decrease in cost to a similar value of under 10 US cent/kilowatt hour. I believe that this model provides a general perspective for acceptable and achievable future energy provision.

The relevance of this model for the real world derives from three facts:

- Solar energy incident per year on Planet Earth exceeds the total present human consumption of energy by a factor of about 10000 (U.S. Department of Energy 2005); available wind energy alone, indirectly also derived from incident sunlight, is, of course, much smaller, but also greatly exceeds total present energy consumption (Archer and Jacobson, 2005).
- The required materials are effectively infinitely abundant: for photovoltaic energy the main material required is silicon; for photothermal energy various effective light-absorbers; and for the capture of wind energy the main material required is steel.
- Although the current contribution of sol-wind energy is still less than 1% of humankind's total energy consumption (REN21, 2008), in recent years production has been growing at the enormous rate of approximately 30% per year (see

above). If extrapolated, this represents growth by a factor of about 200 in two decades and more than a factor of 2000 in three decades. According to this model, the production of sol-wind energy would substantially exceed the current annual production of total energy in a mere 20 to 30 years.

The assumption underlying this model – that sol-wind energy production will continue to grow by about 30% annually for the next 20, or even 30, years – is, of course, extremely optimistic. Nevertheless, it is consistent with the data of the last several years, and also with the effectively unlimited availability of the required materials.

Is this growth also consistent with the availability of labour required to produce, maintain and operate the necessary sol-wind equipment in the short available time-span? I do not have a firm answer to this question; however I can offer the following argument: Today the effective cost of sol-wind power production is about three times higher than the average cost for all forms of energy, including indirect costs due to pollution and global warming. This implies about three times greater labour requirements. Assuming approximately constant future per-capita use of energy, and disregarding possible major scientific-technological advances in energy production, conservation, and efficiency, this suggests that the per-capita labour requirement to produce sol-wind energy would be also about three times greater than today. This substantial load would, of course, be very heavy, but not necessarily prohibitive. Unforeseeable future developments make this estimate very rough and, I believe, probably much too high.

### *Urgency*

The clean and safe sol-wind model described above is emerging from the fossil fuel model of the last two centuries. This new model is the logical consequence of the rapid exhaustion of oil and natural gas over the next 10 to 30 years. Of course, the change of most of the world's energy infrastructure from fossil fuel to sol-wind during this very short time is a huge challenge. Every year waited means a year less before the dreaded global peak-oil year, when uncertainty will begin to morph into a new reality (peak oil happened as predicted in the USA in about 1970; predicted by Hubbert, 1956, and described by Hirsch *et al.,* 2005).

This transformation will be among the greatest challenges ever faced by humankind. We need to do everything in our power – and as rapidly as possible – to stop global warming, including, as previously stated, rapid stabilization, followed by reduction of world population, immediate major per-capita reduction of anthropogenic GHG emissions, dramatic energy conservation, and improved energy efficiency. The time to wait for absolute certainty is far behind us; it has become a time for urgent preventive action.

Energy supply and global warming are make-or-break twin challenges of our times. Unless we put our collective minds to it, the second half of the present century will be a disaster. On the other hand, if we put our minds to it now, I am convinced that we can look forward to a better future, in which solar and wind energy predominate.

# References

Archer, C. L. and Jacobson, M. Z. (2005). Evaluation of global wind power. *Journal of Geophysical Research,* **110,** D12110.

BP (2009). *Solar energy.* http://www.bp.com/sectiongenericarticle.do?categoryId=902378 9&contentId=7044135.

Dell, R. M. and Rand, D. A. J. (2004). *Clean Energy.* Cambridge.

EPIA – European Photovoltaic Industry Association (2008). *Solar Generation V – 2008.* Available at http://www.epia.org/fileadmin/EPIA_docs/documents/EPIA_SG_V_ ENGLISH_FULL_Sept2008.pdf.

Global Wind Energy Council (2008). *Global Wind 2007 Report.* Available at http://www. gwec.net/fileadmin/documents/test2/gwec-08-update_FINAL.pdf.

Hirsch, R. L., Bezdek, R. and Wendling, R. (2005). *Peaking of World Oil Production: Impacts, Mitigation, & Risk Management.* U.S. Government report. Available at http://www.netl.doe.gov/publications/others/pdf/Oil_Peaking_NETL.pdf.

Hofmann, M. and Schellnhuber, H. J. (2009). Oceanic acidification affects marine carbon pump and triggers extended marine oxygen holes. *Proceedings of the National Academy of Sciences,* **106**(9), 3017–22.

Hubbert, M. K. (1956). *Nuclear Energy and the Fossil Fuels.* Houston, Texas. Available at http://www.hubbertpeak.com/hubbert/1956/1956.pdf.

IEA – International Energy Agency (2004). *World Energy Outlook 2004.* Paris.

IEA – International Energy Agency (2007). *World Energy Outlook 2007: China and India Insights, Executive Summary.* Paris.

IEA – International Energy Agency (2008). *World Energy Outlook 2008.* Paris.

IPCC – International Panel on Climate Change (2007). *4th Assessment Report.* http://www.ipcc.ch.

Rahmstorf, S. (2007). A semi-empirical approach to projecting future sea-level rise. *Science,* **315**(5810), 368–70.

REN21 (2008). *Renewables 2007. Global Status Report.* Paris. Available at http://www. ren21.net/pdf/RE2007_Global_Status_Report.pdf.

Sawin, J. L. and Moomaw, W. R. (2008). An enduring energy future. In *State of the World,* ed. L. Starke. New York, 130–50.

U.S. Census Bureau (2009). *World Population: 1950–2050, June 2009 update.* International Data Base. Washington, D. C. Available at http://www.census.gov/ipc/www/ idb/worldpopgraph.php.

U.S. Department of Energy (2005). *Basic Research Needs for Solar Energy Utilization.* Report of the Basic Energy Sciences Workshop on Solar Energy Utilization. Washington, D. C.

WBGU (2006). *The Future Oceans – Warming Up, Rising High, Turning Sour.* Berlin. Available at http://www.wbgu.de/wbgu_sn2006_en.html.

Worldwatch Institute (2007). *Solar Power Set to Shine Brightly.* Available at http://www. worldwatch.org/node/5086.

# Chapter 23

## Low-cost 'plastic' solar cells:
## a dream becoming a reality

---

### Alan Heeger

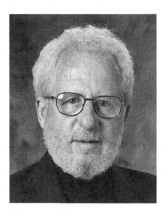

Alan Heeger was born in 1936 in Sioux City, Iowa, USA. He enrolled in studies of physics and mathematics at the University of Nebraska and obtained his PhD at the University of California in Berkeley in 1961. In 2000 he received the Nobel Prize in Chemistry 'for the discovery and development of conductive polymers'. Together with Alan G. MacDiarmid at the University of Pennsylvania, Heeger discovered that plastic can be made electrically conductive. He subsequently helped to develop conductive polymer research into a field of great importance for chemists as well as physicists. As Professor of Physics and Professor of Materials at the University of California at Santa Barbara, Alan Heeger remains active in these research areas. His current research focuses on low-cost plastic solar cells made from semiconducting polymers.

## The problem

It is clear to all of us that we have an energy problem. Luckily, the power from the sun is available to help solve this energy problem.

We installed solar cells on the roof of our house a year ago. It is a wonderful technology: when the sun comes up in the morning my electric meter runs backwards! My electric bill (i.e., my monthly cost of electricity) has dropped to zero. The problem is, however, that the purchase cost (including installation) was much too high. Depending on the details of how the cost of energy increases in the coming years, more than ten years will be required for the savings to repay the installation cost.

There are two problems that we must solve to enable widespread use of photovoltaic solar cell technology. The first is the cost. The second is that we need to produce a lot of area. At noon on a sunny day, we receive one kilowatt per square metre of energy from the sun. This corresponds to sufficient energy received on Earth in one hour to satisfy all of the energy needs for the planet for one year! Thus, the ability to produce low-cost, efficient solar modules in areas sufficiently large to enable significant energy production is a major opportunity.

## The solution: 'plastic' solar cells

An exciting new technology that can produce low-cost solar cells in large quantities uses 'photovoltaic inks'. These inks are organic semiconducting polymers which are in solution with common solvents. Different absorption and transmission associated with the different molecular structures are the reason for different colours of the inks. The unique quality of these coloured liquids is that they have electronic functionality and can be used for printing.

Printing technology was invented by Gutenberg in 1545, more than four hundred and fifty years ago. If printing technology could be used for the fabrication of solar cells, then we could produce low-cost, high efficiency solar cells in large quantities. Indeed, the principles of this old mature technology can be adapted to print solar cells roll-to-roll like newspapers. The potential impact of such printed 'plastic' solar cells on the market for solar technology could be tremendous.

The demonstration sample of a plastic solar cell, shown in Figure 1, has been fabricated by a company called Konarka Technologies. The name of the company stems from a temple dedicated to the Indian Sun God. Initially, this product will be quite expensive, and will therefore only be used by people who can afford it. Possible points of initial use are battery chargers and boats. However, once plastic solar cells are available with high efficiency and printed in large quantities, they will become much more affordable. They could then be given to poor families all over

**Fig. 1.** Solar module, printed on a roll-to-roll tool similar to a printing tool. Its advantages over standard solar cells are flexibility, light weight, low cost, and potential for mass production. (*Source:* A. Heeger)

the world. The access to energy from the sun could then change the lives of millions of people.

## The technology: how to create plastic solar cells

The technology to print plastic solar cells originated from a discovery made in our laboratory at UC Santa Barbara in 1992. We were interested in the potential interaction between our semiconducting polymers with the famous fullerene molecules. We had no concept of solar cells; these initial experiments were motivated purely by curiosity. We discovered that following the absorption of a photon an electron transfer reaction (from polymer to fullerene) occurs on a remarkably short time scale. The rate of this photo-induced electron transfer is two orders of magnitude faster than the first step in photosynthesis. This ultra-fast electron transfer reaction implies that we separate charge (create mobile charge carriers) with a quantum efficiency that approaches unity: every absorbed photon yields a pair of separated charges! This high efficiency of charge separation and mobile carrier generation provides the scientific foundation for creating a technology to produce high efficiency solar cells.

However, our materials, cast from solution into thin films, are very disordered. The analogy would be tangled cooked spaghetti in a bowl rather than rigid straight spaghetti in a box. Because of this disorder, the charges that are separated by photo-induced charge transfer will not travel very far before they recombine. In order to collect these charges, we had to invent a new kind of material comprising charge-separating junctions between two materials – so-called heterojunctions between the donor and the acceptor. Because of the short recombination length, the heterojunction cannot simply be a bi-layer, as is often the case in the semiconductor world. We had to create a nano-morphology with interpenetrating networks of the two components on a length scale of a few nanometres, roughly a hundred angstroms (1 angstrom is equal to 0.1 nanometre). A conceptual sketch of this nano-morphology is shown in Figure 2a.

As demonstrated in Figure 2b, this remarkable nano-structure can already be constructed. How was it formed? The answer is simple, but elegant: we were able to achieve this structure through controlled phase separation of two incompatible components both of which are soluble in the same solvent. When cast as films from solution, the phases of the two components separate as the solvent quickly evaporates. After separation, the two components self-assemble into the material depicted in Figure 2. This so-called bulk heterojunction material has charge-separating junctions everywhere. Each component forms a network that can deliver charges to the electrodes.

By using this bulk heterojunction concept, we can collect photo-generated charge carriers. You might wonder how the electrons know which way to go (for example up and not down). Again this is a simple problem. All one needs to do is to break the symmetry by using two different metals for the electrodes. We were able to control the morphology of the heterojunction material, and are now able to efficiently collect the photo-generated charge carriers. With the specific materials shown in Figure 2, a power conversion efficiency of 5% can be achieved.

The best solar cells fabricated from inorganic semiconductors are triple junction devices that yield power conversion efficiencies in excess of 40%, but because of the high processing costs, these are prohibitively expensive. They can be used in space applications, but not for the kinds of applications we are discussing here. The question is what we can expect to achieve using low-cost plastic solar cells.

### Improving the efficiency of plastic solar cells

The particular material shown in Figure 2, which resulted in solar cells with 5% efficiency, has an absorption spectrum poorly matched to the solar spectrum: the band gap is too large, missing more than half of the solar spectrum (see Fig. 3).

Obviously, there is an opportunity to improve the efficiency of solar energy absorption by doing the proper science. Synthesizing new macromolecules with electronic structures that yield absorption spectra better matched to the solar spectrum could eventually improve the performance of our solar cells by at least a factor of two (see Fig. 4).

Figure 4a depicts such a different molecular structure with a smaller energy gap: the absorption spectrum of the polymers now extends beyond red into the near infrared (see Fig. 4b). Improved performance is achieved through the use of

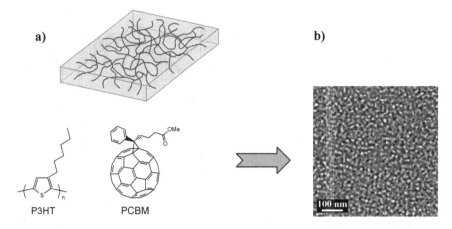

**Fig. 2.** Illustration of the nano-morphology with ubiquitous charge-separating junctions – so-called bulk heterojunction material. a) Conceptual sketch. The black material is an interconnected network of the fullerene (PCBM) and the white material is an interconnected network of a semiconducting polymer (P3HT). Each of the two components is fully interconnected. b) Electron micrograph. The small white bar on the bottom left represents 100 nanometre. (*Source:* Kim *et al.,* 2007; Ma *et al.,* 2007).

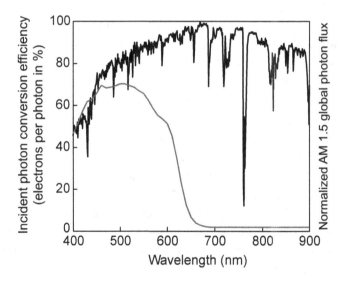

**Fig. 3.** The solar emission spectrum as received on Earth at twelve noon on a sunny day (fluctuating black line) is not well matched by the absorption spectrum of P3HT (see Fig. 2) solar cells (smooth grey line). (*Source:* Peet *et al.,* 2007).

a)

b)

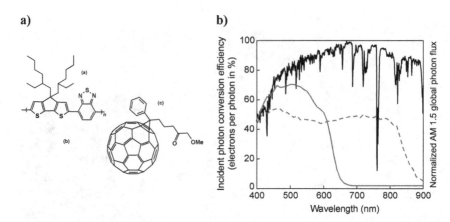

**Fig. 4.** a) Semiconducting polymers with smaller band gaps matching the solar spectrum better than the original polymers used. b) The improved conversion efficiency is shown (dashed line), particularly at wavelengths beyond 650 nm. Wavelengths above 750 nm belong to the infrared spectrum. (*Source:* adapted from Peet *et al.,* 2007)

processing additives (Lee *et al.,* 2007; Peet *et al.,* 2007). While these polymers still do not absorb far enough into the infrared, future synthesis of new molecules with absorption spectra that are even better matched to the solar spectrum will lead to even higher efficiencies.

The next step is to create multi-layer systems. This is possible with the same printing technology, i.e., by processing multi-layers from solution in successive depositions of electronic inks. Multiple layers will further increase the performance of the solar cells. This is because of the simple fact that if two batteries – regular batteries or solar batteries – with voltages $V_1$ and $V_2$ are connected in series ('tandem cells'), then the voltage will be the sum of the two $(V_1 + V_2)$. By connecting batteries in series, we can increase the open circuit voltage, and can take better advantage of the energy delivered in the solar spectrum.

Figure 5 shows that these multi-layer structures can in fact be fabricated. Despite the fact that the depicted films were cast from solution, the interfaces are very well defined – a result that gives us confidence in the success of our approach. By fabricating tandem cells, we have been able to show the expected increase in voltage. So far, we have been able to demonstrate power conversion efficiencies as high as 6.5% (Kim *et al.,* 2007).

While 6.5% represents important progress, it is not high enough. Fortunately, there are many opportunities to further improve the efficiency. A slightly different architecture (Kim *et al.,* 2006) enables us to better harvest the incoming photons and thereby improve efficiency by an additional 25–50%, approaching conversion efficiencies as high as 8–9%. (This architecture adds an 'optical spacer' layer between

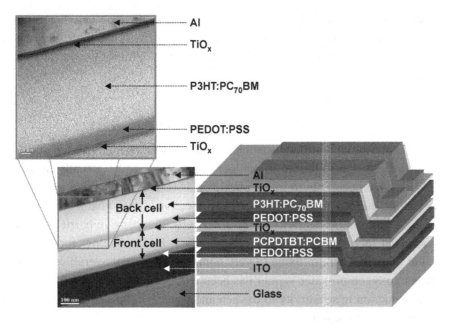

**Fig. 5.** Multi-layer structure of plastic solar cells connected in series (tandem cells). The images on the left are electron micrographs of cross-sections cut through multilayer structure, sliced down like a meat cutter in a delicatessen, turned over and then imaged by electron microscopy. (*Source:* Kim *et al.,* 2007)

the active bulk heterojunction layer and the metal electrode.) Also, we can expect more than a 50% efficiency improvement by creating molecular structures where the energy gap is even better matched to the solar spectrum than our current molecules (see Figs. 3 and 4). It must be emphasized that although we have made some improvements in the charge collection efficiency, we are still collecting only approximately half of the photo-generated carriers. In addition, we foresee optimizing the nano-scale morphology to further improve the charge collection efficiency. By precisely tuning the molecular structure, there is an opportunity to optimize the electrochemistry of semiconducting polymers, and thus to increase the open circuit voltage. It has been demonstrated that in this way power conversion efficiency can be improved by another 50%. The tandem cell configuration offers something between 50% improvement and a doubling of conversion efficiency (Kim *et al.,* 2007).

When the increments for these independent potential improvements are added up, then we could potentially achieve power conversion efficiency in excess of 25% – an efficiency approaching that achieved by existing inorganic solar cells. Each of these separate efficiency improvements have been successfully implemented already. However, realizing all of these improvements at the same time is difficult.

Combining independent improvements is the main challenge we will continue to work on in our laboratories. We are confident that we will reach efficiencies that will enable a major impact on the future solar cell technology, and thus on our future energy system.

## The lifetime of plastic solar cells

One of the questions that people often ask me is whether this 'plastic stuff' will have sufficiently long lifetime in outdoor applications to be actually useful. Although we have been able to make plastic solar cells less sensitive to oxygen or water vapour, they do need barrier films as protective layers. Thanks to the already achieved reduction in sensitivity of the solar cells, inexpensive barrier films such as those used for food packaging can be applied. By depositing, for example, a very thin layer of titanium oxide (a very common material), overall sensitivity of the cells to oxygen or water vapour has been reduced by a factor of 100. We hope that this reduction in sensitivity to oxygen and water will be sufficient to yield the long lifetimes that are required.

Progress on the lifetime issues continues to be promising. The efficiency of plastic solar modules that were on the rooftop for testing over a year (see Fig. 6a) did not decrease; in fact a slight increase was recorded. In the course of November, the efficiency started to fall and people got a little worried. However, it turned out that the temperature coefficient of the efficiency is opposite to that of silicon. When winter came, the efficiency decreased slightly, but it came up again in spring (Hauch *et al.*, 2008). This different temperature coefficient of the efficiency is an advantage, since solar cells increase in temperature when sitting in the heat of the sun. The initial data provide evidence that the lifetime of our solar cells may be sufficient for large-scale applications. Of course, accelerated lifetime testing must continue to provide information on the longer time degradation.

Clearly, plastic solar cells have a very promising future as they are lightweight, portable, and can be produced quickly in large quantities. In addition, their flexibility makes plastic solar cells useful not only for standard areas such as rooftops (see Fig. 6b), but also for a vast number of new applications such as tent and umbrella surfaces, backpacks, or sails. In terms of efficiency of plastic solar cells, improvement efforts have produced some impressive figures of merit. If you evaluate plastic and standard solar cells in terms of watts per gram, plastic solar cells are already more than competitive.

Our goal is to achieve a roll-to-roll manufacturing of low-cost plastic solar cells. With such a production, plastic solar cells could become a very important contribution on our path towards a renewable energy system.

a)

b)

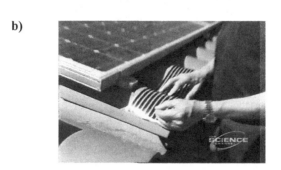

**Fig. 6.** a) Plastic solar cell testing in progress on the rooftop of Konarka Technologies, and b) the author on his own rooftop placing a plastic solar cell next to a conventional silicon solar cell. Although silicon solar cells work well, they have the disadvantage of being heavy and expensive. In contrast, plastic solar cells are lightweight, flexible and potentially produced at very low costs. Building them directly into the roofing tiles is an exciting opportunity. (*Sources:* a) Konarka Technologies, Inc., b) Discovery Channel science)

## References

Hauch, J., Schilinsky, P., Choulis, S. A., Rajoelson, S., Brabec, C. J. (2008). The impact of water vapor transmission rate on the lifetime of flexible polymer solar cells. *Applied Physics Letters,* **93**(10), 103306−9.

Kim, J. Y., Kim, S. H., Lee, H.-H. *et al.* (2006). New architecture for high-efficiency polymer photovoltaic cells using solution-based titanium oxide as an optical spacer. *Advanced Materials,* **18**(5), 572−6.

Kim, J. Y., Lee, K., Coates, N. E. *et al.* (2007). Efficient tandem polymer solar cells fabricated by all-solution processing. *Science,* **317**(5835), 222−5.

Lee, K., Kim, J. Y., Park, S. H. *et al.* (2007). Air-stable polymer electronic devices. *Advanced Materials,* **19**(18), 2445−9.

Ma, W., Yang, C. and Heeger, A. J. (2007). Spatial Fourier-transform analysis of the morphology of bulk heterojunction materials used in 'plastic' solar cells. *Advanced Materials,* **19**(10), 1387−90.

Peet, J., Kim, J. Y., Coates, N. E. *et al.* (2007). Efficieny enhancement in low-bandgap polymer solar cells by processing with alkane dithiols. *Nature Materials,* **6**(7), 497−500.

# Chapter 24

# Smart grids, smart loads, and energy storage

---

## Joachim Luther

Joachim Luther was born 1941 in Hanover, Germany. He obtained his PhD in atomic physics at the University of Hanover in 1970. From 1974–93 he taught at the University of Oldenburg as Professor of Applied Physics. His main research topics were oceanographic laser remote sensing and the physics of renewable energy sources. In 1992 he became a member of the board of directors of the International Solar Energy Society. From 1993–2006 he served as Director of the Fraunhofer Institute for Solar Energy Systems and Professor of Solid-State Physics and Solar Energy at the University of Freiburg. Luther was a member of the German Advisory Council on Global Change, and joined the Expert Commission of the German Government on Science and Innovation in 2007. Since 2008 he has served as Chief Executive Officer at the Solar Energy Research Institute in Singapore (SERIS).

The energy system of the future must be sustainable and must, therefore, be largely based on renewable energy sources. Solar energy will play by far the most important role, but wind energy, biomass, hydro energy, geothermal energy, ocean energy, and others will also contribute to a sustainable energy supply. The use of fossil fuels will remain essential in the next few decades, during which a sustainable energy supply system must be established (WBGU, 2003). However, during this transition period the carbon-dioxide emission rates of fossil fuels must be greatly reduced, for example through technological advances and the large-scale introduction of carbon-capture technologies such as sequestration. In this context it is essential that reliable and cost-effective carbon dioxide sequestration technologies become available quickly.

Due to energy scarcity, the rising cost of energy, and the fact that carbon dioxide emissions must be greatly reduced, efficiency of energy use will become increasingly important. In sustainable energy systems this will lead to buildings characterized by an extremely low demand for external energy input for heating and air-conditioning, and to a highly efficient transport system based largely on electric batteries, biofuels, and novel fuels like hydrogen (generated using electricity from renewable sources) or hydrogen derivatives. Simultaneously, electricity will become by far the most important form of distributed and traded energy. The question this raises is how to implement a reliable electricity supply system that distributes the required energy and that is powered to a large extent by fluctuating energies from solar and wind resources.

The answer has several principal components:

1. distributed energy generation and smart grids;
2. energy meteorology;
3. smart loads;
4. careful use of dispatchable sources for electricity generation; and
5. energy storage systems (both centralized and decentralized).

The balance in electricity supply will be provided by a mix of electricity generating systems powered by fossil fuels, biomass, or hydro energy (point 4 from above). This point will be addressed in combination with the discussion of point 1. In general, it will be essential to merge information technologies, power generation, power distribution, energy storage, and demand-side management in an optimal way.

## Distributed energy generation and smart grids

In contrast to fossil-fuel and nuclear systems, renewable energy sources – particularly solar, wind and biomass – are characterized by a relatively low spatial power density ($W/m^2$). Thus, these technologies will necessarily be large-area technologies (several percent of the global land surface will be required), and they will be applied in a highly distributed manner (to use as far as possible existing anthropogenic structures such as buildings as installation sites and in order to avoid unacceptable environmental effects). The low power density leads, on the one hand, to relatively high initial investment costs (however, the 'fuel' for operation is free); on the other hand, large-area statistical effects even out the characteristic fluctuations of solar and wind energy availability.

The temporal variations in solar and wind energy fluxes have two components: a trend pattern (daily and seasonal), and a random (or stochastic) component. The stochastic component is characterized by a spatial coherence that decays approximately exponentially with increasing distance between the sites. That is, the power fluctuations of two wind turbines situated at the same site are considerably larger than the fluctuations of the lumped power output of two turbines installed much further (e.g., 100 km) apart. The 'decay constant' mentioned above is roughly inversely proportional to the frequency of the power fluctuations. In other words, high-frequency fluctuations (in the range of seconds to minutes) are evened out much more effectively than low-frequency fluctuations (in the range of hours) (Beyer *et al.*, 1993). By means of computer simulation it has been shown that for large-area grids (with spatial dimensions exceeding 1000 km) and distributed generation of wind and solar electricity, the stochastic fluctuations with frequencies higher than 30 minutes are almost completely eliminated (Bubenzer and Luther, 2003).

Thus, if very low-frequency fluctuations are compensated by dispatchable electricity generators connected to the grid (powered, for example, by fossil fuels, hydro power or biomass) a reliable electricity supply can be guaranteed. In order to implement such an electricity supply scheme, two prerequisites have to be met: strong bidirectional grids, and the availability of sufficient and suitable dispatchable power generation capacity. This will, of course, require investments in grids as well as appropriate power plants.

As part of this infrastructure build-up, electricity generation from fossil fuels will also be decentralized to a certain extent. This will have the advantage that the locally generated waste heat from power plants can be used, for example, for district heating, dehumidification of air, and/or cooling of air, thereby increasing the overall efficiency of the energy supply system.

The effectiveness of evening out the stochastic fluctuations depends greatly on the spatial extension of the grid. By using an intercontinental grid (e.g., from western France to eastern Russia) that spans several time zones, even the daily trend component of the solar energy flux can be significantly evened out. Such long distance electricity transport (e.g., by means of high-voltage, direct current links) is technically state of the art. Thus, in particular a large-area network of solar and wind power plants can produce a considerable amount of base power with the same reliability as conventional power plants. The fraction of this base power component compared to the peak power of the whole installation (the capacity credit) depends greatly, of course, on the spatial extension of the network. Naturally, strong bidirectional electricity transport over long distances is essential for this scheme. Today, high-voltage direct-current technology would be the technology of choice. In the future this may be complemented by transmission lines based on high-temperature superconductors.

In distributed power generation schemes, a very large number of power generators will be connected to the distribution grid. Each generator will have its own power electronics unit that serves as the interface with the grid. If these units are designed properly, and if they are connected via information technology links, several additional benefits can be realized in future smart grids: (i) increase of power quality in the grid by means of local suppression of harmonics, local provision of reactive power and local voltage control, and (ii) increase of power supply security (e.g., in the case of natural disasters or terrorism) by forming island grids that guarantee at least a basic electricity supply. In such cases the cold-start capability of grids has to be addressed carefully in the design of the networks.

## Energy meteorology

In order to assess and predict the behaviour of smart distributed electricity grids that are largely powered by solar and wind energy, the temporal and spatial behaviour of the solar and wind energy fluxes must be known with high precision. A combination of distributed ground-based measurements and satellite information (most likely special sensor systems will be needed) seems to be the best way to collect the required data. Using elaborated meteorological models and suitable data distribution systems (e.g., the Internet), essential information concerning the meteorological energy fluxes and the status of the grid will be readily available whenever and wherever it is needed.

Statistical information is essential to optimally design (i) the spatial distribution of solar and wind electricity generators, (ii) the optimal fraction of solar- and wind-generated electricity in the grid (taking into account their partly complementary behaviour), (iii) the structure of the grid, and (iv) the information and control system of the entire network. Real-time information on the energy production of the individual electricity generators is essential for operating the smart distributed electricity supply system in an optimal way. This includes control of solar and wind power stations (in the event of electricity surpluses), control of smart loads (see below), control of dispatchable power plants (including distributed fossil fuel-powered combined heat/cold power units), and optimization of the power quality in the grid.

'Energy weather forecasts' for several days will be highly useful in efficiently operating dispatchable power plants and storage systems. All this information will also be essential for the electricity stock markets.

With the help of the information supplied by the required advanced energy meteorology systems, the hardware requirements of smart distributed energy supply systems can be considerably reduced; information in this case would substitute for hardware.

## Smart loads

Today's electricity supply systems are designed to ensure that most of the power plants can operate for the longest possible amount of time during a year. Economically this is sensible given the high investments in the power supply system. Since generation and load must match at any point in time, this means that the temporal variations of the lumped load must be smoothened as far as possible. Generally, this is achieved via sophisticated tariff structures such as penalizing peak loads, favouring electric night-time heating, and switching off large loads (e.g., refrigerating units, air-conditioners, etc) by the utility companies.

The same set of tools will also be applied in solar-dominated electricity supply systems. However, in contrast to today's approach, the lumped load pattern will be shaped such that there is a peak around noon times. A prerequisite for this are smart loads; loads that can easily react to external tariff signals (e.g., washing machines, heating units, etc.) and/or loads that can be externally switched off or on by the utility companies. In all these cases the quality of energy services has, of course, to be maintained.

The realization of smart loads generally includes two components: a certain 'technical intelligence' within the load combined with connection to an information network (in this case the issue of data security will have to be addressed carefully), and a certain storage capacity for energy in various forms. Examples of the latter include batteries in electronic devices, the heat capacity of buildings, the heat capacity of cooling units, compressed air, and process heat storage systems (for heat temperatures greater than 100 °C). It is not necessary that all the switching or control of loads will be done automatically (locally or remotely); the consumer who reacts to tariff signals will also create smart loads.

The largest single type of smart load will very likely be electric cars. The car batteries can be charged according to current demand (priority charging), the prevailing electricity tariff (via smart electronics), or remotely controlled by a utility company. If more than 50% of local transport needs is met by battery-powered electric vehicles, this will constitute a smart load representing 10–20% of total energy demand in Western Europe (Langniß *et al.*, 1998).

## Energy storage systems

By applying the above-mentioned schemes to large-area electricity grids, a high penetration of the grids with fluctuating energy inputs is feasible without the need for large energy-storage systems. This has been shown both by computer simulations and in practice (e.g., in Denmark and in Northern Germany). A large-area smart electricity supply system in Europe can handle a penetration with fluctuating energy inputs of at least 30 % without applying bulk energy storage (Langniß *et al.*, 1998). A prerequisite for this are, of course, targeted investments in grids, loads and information technology.

A higher penetration of grids with fluctuating energy inputs will require increasingly large energy-storage systems. Today, the main options for high-capacity storage are electrochemical systems and hydro power, although the latter has only a limited capacity on a global scale (WBGU, 2003).

Among the electrochemical storage systems, hydrogen-based systems have in principle an unlimited capacity; using electricity-powered electrolysers, water is split into hydrogen and oxygen (Luque and Hegedus, 2003). These gases are stored and later recombined in a fuel cell to generate electricity. The main disadvantage of this process is its low energy-efficiency. Even in future optimized systems, the overall efficiency will not be much higher than 50 %. Other storage options include advanced batteries (in particular for cars and other smart loads), redox systems (e.g., on the basis of vanadium compounds), supercapacitors, compressed air systems, and superconducting units. In solar thermal-power plants the possibility exists to store thermal energy at a high temperature, enabling an extension of the daily operating time by several hours. All of these technologies provide the basis for an appropriate storage of electricity on different time scales and with different capacities per unit; some are suitable to stabilize the grid on a short time-scale (seconds), while others may be utilized for bulk electricity storage. Some of the technologies mentioned above are not yet available for use in electricity supply systems. Further targeted research and development is needed.

## Conclusion

By applying the concepts of smart grids, smart loads and energy storage, grids with a high penetration of fluctuating energy inputs from solar and wind sources can be designed and operated reliably, while at the same time maintaining a high degree of energy security and power quality. The three concepts have to be viewed and optimized as a whole (this is why, from a technical point of view, 'unbundling' of power supply systems does not seem to be the best path towards a sustainable energy system).

As a rule of thumb, the larger the spatial extension of such a grid, the smaller the (relative) investment needed to construct and to operate the energy supply system.

Investments in storage systems, in transmission lines, and in smart control technologies have to be seen as three necessary steps that complement each other. Given today's penetration levels, there is presently no urgent technological need for large centralized bulk energy storage systems (e.g., on the basis of hydrogen), provided that proper investment is made in enhancing grid capacity (including smart loads, etc.). Bulk storage capacity will become important once the penetration of fluctuating energy inputs in large-area grids exceeds 20–30%.

The latter statement does not apply to small-area systems such as remote power systems or village and island power supplies. In these cases storage demand will become important much earlier, because of the inability to even out the fluctuations in energy input through statistical effects, and because of the relatively small number of (smart) loads. From this it follows that, if economically and politically feasible, such units should be electrically linked and operated as larger-area smart systems.

The world-wide installation of a sustainable electricity supply system based to a large extent on solar and wind energy sources is not, fundamentally, a technological problem. A basic set of proven energy conversion and distribution technologies already exists and will be further developed. This will lead to a considerable reduction of the cost of energy from renewable sources. Political will, coherence and a suitable global financing scheme are required to transform today's energy supply system towards sustainability.

## References

Beyer, H. G., Luther, J. and Steinberger-Willms, R. (1993). Power fluctuations in spatially dispersed wind turbine systems. *Solar Energy,* **50**(4), 297–305.

Bubenzer, A. and Luther, J., eds. (2003). *Photovoltaics Guidebook for Decision-Makers.* Berlin.

Langniß, O., Luther, J., Nitsch, J. and Wiemken, E. (1998). Strategien für eine nachhaltige Energieversorgung. In H. P. Hertlein, P. Tolksdorf, eds., *Workshop des Forschungsverbunds Sonnenenergie, Freiburg, 12. Dezember 1997.* Köln.

Luque, A. and Hegedus, S., eds. (2003). *Handbook of Photovoltaic Science and Engineering.* Chichester.

WBGU – German Advisory Council on Global Change (2003). *World in Transition – Towards Sustainable Energy Systems.* London.

# Chapter 25

# The SuperSmart Grid – paving the way for a completely renewable power system

## Antonella Battaglini, Johan Lilliestam, and Gerhard Knies

Antonella Battaglini is a senior scientist at the Potsdam Institute for Climate Impact Research (PIK). Within PIK and the European Climate Forum (ECF) she leads the SuperSmart Grid (SSG) process, which explores investment and technology options for transition to a decarbonized economy. The SSG concept was first developed by Battaglini in 2007.

*Note:* Photos and biographies of co-authors can be found in the appendix.

Renewable energy resources are abundant in all of Europe and neighbouring countries. Nonetheless, the current share of modern renewable energy sources in the European energy mix is very low, due to past political and technological decisions. If we are to prevent dangerous climate change, the political decisions that have shaped the current energy mix need to be revised to fully recognize the role of renewables in the immediate future, and to create a suitable environment for a sustainable energy system.

The idea of using solar energy for mechanical operations is very old, and its 'development across the centuries has given birth to various curious devices', as Augustin Mouchot stated as early as 1878 at the Universal Exposition in Paris. In 1861 Mouchot developed a steam engine powered entirely by the sun. But its high cost, coupled with the falling price of coal, doomed his invention to become a footnote in energy history. Since then, due to strong belief in the overarching advantages of fossil energy sources, investment and research in renewable energy technologies have comprised a negligible fraction of the funds provided for fossil and nuclear energy sources. Things slowly began to change during the energy crises in the 1970s, and gained momentum in recent years due to high energy prices and price volatility, and due to the threats posed by climate change.

It is the common view that the long-term climate target for Europe is an 80% reduction of greenhouse gas emissions by 2050 (see, for example, 2009/29/EC, 2009, p. 8). Reducing EU emissions by 80% in 40 years is a huge challenge and will require a transformation of the entire energy system, with great implications for societies and economies. In some sectors – such as the power sector – technological solutions that could enable significant emissions reductions already exist and many new technologies are being developed. Other sectors – such as agriculture or transport – could have a more difficult time reducing emissions at the required magnitude. For these reasons, we believe that the European power sector will have to be the first sector to be fully decarbonized by 2050. This paper discusses the European power sector and how to achieve its decarbonization.

## The SuperSmart Grid

There are several options for decarbonizing the power sector. None of them is easy, most require new mental models and political reform, but many are feasible. Among the main options discussed today are energy efficiency, carbon capture and storage (CCS), nuclear power, and renewables (see Bruckner *et al.*, this volume). Demand-side action such as energy efficiency measures will be increasingly important, but this point is not discussed in detail in this paper. On the supply side, the renewable energy option is the only truly sustainable solution, and therefore the less risky option, regardless of whether long-term investment risks, environmental risks, policy risks or other related risks are in focus. This option comprises a variety of technologies, some of them already mature (like hydropower, onshore wind, and biomass) and others in different development phases (offshore wind, photovoltaic (PV), and concentrating solar power (CSP)).

Within the broad field of renewable options, there are two main approaches. The first approach involves centralized, utility-scale power generation spread over a wide area. It requires electricity to be transported over long distances, from generation sites to load and storage areas. This is possible with minimum losses by using high-voltage direct current (HVDC) transmission technologies, which have been in use for decades on all continents. This approach is widely known as *Super-Grid*. The second approach – the virtual power plant – consists of a multitude of scattered generation sources which are aggregated together and managed by intelligent technologies (such as two-way communication between consumers and producers, as well as between producers), and Smart Meters, which enable consumers to manage their load – and thus their electricity cost – automatically. This intelligent operation of decentralized renewable power production, combined with demand-side management measures to better match the volatile supply with the demand are commonly known as *SmartGrid*. These two approaches are often perceived as exclusive alternatives, but it is conceptually necessary and technically possible to merge them. The combination of these two approaches is what we call the *SuperSmart Grid (SSG)*. We strongly believe that by combining them we can not only speed up the decarbonization process, but also open the way to further develop technologies that can address very different energy needs in Europe, in neighbouring countries, and elsewhere in the world.

A first step towards a Northern European SuperSmart Grid was recently taken by the Swedish EU presidency. One of its main objectives is the creation of an interconnected power grid in and around the Baltic sea (known as the Baltic sea power ring) and a joint Baltic power market. A Northern European power market, if successful, could form the nucleus of a pan-European energy market. The Nordic experience with running an international power market could also strongly contribute to the success of the Baltic market. Another example of development towards a SuperSmart Grid is the proposed North Sea grid for integration of wind power – supplying almost 15 % of the electricity needs of the seven North Sea countries by 2020 – and, as a positive side-effect, the physical unification of the North Sea power markets (Woyne *et al.,* 2008). Gregor Czisch, a SuperGrid energy expert at Kassel University, states that the potential for offshore wind in the North Sea is 6600 TWh/a, or almost twice the current EU-27 electricity consumption (Czisch, 2005). While this figure may be contested, it nonetheless indicates the potential that can be harnessed by grid expansion projects. A similar approach to the SuperSmart Grid is currently being implemented in the United States, although there it is only called Smart Grid, and is supported by the American stimulus package. Theoretically, the SuperSmart Grid approach could be applied to every power system currently in place.

In Europe, an important first step towards a large-scale, trans-continental Super-Grid for Europe and North Africa was recently taken by the Desertec Industrial Initiative, based on the work of Franz Trieb at the German Aerospace Center and the Desertec Foundation (Club of Rome, 2008; DLR, 2005; DLR, 2006; DLR, 2007). In this initiative, 12 companies – among them Munich Re, Deutsche Bank, E.ON and Siemens – have agreed to 'analyse and develop the technical, economic, political, social and ecological framework for carbon-free power generation in the deserts of North Africa'. The long-term goal of the Desertec consortium is to produce approximately 15% of European electricity requirements from renewable sources as dispatchable power, mainly from thermal solar power plants, in the Sahara desert, and to transport this electricity into the European power grid (DII, 2009).

### *Reducing costs through learning effects*

The cost of almost any technology starts off high and decreases over time, as increasing cumulative production triggers learning effects (costs are reduced through 'learning by doing') and economies of scale; each piece becomes cheaper as total production increases, since the costs of machines, for example, can be distributed over greater production (Coulomb and Neuhoff, 2006). Today, the costs of CSP are about EUR 0.25 per kWh in Spain and some EUR 0.15 per kWh in southern USA and in the desert of North Africa. These costs are expected to decrease by at least 20–40% in the next decade if 20 GW of new capacity goes online (Munich Re, 2009; Club of Rome, 2008; DLR, 2006; Ummel and Wheeler, 2008). CSP technology is still far from mass production and it remains to be seen how quickly these learning rates can indeed be achieved or even exceeded.

For wind, the principle is similar, although onshore wind technology has already passed through a large part of its learning curve, limiting the potential for further cost reductions. Nonetheless, onshore wind power can be expected to become some 10% cheaper per doubling of the cumulated capacity[1] and can at normal sites (see section on quality of sites below) asymptotically reach about EUR 0.06 per kWh in the long run (Krohn *et al.*, 2009; GWEC, 2009; Neij, 2008; Nitsch, 2008). Offshore wind technology is still rather expensive and has only been installed at relatively small scales. The production costs today are about EUR 0.15 per kWh, but are expected to be half of that – EUR 0.075 per kWh – in 2020 (Nitsch, 2008).

---

[1] In the decade up to 2009, the cumulated global wind capacity has doubled about every three years. This trend is expected to remain the same in the medium-term future.

By comparison, the production costs of new nuclear and coal power based on current world market fuel prices, a carbon dioxide cost of EUR 20 per tonne, and investment costs as provided by the companies constructing new power stations in Europe – excluding costs of all insurances, decommissioning, final storage, interest fees for capital invested during the construction time and all external costs – are between EUR 0.055 and 0.075 per kWh (Olkiluoto 3, nuclear) and EUR 0.045–0.055 per kWh (Neurath 2 and 3, lignite), depending on interest rate and economic lifetime (AFP, 2008; Ernst & Young, 2006; RWE, 2009). A recent meta-study of the costs of new nuclear power stations puts the costs at EUR 0.085–0.145 per kWh (Cooper, 2009). It should be noted that nuclear power is one of the few technologies that is not getting cheaper with time; instead, new nuclear power tends to become more expensive with time (Cooper, 2009; Neij, 2008).

A sustained level of wind and solar power expansion is therefore the key to making renewable technologies competitive, reaching grid parity[2] and eventually becoming the cheapest option for new power stations. Such large cost reductions are not only important for the cost-efficient implementation of renewable electricity in Europe, but also extremely relevant for investments in developing countries where resources are limited and investment competition among different sectors is high. Developing countries today simply cannot invest in the still much more expensive renewable technologies. Today, the upfront investment for electricity generation is substantially lower for old fossil-fuel-based technologies than for renewable energy technologies. That is a major reason why large amounts continue to be invested in old technologies, even in developed countries,[3] despite the threats posed by climate change, the risks of increasing fuel costs, and the risk of an increasing carbon price. Strong European investment in renewable power generation technologies will bring the costs of these technologies down, which will make the renewable option the cheapest and least risky solution to satisfy the rapidly increasing electricity demand in developing countries. Therefore, the impact of European leadership in renewables expansion will extend far beyond the immediate emissions reductions in the European power system since, together with the considerable US efforts in green investments, it can pave the way for even greater reductions globally.

---

[2] There are numerous definitions of grid parity. Here, we refer to the break-even point of the costs of producing your own electricity and the price of electricity from the grid, including taxes and grid fees.
[3] Between 2007 and 2012 RWE plans to invest EUR 12 billion in power plants, lines and open-pit mines; E.ON even plans to invest EUR 30 billion between 2009 and 2011, mainly in 'renewing and maintaining ... and expanding our conventional generation capacity' (E.ON, 2009; RWE, 2007).

## *Reducing costs by choosing only the best sites*

Today, the generation costs of most forms of new renewable electricity are still higher than those of fossil-fuelled electricity, with the possible exception of on-shore wind power on good sites (see cost estimates above). The renewables – except for biomass and biogas – have a completely different cost structure to fossil power; investment accounts for by far the greatest share of generation costs, as the fuel costs are zero. Instead of fuel price, the quality of the production site – for example as measured by average wind speed and direct solar insolation[4] – becomes the main variable for determining the production costs. That means that good sites have much lower production costs than marginal sites.

In Europe, the renewable energy potential is high and is probably sufficient to satisfy the current levels of electricity demand (see Fig. 1). However, resources are not evenly distributed and in some countries the renewable potentials exceed the national demand. For example, Sweden and Spain with their extensive renewable resources (biomass, wind and hydro for the former, and solar for the latter) could achieve a 100% renewable power system if they decided to. Other countries, such as France, Germany, Italy and the Benelux countries, are not as rich in renewable resources, mainly due to high population density and geography. The economic potentials in these countries are much too low for a completely carbon-neutral power system based on renewables and they would have to utilize bad, and thus expensive, production sites in order to achieve very high shares of renewable power.

In the event that European electricity demand should increase significantly in the future, for example by the widespread introduction of electromobility, even the combined and optimally interconnected domestic EU potentials, at reasonable economic cost, may not be large enough.

Enormous potentials for renewable power are found just outside of Europe, for example in the neighbouring North African countries. The solar energy potential is immense all across the Sahara Desert and there is a multitude of very good wind sites, for example along the Red Sea and the coasts of Morocco. The economic solar and wind power potentials of the five countries on the southern Mediterranean rim is two orders of magnitude larger than the combined electricity demand of Europe and North Africa in any realistic scenario (see Fig. 2). Utilizing these resources would allow 'cherry-picking' of production sites. Marginal sites could be completely discarded and only the best ones utilized for electricity production, which would allow for high economic efficiency of the transformed renewable power system by providing dispatchable and controllable capacity (for explanations of these terms, see below). Moreover, if electromobility or other large new

---

[4] incident solar radiation

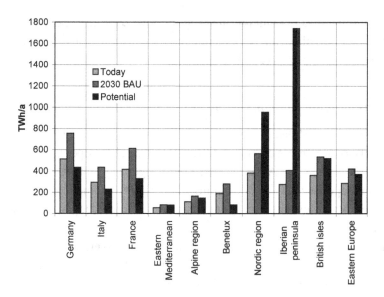

**Fig. 1.** Electricity consumption today and in 2030 (EU 'business as usual' case) and the economic potentials for all renewable electricity sources in different regions of the EU-27.[5] (*Sources:* DLR, 2005; DLR, 2006; Eurostat, 2009; Resch *et al.,* 2006)

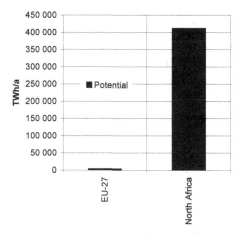

**Fig. 2.** The economic potential for renewable electricity in the EU-27 and North Africa. Note that this scale is 100 times larger than the scale in Figure 1. Graphically comparing the consumption of Europe (approximately 3000 TWh/a) or North Africa (approximately 200 TWh/a) is not useful, since the potentials are so much larger than any realistic consumption. (*Sources:* DLR, 2005; DLR, 2006)

---

[5] It should be noted that the potentials in the figure above are averages. As most renewable sources are intermittent, these numbers only indicate that the potentials are, on average, sufficient to decarbonize the power system, but do not indicate that sufficient production will be available at any given time (see section on generation intermittency).

power consuming systems emerge in the future, utilizing the resources in North Africa and other neighbouring regions may be the only way for Europe to sustainably decarbonize its power system at reasonable costs.

### *Maintaining and improving geopolitical security of supply*

The transformation and decarbonization of the power system can only succeed if energy supply is secured at all times. Often, the idea of Europe importing renewable electricity from North Africa is criticized for getting Europe into yet another energy import dependency (see for example Zeller, 2009), adding to Europe's already high import dependency (see Fig. 3) and, as a consequence, jeopardizing European security of supply.

The main option to increase security of supply is to diversify sources, increase the share of domestic fuels and make the power system more flexible.[6] A well developed SmartGrid with a large share of decentralized and distributed renewables generation, linked into a highly flexible grid capable of transporting electricity over vast distances and in all directions, would greatly improve Europe's security of electricity supply (EC, 2006; Jansen *et al.*, 2004; Ocaña and Hariton, 2002; Ötz *et al.*, 2007; Scheepers *et al.*, 2006). Including North Africa in the European power system can lead to further diversification of source countries, fuels and technologies, and reduce import dependency on fossil fuels even in the transition phase to a completely renewable power system, thus improving overall security of electricity supply (Ötz *et al.*, 2007; DII, 2009). In the long run, imports of renewables will be the only imports to the electricity sector, and the total import dependency will be much lower than it is today.

It is a matter of good governance to ensure that these imports are secure and beneficial for both sides. The twin objectives of guaranteeing European electricity supply while avoiding colonial tendencies – real or perceived – and the resource curse[7] for the exporting countries are equally important and should be pursued in tandem. Good governance is not usually addressed in today's world energy market, but it can be. Norway, for example, which today exports large amounts of gas and oil to other European countries (see Fig. 3), is considered at least as secure as any EU member state, it does not suffer from the resource curse, and does not feel colonized or exploited by its energy customers.

An electricity relationship between the EU and its different North African partners

---

[6] Or, to put it in the almost 100-year-old words of Winston Churchill, 'safety and certainty in oil lie in variety, and variety alone' (Ladoucette, 2002)

[7] The resource curse refers to the paradox that countries with large exports of unrefined natural resources tend to have a slower economic and social development, suffer more corruption and are less democratic than countries with only small exports of natural resources

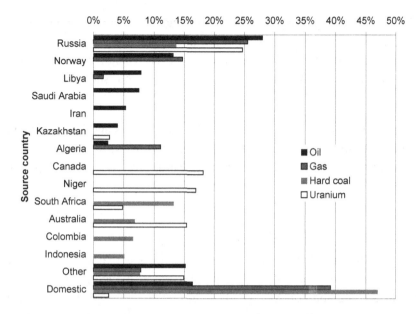

**Fig. 3.** Import shares to the EU-27 in 2006 from different source countries, as share of total consumption of each fuel. Hard coal imports calculated on the basis the 'solid fuel' category of Eurostat statistics, based on average heating values and with the assumption that all lignite is domestic. (*Sources:* BP, 2008; DG TREN, 2008; Euracoal, 2008 a; Euracoal 2008 b; Euratom, 2008; Eurostat, 2009)

will create interdependency and must be based on clear and stable treaties, as well as the economic and development needs of both sides. A number of specific issues must be explicitly addressed:

• A business model that takes into consideration and satisfies North African power demand and expectations should be developed to guarantee stable and long-lasting cooperation.

• The desert land that, from a European perspective, seems empty and worthless is in fact inhabited by different peoples, such as numerous Bedouin tribes. Although only a small fraction of the Sahara Desert will be exploited, power stations and power lines will be an intrusion into these lands, and the people living there must be included in the planning of CSP and wind capacities.

• If the rights of the desert peoples, as well as local populations in general, are not recognized, European and North African security of supply may be at greater risk and the threat of attacks against power plants and lines may increase. Terrorist attacks against the long power lines through the desert will be a real threat to both North African and European security of supply and measures to minimize this risk will be required. However, a comparison with the gas sector may prove useful: already today, long pipelines stretching from southern and central Algeria and Libya to the coast exist and are not targets for terrorists, despite their exposed situation. The terrorist threat against energy installations will always be present and must be taken seriously in any location, but should not be exaggerated.

A fair and well thought-out deal between North Africans and Europeans will set the fundaments for a reliable electricity supply and avoid the sort of disruptions or blackmail seen in the Russian-Ukrainian-European gas relationship in recent years.

### *Handling generation intermittency*

The greatest difficulty with renewable energy sources is that they are intermittent and supply-controlled (see Luther, this volume). Fossil-fuelled power plants, on the other hand, are demand-controlled and can be operated whenever there is demand, which is one of their major advantages. A wind power plant can only produce electricity at times when there is wind, and these times may or may not coincide with the times of consumption in the surroundings. CSP production is not necessarily as intermittent as wind, due to the possibility of thermal storage directly in the power station. Some of the heat generated during the day can be stored and used, for example, at night. If the storage and the mirror fields are large enough, a CSP station at a very good site, for example in the desert, can provide firm capacity most of the time (Trieb *et al.*, 2009; DLR, 2006). By adding back-up systems, such as a fossil- or biomass-fuelled combustion chamber to replace or support the solar field during longer periods with little or no sun, firm capacity can be guaranteed at any time. The 'intelligent' operation of CSP plants with intrinsic thermal storage, combined with other, entirely supply-controlled power sources in a broad electricity mix, could be one of the easiest ways – and therefore also one of the cheapest ways – to handle intermittency. In principal, however, all existing renewable power options, except biomass-based ones, are, to different degrees, supply-controlled and intermittent. As electricity has to be consumed instantly, this stochastic behaviour of renewable electricity production has to be managed.

Electricity storage and back-up capacities are often mentioned as necessary tools to maintain stability in power systems with high penetrations of renewables. These options have the advantages that they are easily controllable and fit well into the paradigm of the current system. The main disadvantages are their high costs: long-term storage (on the scale of weeks) and short-term storage ('peak shaving', on the scale of a few hours or up to a day) cost from EUR 0.37 per kWh (long-term pressurized air) – or even EUR 0.5 per kWh for lithium-ion batteries – to EUR

0.1–0.2 per kWh for most short-term technologies, with a singular minimum cost of EUR 0.05 per kWh for short-term pressurized air storage (Leonhard *et al.*, 2008). Even if the costs were to decrease by 50%, most electricity storage technologies would still be too expensive. They would be uneconomical compared to fossil power and CSP with thermal storage even if the electricity generation were cost-free.[8] The costs of back-up generation vary greatly depending on the power system configuration and electricity mix, but some EUR 0.02–0.04/kWh are realistic wind back-up costs at current wind penetrations. Due to the low load factor[9] of photo-voltaic power, the back-up costs can be expected to be higher than this. The back-up costs tend to increase with higher penetrations of intermittent renewables (RAENG, 2004; IEA *et al.*, 2005), and will probably not be viable on a very large scale. Depending on the configuration and size of thermal storage, the need for and costs of back-up and electrical storage for CSP electricity could be significantly lower than for wind; if the thermal storage were large enough, no electrical storage outside of the power plant would be needed.

Another way to deal with intermittency, and the one advocated in the SuperSmart Grid concept, is a mix of different generation technologies in a SmartGrid virtual power plant approach, as well as stochastic smoothing over vast distances. A virtual power plant consists of a number of renewable power stations of different kinds and with different fuels – a broad fuel mix of wind, solar, bio and hydro is the key – that are operated as an aggregate power plant. The combination of supply-controlled technologies (such as wind and PV) and demand-controlled technologies (such as biogas or hydropower with dams, and CSP with thermal storage) makes it possible to operate the aggregate of supply-controlled renewable power stations in a demand-controlled way, offering dispatchable capacity[10] or – in the future – even base-load[11] generation (Mackensen *et al.*, 2008). If the power system is geographically larger than a weather system, which it would be in a SuperGrid Europe, there will always be wind somewhere and sun somewhere else within the area (alternatively, at night electricity from CSP storage plants can be used). If the grid is efficient, densely meshed and flexible, electricity can flow from A (with high production) to B (with low production) at one moment and from B to A in the

---

[8] This is true only in the current accounting system, which does not include environmental externalities

[9] The load factor is a measurement of the utilization of a power plant and is defined as the quota of the actual yearly production divided by the maximum potential yearly production. Wind power plants typically have load factors of 25–30%, whereas photovoltaics usually lie around 10–15%. Baseload power stations, such as lignite and nuclear power plants, typically have load factors between 80 and 90%. CSP equipped with storage capacity can provide a similar base load to fossil fuels.

[10] The term 'dispatchable capacity' refers to power stations which can provide capacity on demand and when electricity is needed.

[11] Base-load generation involves power plants that operate permanently at full or almost full capacity and are characterized by very high load factors (see above). In the current system design, these power stations provide the base of the supply system, whereas peak-load power plants handle fluctuations and provide electricity during times of high or volatile consumption.

next when the weather system has moved and the production pattern has changed, even if the two points are thousands of kilometres apart.

According to Gregor Czisch (2005), this *correlated* and *stochastic* smoothing[12] over all of Europe as well as North Africa and the Middle East is enough to satisfy power demand at any given time, completely without electric storage and back-up. Even if the power system, for example during the transition phase, does not allow for sufficient smoothing to meet demand at exactly all times, correlated and stochastic smoothing will greatly reduce the need for back-up or storage. The costs of transmitting the renewable electricity to just about any point in Europe, which would be up to about 3000 km, with high-voltage direct current (HVDC, see below) power lines is about EUR 0.01–0.02/kWh (Czisch, 2005; DLR, 2006; Jochem *et al.*, 2008; May, 2005), which makes correlated and stochastic smoothing in combination with dispatchable CSP power by far the cheapest option for handling intermittent renewable energy resources.

### *Expanding power grids and generation capacities: a policy matter*

Today, the power grid is a major bottleneck for a further large-scale expansion of renewable electricity production. Already today, many power lines – especially cross-border interconnectors – are congested and overloaded (Battaglini *et al.*, 2009; DENA, 2005). Long-term strategic planning for a truly European power grid, also recognising the benefits of stochastic smoothing, is urgently required. Currently, expansion plans are made nationally on the basis of 'business as usual' developments and with a time-frame of about ten years. The implementation of such plans generally takes much longer than that, due to bureaucracy and strong opposition by the public. By the time the planned lines are finished, the 2020 renewables targets – and possibly the climate target as well – will no longer be reachable.

The transmission system operators (TSOs) are today neither requested nor paid to have a vision for the future power system. Therefore, they generally do not engage in investigating different development scenarios for future required European transmission capacity. They are not encouraged to have an international, not to mention pan-European, approach to grid expansions, but rather to optimize the national system in the short- to medium term, which is often not the best solution from a long-term perspective. For the integration of electricity produced in North Africa and offshore in the North and Baltic Sea regions into the European power

---

[12] Correlated smoothing refers to smoothing effects emerging from weather correlations over large distances and among energy generation technologies in a broad technology mix, whereas stochastic smoothing is an effect caused by a random input of wind and solar power (mainly PV) over the wide geographical spread of the power grid.

system, HVDC lines will be required. Such power lines have much lower losses than conventional AC (alternating current) power lines, and are cheaper to build over long distances.[13] On short-distance lines, including almost all national power lines, HVDC is, however, more expensive than AC; the break-even point is about 800–1000 km (DLR, 2006). Therefore, HVDC lines are economically suboptimal in a national, short-term perspective, and TSOs are today de facto not allowed to build these. Thus, the nationally limited grid regulations based on short-term economic efficiency prevent Europe from reaching the longer-term renewables and climate targets in an economically sound way.

Due to these obstacles (national borders, focus on short-term economic efficiency), there is a risk that Europe will build itself into a situation far from a comfortable pathway to the 2050 emission reduction targets. A first step to tackling this problem would be to give the newly created European Network of Transmission System Operators for Electricity (ENTSO-E) the mandate to develop expansion plans for different carbon-neutral electricity mixes until 2050, including different scenarios of entirely renewable power systems, and allowing for imports of electricity from outside the European borders. The short-term and nationally limited perspective on regulation must be abandoned, and long-term, pan-European regulations introduced, which would allow the financing and the construction of the required HVDC lines. The process of restructuring and expanding the transmission grid must be inclusive and involve NGOs and affected communities. The communication and discussions of the grid expansion issue must be far more holistic than is currently the case – especially from the side of green NGOs – and the focus must be expanded to include both generation and transmission. The Renewables Grid Initiative,[14] bringing together NGOs and TSOs, is a first step in this direction.

Moreover, long-term targets and planning are fundamental to building up the supply chains of new renewable generation capacities, which today are not sufficient to realize the required transformation of the electricity system at the required pace. The capacities for the production of new renewable power stations are growing fast, but demand is growing even faster in some regions. Limitations in the renewable power station supply chains already hamper renewables expansion in some areas, especially in the wind power sector, and these supply chain constraints are an important determinant of how fast the transformation of the system can be (see, for example, EWEA, 2009; Krohn *et al.,* 2009). It is the task of policymakers to clearly define the long-term direction and create confidence for investors to channel funds into expanding the supply chains, in order to ensure a faster pace in the

---

[13] HVDC lines have full load losses of about 2–3% per 1000 km, whereas conventional high-voltage alternating current (HVAC) lines have losses of 7–10% per 1000 km (Battaglini *et al.,* 2009; Czisch, 2005; DLR, 2006).
[14] http://www.renewables-grid.eu

transition towards a renewable power system. It is important to note that serious supply chain bottlenecks are present for new renewable generation technologies and transmission lines, but also for other potential options such as CCS and, most significantly, nuclear energy. Although these technologies are well established, the power plant construction capacity at present is limited and the supply chain would need to be expanded to ensure power supply even in a fossil-fuel-based future power system (DG TREN, 2008).

## Conclusion and outlook

The potential for renewables in Europe and North Africa is sufficient to entirely decarbonize the power system. However, this can only be achieved through a co-ordinated pan-European and trans-Mediterranean approach and not by single countries autonomously, as the renewable electricity potentials for most countries are simply not large enough. For some countries costs will be too high and intermittency of supply will cause serious trouble, adding to the cost problem. These problems could be addressed and eased by developing a unified European power market, equipped with smart technologies, and by unifying the European and North African markets into a pan-European, trans-Mediterranean SuperGrid. Such a SuperSmart Grid has the potential to satisfy any electricity needs of the future, to minimize costs by enabling cherry-picking of sites, and manage intermittency problems.

Strong political leadership is required to foster and promote the transition to a largely renewable-based power system. European and American efforts to develop renewable technologies will generate a lot of synergies and accelerate economies of scale. Reduced investment costs and the expected increase in fossil fuel energy prices will provide the economic stimulus to channel investments into renewable technologies, not only in Europe and other developed countries, but also in developing countries. This will contribute greatly to reducing emissions worldwide, and at the same time help guarantee developing countries' right to economic development. It is a difficult process, but achievable nonetheless.

During the second half of the twentieth century, Europe was divided. Most people thought that this division was impossible to overcome, but the vision of reunification was still in the minds of people on both sides. In early 1989, the East German leader Erich Honecker stated that the Berlin Wall would endure for another 100 years. Just a few months later the wall fell. The vision of a power system based entirely on renewable energy sources is not new; it has been discussed for decades, with Desertec in recent years taking the lead in advocating energy cooperation with North Africa to meet Europe's and North Africa's energy needs. The interest among politicians and the business community in the renewable energy option has never been greater than today, and that interest keeps on growing. Nonetheless, the dominance of fossil fuels seems insurmountable. However, sooner or later, just like the Berlin Wall, the fossil-based energy system will crumble, and the time of renewables will come.

# References

2009/29/EC (2009). *Directive 2009/29/EC of the European Parliament and of the Council of 23 April 2009 Amending Directive 2003/87/EC so as to Improve and Extend the Greenhouse Gas Emission Allowance Trading Scheme of the Community.* Available at http://eur-lex.europa.eu/LexUriServ/LexUriServ.do?uri=OJ:L:2009:140:0063:008 7:EN:PDF, accessed 2 September 2009.

AFP – Agence France Presse (2008). *Areva Faces 50 Pct Cost Rise for Finnish Nuclear Reactor: Report.* Paris. Available at http://afp.google.com/article/ALeqM5h-habwlIEsV9fgMzGXgFKDzg70KMw, accessed 5 August 2009.

Battaglini, A., Lilliestam, J., Haas, A. and Patt, A. (2009). Development of SuperSmart Grids for a more efficient utilisation of electricity from renewable sources. *Journal of Cleaner Production,* **17,** 911–18.

BP (2008). *BP Statistical Review of World Energy.* London.

Club of Rome (2008). *Clean Power from Deserts.* Hamburg.

Cooper, M. (2009). *The Economics of Nuclear Reactors.* South Royalton.

Coulomb, L. and Neuhoff, K. (2006). *Learning Curves and Changing Product Attributes: the Case of Wind Turbines.* Cambridge.

Czisch, G. (2005). *Szenarien zur zukünftigen Stromversorgung. Kostenoptimierte Variationen zur Versorgung Europas und seiner Nachbarn mit Strom aus erneuer-baren Energien.* Kassel.

DENA – Deutsche Energieagentur (2005). *Energiewirtschaftliche Planung für die Netz-integration von Windenergie in Deutschland an Land und Offshore bis zum Jahr 2020.* Köln.

DG TREN – Directorate-General Transport and Energy, European Commission (2008). *European Energy and Transport. Trends to 2030 – Update 2007.* Brussels.

DII (2009). *12 Companies Plan Establishment of a Desertec Industrial Initiative.* Press release of Munich Re Group. Available at http://www.munichre.com/en/press/press_releases/2009/2009_07_13_press_release.aspx, accessed 13 July 2009.

DLR – German Aerospace Center (2005). *Concentrating Solar Power for the Mediterra-nean Region.* Stuttgart.

DLR – German Aerospace Center (2006). *Trans-Mediterranean Interconnection for Concentrating Solar Power.* Stuttgart.

DLR – German Aerospace Center (2007). *Concentrating Power for Seawater Desalini-sation.* Stuttgart.

EC – European Commission (2006). *Green Paper. A European Strategy for Sustainable, Competitive and Secure Energy.* COM(2006)105 final. Brussels.

Eon (2009). Investments for energy security and growth. Düsseldorf. Available at http://www.eon.com/en/unternehmen/12513.jsp, accessed 06 August 2009.

Ernst & Young (2006). *Energiemix 2020. Szenarien für den deutschen Stromerzeugungs-markt bis zum Jahr 2020.* Düsseldorf.

Euracoal – European Association for Coal and Lignite (2008a). *Coal Production and Imports in EU-27 in 2007 in Mt.* Brussels. Available at http://www.euracoal.org/pages/medien.php?idpage=400, accessed 07 May 2009.

Euracoal – European Association for Coal and Lignite (2008 b). *Trade with Hard Coal in EU-27 in 2007 (in Mt)*. Brussels. Available at http://www.euracoal.be/pages/medien. php?idpage=397, accessed 07 May 2009.

Euratom (2008). *Euratom Supply Agency. Annual Report 2007*. Luxemburg.

Eurostat (2009). *Energy and Transport in Figures. Statistical Pocketbook 2009*. Brussels.

EWEA – European Wind Energy Association (2009). *Wind Energy – The Facts*. Brussels.

GWEC – Global Wind Energy Council (2009). *Global Cumulative Installed Capacity 1996–2007*. Brussels. Available at http://www.gwec.net/uploads/media/chartes08_ EN_UPD_01.pdf, accessed 18 June 2009.

IEA – International Energy Agency, NEA – Nuclear Energy Agency, OECD Organisation for Economic Co-Operation and Development (2005). *Projected Costs of Generating Electricity*. 2005 Update. Paris.

Jansen, J. C., van Arkel, W. G. and Boots, M. G. (2004). *Designing Indicators of Long-Term Energy Security of Supply*. ECN-C-04-007. Petten.

Jochem, E., Jaeger, C. C., Battaglini, A. *et al.* (2008). *Investitionen für ein klimafreundliches Deutschland*. Potsdam and Karlsruhe.

Krohn, S., Morthorst, P.-E. and Awerbuch, S. (2009). *The Economics of Wind Energy*. Brussels.

Ladoucette de, V. (2002). Security of supply is back on the agenda. *The Middle East Economic Survey*. Available at http://www.mafhoum.com/press4/121E17.htm.

Leonhard, W., Buenger, U., Crotogino, F. *et al.* (2008). *Energiespeicher in Stromversorgungssystemen mit hohem Anteil erneuerbarer Energieträger. Bedeutung, Stand der Technik, Handlungsbedarf*. Frankfurt/Main.

Mackensen, R., Rohrig, K. and Emanuel, H. (2008). *Das regenerative Kombikraftwerk*. Kassel.

May, N. (2005). *Eco-Balance of a Solar Electricity Transmisison from North Africa to Europe*. Braunschweig.

Munich Re – Münchener Rückversicherungs-Gesellschaft (2009). *Munich Re Newables. Our Contribution to a Low-Carbon Energy Supply*. München.

Neij, L. (2008). Cost development of future technologies for power generation: a study based on experience curves and complementary bottom-up assessments. *Energy Policy*, **36**(2008), 2200–211.

Nitsch, J. (2008). *Leitstudie 2008. Weiterentwicklung der "Ausbaustrategie Erneuerbare Energien" vor dem Hintergrund der aktuellen Klimaschutzziele Deutschlands und Europas*. Berlin.

Ocaña, C. and Hariton, A. (2002). *Security of Supply in Electricity Markets*. Paris.

Ötz, S., Sims, R. and Kirchner, N. (2007). *Contribution of Renewables to Energy Security*. Paris.

RAENG – The Royal Academy of Engineering (2004). *The Cost of Generating Electricity*. London.

Resch, G., Obersteiner, C., Auer, H. *et al.* (2006). *Guiding a Least Cost Grid Integration of RES-Electricity in an Extended Europe. Database on Potentials and Cost of RES-E and Energy Efficiency for the EU-25 Member States (+NO, CH, RO, BG, HR)*. Vienna.

RWE (2007). *RWE investiert Milliardenbetrag in den europäischen Energiemarkt*. Essen. Available at http://www.rwe.com/web/cms/de/110504/rwe/investor-relations/ nachrichten/news-ad-hoc-mitteilungen/?pmid=4001505, accessed 5 August 2009.

RWE (2009). *Hier investiert RWE 2,2 Mrd. Euro in die Zukunft der Region*. Essen. Available at http://www.rwe.com/web/cms/de/12068/rwe-power-ag/kraftwerksneu bau/boa-2-3/, accessed 5 August 2009.

Scheepers, M., Seebregts, A., de Jong, J. and Maters, H. (2006). *EU Standards for Energy Security of Supply.* ECN-C-06-039/CIEP. Petten.

Trieb, F., O'Sullivan, M., Pregger, T., Schillings, C. and Krewitt, W. (2009). *Characterisation of Solar Electricity Import Corridors from MENA to Europe. Potential, infrastructure and cost.* Stuttgart.

Ummel, K. and Wheeler, D. (2008). *Desert Power: The Economics of Solar Thermal Electricity for Europe, North Africa, and the Middle East.* Working paper 156 of the Center for Global Development. Washington D. C.

Woyne, A., de Decker, J. and van Thong, V. (2008). *A North Sea Electricity Grid [R] evolutioni. Electricity Output of Interconnected Wind Power.* Brussels.

Zeller, T. (2009). Europe looks to Africa for solar power. *The New York Times,* 21 June 2009. Available at http://www.nytimes.com/2009/06/22/business/energy-environment/22iht-green22.html?_r=4&pagewanted=1&partner=rss&emc=rss, accessed 13 July 2009.

# Chapter 26

# Getting the carbon out of transportation fuels

---

## Felix S. Creutzig and Daniel M. Kammen

Felix S. Creutzig is a postdoctoral fellow at the Technical University of Berlin. Previously, he worked with Daniel Kammen at the Berkeley Institute of the Environment on various aspects of transportation, ranging from transport policy to technology evaluation of the compressed-air car. He has also worked for the Energy Foundation China in Beijing. Creutzig obtained his PhD in Computational Neuroscience at the Humboldt Universität in Berlin after graduating in Theoretical Physics from the University of Cambridge.

Daniel M. Kammen received his PhD in physics from Harvard in 1988 for his work on theoretical solid-state physics and computational biophysics. From 1993–9 he was Professor and then Chair of the Science, Technology and Environmental Policy Programme at Princeton University. He then moved to the Energy and Resources Group and the Goldman School of Public Policy at the University of California, Berkeley. Kammen is also the founding director of the Renewable and Appropriate Energy Laboratory and a Coordinating Lead Author for the Intergovernmental Panel on Climate Change, which shared the 2007 Nobel Peace Prize.

Transport is currently responsible for 13% of global greenhouse gas (GHG) emissions and it contributes 23% of global carbon dioxide emissions from fuel combustion (International Energy Agency, 2008). Global transport-related carbon dioxide emissions are expected to increase by 57% in the period 2005–2030, making this the fastest growing sector globally. At the same time, there is broad consensus in science and politics that global GHG emissions must be reduced by more than 80% from 1990 levels by 2050 to avoid perilous global warming. It is clear that the transport sector will need to be central to mitigation efforts. One important contribution towards this goal can be to reduce the carbon content of fuels or, more generally, vehicle propellants. In this essay, we investigate the potential of biofuels and electric mobility to decarbonize car transportation. As with most areas of a sustainable energy economy, large improvements are possible, but they require a 'systems science' approach that works across disciplines and considers traditional vehicles approaches and stationary power. Science, technology, policy, economics, and cultural awareness must be utilized in concert.

## Innovations in response to challenges: from lead to carbon

During the late 1970s and early 1980s, the highest-profile environmental issue in the vehicle and fuel industries was the establishment of a ban on lead additives in petrol – encapsulated by the slogan *get the lead out*. After initial uncertainty and some opposition based on the fear that prices would rise and vehicle performance would suffer, the transition to unleaded fuels proved remarkably easy and effective. Between 1970 and 1987 the average blood-lead level in the US population dropped by 75%, and the blood-lead levels of up to two million children were reduced to below toxic levels every year as leaded petrol use was curtailed.[1] In direct response to the reduction in atmospheric lead the IQ levels of previously lead-exposed urban children increased (Thomas, 1995).

The US Congress also enacted the Corporate Average Fuel Economy (CAFE)[2] regulations, a sustained effort to raise average vehicle efficiency standards in response to the 1973 Arab oil embargo. This measure increased vehicle mileage standards by more than 25%. Such examples demonstrate that ambitious, yet achievable, targets can be codified, enforced, and adjusted as technological, economic, and environmental needs change. These targets set a precedent for what is possible. In other words, technological innovation combined with economic and environmental necessity is altering the landscape of vehicle efficiency. Today's innovation is reminiscent of the effort to *get the lead out,* only this time the goal is to *get the*

---

[1] http://www.thenation.com/doc/20000320/kitman
[2] http://www.nhtsa.dot.gov/cars/rules/cafe/overview.htm

*carbon out* of transportation fuels. One policy measure that supports ambitious emission-reduction targets is the low-carbon fuel standard.

The low-carbon fuel standard is a simple and elegant concept that targets the amount of GHGs produced per unit of energy delivered to the vehicle; i.e. the vehicle's so-called 'carbon intensity'[3]. In January 2007, California Governor Arnold Schwarzenegger signed Executive Order S-1-07,[4] which called for a 10% reduction in the carbon intensity of his state's transportation fuels by 2020. Eight months later, a coalition that included one of the authors (DMK) and other researchers at the University of California and non-governmental groups responded with a technical analysis[5] of low-carbon fuels that could be used to meet that mandate. The report relies upon life-cycle analysis of different fuel types, taking into consideration the ecological footprint of all activities included in the production, transport, storage, and use of the fuel.

If a low-carbon fuel standard were established, fuel providers would track the 'global warming intensity' (GWI) of their products and express it as a standardized unit of measure – the grams of carbon dioxide equivalent per megajoule ($gCO_2e$/MJ) of fuel delivered to the vehicle. This value measures not only direct vehicle emissions but also indirect emissions, such as those induced by land-use changes related to biofuel production. The global warming intensity also provides a common frame of reference to compare propellants as diverse as petrol and electricity. Before discussing the GWI of biofuels and 'electromobility', let us contrast the low-carbon fuel standard with current policies on biofuels.

## Problematic biofuel policies

Unfortunately, the first biofuel policies were developed before the true impact of global warming was known, with the main examples coming from the USA and EU. In the USA, two current policies promote biofuels: a USD 0.51 tax credit per gallon of ethanol used as motor fuel, and a mandate that up to 7.5 billion gallons (5–6% of total US fuel demand) of 'renewable fuel' be available at US petrol stations by 2012. The EU aims that by 2020 biofuels will account for 10% of fuels used in the transport sector.[6]

Government policies to promote biofuels intend to improve environmental quality (for example, to reduce the impact of global warming) and aim to support agriculture and to reduce petroleum imports. In practice, however, current government

---

[3] Our team published a paper and an open-access life-cycle model, called 'EBAMM', which has been widely used to assess the carbon impacts of a broad range of fuels (Farrell *et al.*, 2006).
[4] http://gov.ca.gov/executive-order/5172
[5] http://www.energy.ca.gov/low_carbon_fuel_standard/UC-1000-2007-002-PT1.PDF
[6] http://www.euractiv.com/en/transport/biofuels-transport/article-152282

biofuel policies tend to function most directly as agricultural support mechanisms, involving measures such as subsidies or mandates for the consumption of biofuels. By contrast, the environmental impacts of biofuels, and more specifically the GHG emissions related to fuel production, are often not measured, let alone used to adapt financial incentives or to guide government regulation. Yield maximization for a number of agricultural staple crops often involves high levels of fossil-fuel inputs (e.g., for fertilizers), further complicating the mix of rationales for biofuel support programmes. It is important to apply a fairly broad framework on biofuel policies to avoid repeating past mistakes.

**Sustainability and economic path dependency.** The biofuel industry has been growing rapidly and can be very profitable when world oil prices are high. Government policies to further subsidize, mandate, and otherwise promote biofuels are being implemented, and more are proposed. Given the large investments in research and capital that continue to flow into the biofuels sector, it is time to carefully assess the types and magnitudes of the incentives that are meant to mitigate global warming. By engaging in this analysis, we can reward sustainable biofuel efforts, and avoid the very real possibility that the economy could be further saddled with the legacy costs of short-sighted investments.

**Global warming impact.** Biofuels are often proposed as a solution to environmental problems, especially climate change. However, biofuels can have a positive or negative global warming impact relative to petrol, depending on the precise production pathway (Farrell *et al.*, 2006), as we will discuss in the next section. To distinguish between these two cases, and the myriad of other feedstock-to-fuel pathways, as illustrated in Figure 1, clear standards, guidelines, and models are needed.

**Development of novel biofuels.** Many new fuels, feedstocks, and processing technologies are now emerging, and numerous others are under consideration (Tilman *et al.*, 2006; Gray, 2007; Stephanopoulos, 2007). These are being developed as biofuel technologies per se; they are not merely adaptations of pre-existing agricultural production methods. If these developments can be managed to achieve high productivity while minimizing negative environmental and social impacts, the next generation of biofuels could avoid the disadvantageous properties of a number of current biofuels (e.g., low energy-density, corrosiveness, and poor performance at low temperatures).

A transparent set of data on what we wish biofuels to provide, as well as clear and accessible analytic tools to assess different fuels and pathways, are critical to efforts aimed at providing appropriate incentives for the commercialization of cleaner fuels. This entire analysis, however, needs further elaboration.

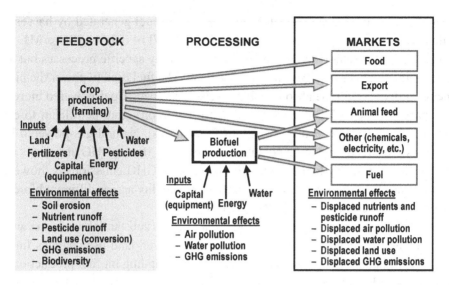

**Fig. 1.** Simplified general biofuel pathway with inputs and environmental impacts. Many effects are displaced; i.e., they occur at different locations due to market-mediated forces. (*Source:* adapted from Kammen *et al.*, 2008).

### What is the carbon impact of biofuels ... and of other new fuels?

Biofuels and related GHG emissions are a contentious issue, both in the political and research arenas. A variety of different GHG emission values have been reported, ranging from a 20% increase to a 32% decrease when switching from petrol to ethanol in the United States (Farrell *et al.*, 2006). Our group developed EBAMM (The ERG Biofuel Meta-Analysis Model; Farrell *et al.*, 2006; Kammen *et al.*, 2008) to compare and reconcile these different values. A major reason for inconsistencies was the choice of different system boundaries; i.e., the choice of which processes to include in biofuel GHG emission accounting, and which to exclude. Harmonization of boundaries – for example, excluding emissions induced by human labour but including the displacement of GHG emissions by energy-valuable co-products of ethanol – brings the GWI of the different processes closer together. Any significant remaining uncertainty is mostly due to the unknown and not-well-studied effect of lime application (lime is added to correct the pH of acidic soils; it is applied only once, and it is crucial to account for GHG emissions over the full yield period). According to the updated EBAMM,[7] ethanol produced using a carbon-dioxide-intensive refining process (e.g., a lignite-powered ethanol plant) has a marginally better GWI than petrol (i.e., 91 $gCO_2e/MJ$ instead of 94 $gCO_2e/MJ$), while average

[7] http://rael.berkeley.edu/ebamm

ethanol production has a GWI of 77 $gCO_2e/MJ$. Biofuel generated by harvesting cellulose from switch grass is projected to have a GWI of only 11 $gCO_2e/MJ$.

The EBAMM meta-analysis points out that not only specific processes but also agricultural practices largely determine the GWI. The fuel used to power the biorefineries is decisive for the absolute climate change impact. Coal-powered biorefineries barely reduce GHG emissions (but shift emissions from petroleum to coal, thus reducing energy dependency in OECD countries). Natural-gas-powered biorefineries are already having a positive net effect; i.e., fewer GHG emissions than when using petrol. The highest potential in terms of GHG emissions is, however, in cellulosic ethanol. Figure 2 summarizes the variability across different biorefinery processing scenarios (Wang *et al.*, 2007).

From this discussion, it is already clear that there is substantial need to evaluate each fuel using a detailed life-cycle analysis. However, land-use changes further complicate matters. Recent studies indicate that expanding biofuel production induces large GHG emissions from land-use change for biofuels, in particular when biofuel production competes with other land uses such as the production of food. Indirect effects are difficult to evaluate but highly significant. Commodity substitutability and competition for land transmit land-use change across global markets; for example, when US ethanol production increases the global corn price, making it profitable to clear rainforests for additional corn or crop production in Brazil. These market-mediated land-use change emissions are separated from the biofuel production process by several economic links, as well as by physical distance.

A critically important new study finds that such indirect land-use changes induce GWI above petrol emissions on a century time-span (Searchinger *et al.*, 2008). If grassland is converted to crops, both land conversion (e.g., by fire) and land cultivation cause significant emissions. For example, if one acre of land is devoted to bioethanol production, which involves the conversion of 0.6 acres of forest and 0.24 acres of grassland to agricultural land, then 30 metric tonnes of carbon dioxide are released. One acre produces approximately 400 gallons of ethanol per year, saving one tonne of carbon dioxide annually. Hence, the GHG payback time is 30 years (CARB, 2009). Searchinger *et al.* (2008) estimate that GHG payback time is over 150 years in some cases. In particular, expansion of US bioethanol production will cause previously uncultivated land to be utilized for crop production, both in the USA and elsewhere (primarily in Brazil, China and India). Hence, there will be significant loss of pristine grasslands and forests, as well as lost opportunities for carbon sequestration on idle arable land. It is generally recognized that there are significant GHG emissions related to indirect land-use changes. While the extent of this effect is disputed, as 1) model assumptions cannot easily be verified, and 2) the system is highly complex; deforestation, for example, is multi-causal (there are also local drivers of deforestation). The following factors produce major uncertainty:

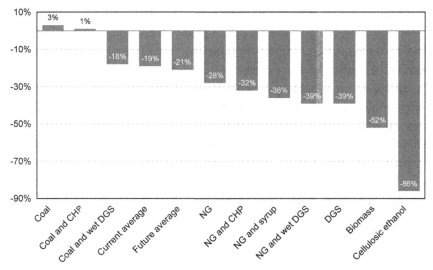

CHP = combined heat and power; DGS = distiller grains and solubles; NG = natural gas

**Fig. 2.** Comparison of the GWI of different biofuel refineries compared to petrol. Note that land-use effects are not part of this analysis. Taking resource supply (cellulosic biomass versus corn) into account, cellulosic ethanol appears as the ultimate ethanol option, reducing GHG emissions by 86% in comparison to petrol – if it can be produced for a competitive market. (*Source:* adapted from Wang *et al.,* 2007)[8]

- Carbon emission factors related to agro-ecological zones and new land (i.e., the precise location of biofuel production, and the carbon content of the land prior to conversion to biofuel plants);
- Future land-use trends, such as the total global demand on food production, which itself depends on population growth;
- Policies and competition for different land-use types (e.g., the existence and effectiveness of rainforest protection measures).

Another issue is the accounting of time. To obtain a GWI, most studies averaged the total indirect emissions over the total fuel produced during a production period and add these to the direct emissions. This straight-line amortization has been proposed for the Californian LCSF (Arons *et al.,* 2007; CARB, 2009). Hence, a unit of GHG emissions released today is treated as though it had the same consequences as one released decades in the future. Annual GHG flows are, in general, a poor proxy for economic costs; most climate change costs are imposed by GHG stocks in the atmosphere. Furthermore, consideration of long timeframes involves assumptions

---

[8] These assessments from 2007 do not fully capture the concerns raised by Searchinger *et al.* (2009) about the generally far smaller than previously thought benefits of many biofuel-to-transportation fuel (liquid or via electricity) pathways. Further analysis is needed to chart the actual benefits of these technology/fuel systems.

about technological innovation and land-use changes over that timeframe, including post-cultivation changes in land use. A proper accounting of time, recognizing the physics of atmospheric carbon dioxide decay, significantly worsens the GWI of any biofuel that causes land-use change in comparison to fossil fuels (O'Hare *et al.,* 2009). The key point is that a lot of emissions appear due to land-use changes at the beginning of biofuel cultivation, while emission savings occur later. Emissions occur front up, and as a result, cumulative warming – global warming produced by emissions within a fixed analytical horizon (e.g. 50 years) – and associated damages in the near-term are more severe than future ones.

Biofuel production has also been criticized for competing with global food supply (Runge and Senauer, 2007), and for raising global corn prices as a consequence. For the world's poor a marginal price increase can have devastating effects. The corn required to fill the fuel tank of a SUV with bioethanol contains enough calories to feed one person for a year; the SUV driver will often pay more for the corn (indirectly as fuel) than people in poor countries can afford. From a narrow market perspective, the starvation of the poor can in fact be an efficient market outcome, making bioethanol policies in the USA and EU even more questionable. To understand the relevance of policies in specific world regions, we should note that, for example, 40 % of global corn (maize) production is in the USA.

One way out of this problem is to decouple biofuel cultivation, first from food production by using waste products (second generation) and, in the long run, from land-use; for example, by relying on biofuels produced from algae (third generation). Currently, these technologies are not cost-effective, but significant research and money is being invested.

Overall, major uncertainties about the sustainability of current biofuel production persist. Indirect land-use change effects are too diffuse and subject to too many arbitrary assumptions to be useful for rule-making. To ascertain a minimum environmental quality of biofuels, a suggested low-carbon fuel standard can include evolving minimum criteria related to GHG emissions, for example as identified by Börjesson (2009). One could start by placing restrictions on biorefineries, requiring improved agricultural practices, such as conservation tillage, and in a few year's time allow only biodiesel and biofuels of the second generation. The Roundtable on Sustainable Biofuels[9] develops criteria according to which a third party could perform a life-cycle assessment of biofuels and certify the fuels according to established standards.[10]

---

[9] The Roundtable on Sustainable Biofuels is an international initiative that brings together farmers, companies, governments, non-governmental organizations, and scientists who are interested in the sustainability of biofuel production and distribution.
[10] http://cgse.epfl.ch/page65660.html

## Electromobility

Biofuels represent a minor modification in vehicle propulsion. Electromobility is a more radical and rapidly evolving technological change that dates back to the nineteenth century. Electromobility not only requires a different propellant but also different vehicle technology (an electric motor) and storage system (for example, a battery). There are two main advantages of electromobility:

1. An electric motor has 70–80% well-to-wheel efficiency[11] and, hence, is far superior to the combustion engine (with 15–25% well-to-wheel efficiency).
2. In principle, it is a straightforward process to get the carbon out of electromobility by increasing the deployment of renewable energies for electricity generation.

A significant challenge for large-scale electromobility is battery technology. Current batteries need to be improved in terms of storage capacity but also in terms of cost. All-electric cars must be relatively light in order to reduce overall energy demand. Altogether, the electricity used by a battery-powered electric vehicle in California has a GWI value of 27 $gCO_2e/MJ$ (Lemoine *et al.*, 2008; Kammen *et al.*, 2009), a considerable improvement on petrol and ethanol. Other comparable technologies, based on the current electricity mix and different storage media – such as compressed air or hydrogen – have at present a worse GWI than petrol (Creutzig *et al.*, 2009).

The evaluation of the GWI of electric cars is not a trivial issue. Rather than the GHG emissions of the average power plant, it is the marginal power plant (added when there is additional electricity demand) that must be evaluated in terms of climate change impact. Potentially, car batteries can be used for demand management (for example, cars can be charged by wind energy at night, when there is no other electricity demand; see also the chapter by Joachim Luther on smart loads, this volume). Electromobility is not merely synonymous with electric cars, but also includes smaller vehicles such as electric bikes. For OECD countries, electric bikes are still relatively exotic. However, in China – by 2009 the world's largest market for cars – more electric bikes than conventional cars are sold.

It is important to consider the full spectrum that lies between conventional petrol-operated cars and all-electric cars. For example, average fuel savings in the USA can easily be doubled (and fleet emissions halved) by deployment of existing technological advances, weight reductions and a reasonable market penetration of hybrid vehicles (American Physical Society, 2008). In contrast, plug-in (hybrid) electric vehicles (relying on battery for short distances and petrol for longer distances) are

---

[11] Well-to-wheel efficiency is the percentage of the primary energy that is used for powering the car.

expected to contribute little to total emission savings until 2030. In the case of urban transportation, even more can be gained. If inner-city transport switches from cars to non-motorized transport and electromobility, urban transportation can be effectively decarbonized.

---

### Beyond fuels

Car transportation emissions can be factorized into vehicle distance travelled, fuel efficiency, and carbon content. In this chapter, we mostly discuss the carbon content of fuels. There is, however, a need to reduce transportation emissions drastically, and both other factors will have to contribute. Fuel efficiency can be increased through better technologies and by reducing the weight of vehicles. There is huge potential to decrease average vehicle weight, particularly in the USA (Schipper, 2007). Vehicle distance travelled can be reduced by appropriate land-use policies (e.g., transit-oriented development), and by demand management (e.g., by parking management and city tolls). Pricing mechanisms, such as city tolls, are efficient ways of addressing all social costs of motorized transportation (both those internal to the transportation system such as congestion, and environmental costs such as air pollution and GHG emissions), and are most effective in joint extension of public transit (Creutzig and He, 2009). The greatest GHG mitigation potential lies in policies that address vehicle distance travelled.

---

### Outlook on international carbon fuel measures

Equipped with detailed measurements that relate directly to the objectives of a low-carbon fuel standard, policymakers can set standards for a state or nation, and then strengthen them over time. The standard applies to the mix of fuels sold in the region, so aggressively pursuing cleaner fuels permits a certain percentage of more traditional, dirtier fuels to remain, a flexibility that can facilitate the introduction and enforcement of a new standard.

California introduced a low-carbon fuel provision (specifying the low-carbon fuel standard from 2007) in April 2009, mandating emission reduction of 10% from the entire fuel mix by 2020 (CARB, 2009). The regulation also requests lifecycle emissions scores for biofuels that include indirect pollution from the conversion of forests to farm land for cultivation of corn and other fuel-feedstock crops. The US Environmental Protection Agency (EPA) proposed a revised Renewable Fuel Standard in May 2009,[12] mandating total renewable fuel volume requirements and GHG

---

[12] http://www.epa.gov/otaq/renewablefuels/#regulations

emission reduction targets for different biofuel categories ranging from 20% to 60%. An evaluation of full lifecycle emissions was also proposed. The American Clean Energy and Security Act of 2009 (ACES, also known as the Waxman-Markey Act), which was approved by the US House of Representatives but is still up for debate by the Senate, includes a mandate for the EPA to *exclude* any estimation of international indirect land-use changes due to biofuels for a five-year period.

The EU acknowledges criticism of its biofuel targets. It has confirmed its 10% 'green fuel' target by 2020, but this includes not only biofuels but all renewable energy used in transport, such as electric vehicles powered by renewable sources. Furthermore, it has clarified that biofuels must offer at least 35% GHG emission savings, a value that will be incrementally increased to 60% by 2017. Indirect land-use emissions, however, are not included in the formula to calculate overall GHG performance.

The appeal of a low-carbon fuel standard is that it establishes performance levels and opens the transportation fuels market to new competitors, not allowing the government to lock in on preferred programmes (such as biofuel subsidies) or technologies. Liquid fuel providers who produce and sell diesel fuel, petrol, or biofuels – as well as electricity providers who 'fuel' plug-in hybrid vehicles with electricity generated by renewables – can all now compete equally for transportation spending. Competition and market forces are tremendously useful in encouraging innovation that brings down costs.

All of this momentum is pushing a steady evolution to cleaner fuels, but there is no reason to stop at eliminating GHG emissions. As described above, there are other ramifications of fuel usage that we can measure and need to improve. The impacts of biofuel production, for example, range from excessive water use to erosion of formerly fallow land, to competition with food production. A natural next step is to evolve from a low-carbon fuel standard to a *sustainable* fuel standard.

Finally, a lurking issue is how fuel standards will more generally interact with the prices for carbon emissions that are likely to be established in a number of regions. Europe has already enacted a carbon trading scheme. California and the New England/mid-Atlantic region of the USA have begun to work out regional frameworks, likely based around a 'cap and trade' system, and several other regional markets may evolve in the USA. The Waxman-Markey Act aims to introduce US-wide cap-and-trade. If these carbon pricing projects are successful, the use of sector-specific regulations will likely need to evolve, both to address areas where the carbon price is too low to induce real change, and to focus on ecological and cultural sustainability issues, as the idea of a 'sustainable fuel standard' implies.

# References

American Physical Society (2008). *Energy Future: Think Efficiency Report.* Available at http://www.aps.org/energyefficiencyreport/report/index.cfm.

Arons, S. R., Brandt, A. R., Delucchi, M. *et al.* (2007). *A Low-Carbon Fuel Standard for California Part 1: Technical Analysis.* Available at energy.ca.gov/low_carbon_fuel_standard/UC_LCFS_study_Part_1-FINAL.pdf.

Börjesson, P. (2009). Good or bad bioethanol from a greenhouse gas perspective: what determines this? *Applied Energy,* **86**(5), 589–94.

CARB – California Air Resources Board (2009). *Low Carbon Fuel Standard Program.* Available at http://www.arb.ca.gov/fuels/lcfs/lcfs.htm.

Creutzig, F. and He, D. (2009). Climate change mitigation and co-benefits of feasible transport demand policies in Beijing. *Transportation Research D,* **14**, 120–31.

Creutzig, F., Papson, A., Schipper, L. and Kammen, D. M. (2009). Economic and environmental evaluation of compressed-air cars. *Environmental Research Letter,* submitted.

Farrell, A. E., Plevin, R. J., Turner, B. T. *et al.* (2006). Ethanol can contribute to energy and environmental goals. *Science,* **311**, 506–8.

Gray, K. A. (2007). Cellulosic ethanol: state of the technology. *International Sugar Journal,* **109**(1299), 145–51.

International Energy Agency (2008). *World Energy Outlook.* Paris.

Kammen, D., Farrell, A. E., Plevin, R. J. *et al.* (2008). Energy and greenhouse impacts of biofuels: a framework for analysis. In OECD/ITF, *Biofuels: Linking Support to Performance.* Paris, 41–74.

Kammen, D. M., Arons, S. R., Lemoine, D. and Hummel, H. (2009). Cost-effectiveness of greenhouse gas emission reductions from plug-in hybrid electric vehicles. *Plug-in Electric Vehicles: What Role for Washington?,* chapter 9, 170–91, Washington, D. C.

Lemoine, D. M., Kammen, D. M. and Farrell, A. E. (2008). An innovation and policy agenda for commercially competitive plug-in hybrid electric vehicles. *Environmental Research Letter,* **3**, 014003.

O'Hare, M., Plevin, R. J., Martin, J. I. *et al.* (2009). Proper accounting for time increases crop-based biofuels' greenhouse gas deficit versus petroleum. *Environmental Research Letter,* **4**, 024001.

Runge, C. F. and Senauer, B. (2007). How biofuels could starve the poor. *Foreign affairs.* Available at http://www.foreignaffairs.com/articles/62609/c-ford-runge-and-benjamin-senauer/how-biofuels-could-starve-the-poor.

Schipper, L. (2008). *Automobile Fuel, Economy and CO_2 Emissions in Industrialized Countries: Troubling Trends through 2005/6.* Washington, D. C.

Searchinger, T., Hamburg, S., Melillo, J. *et al.* (2009). Fixing a critical climate accounting error. *Science,* in press.

Searchinger, T., Heimlich, R., Houghton, R. A. *et al.* (2008). Use of U.S. croplands for biofuels increases greenhouse gases through emissions from land use change. *Science,* **319**, 1238–40.

Stephanopoulos, G. (2007). Challenges in engineering microbes for biofuels production. *Science,* **315**, 801–4.

Thomas, V. M. (1995). The elimination of lead in gasoline. *Annual Review of Energy and the Environment,* **20**, 301–24.

Tilman, D. A., Hill, J. and Lehman, C. (2006). Carbon-negative biofuels from low-input high-diversity grassland biomass. *Science,* **314**, 1598–1600.

Wang, M., Wu, M. and Huo, H. (2007). Life-cycle energy and greenhouse gas emission impacts of different corn ethanol plant types. *Environmental Research Letter,* **2**, 024001.

# Chapter 27

# Opportunities for technological transformations: from climate change to climate management?

———

## Maria Magdalena Titirici, Dieter Murach, and Markus Antonietti

Maria Magdalena Titirici was born in Bucharest, Romania, where she began her studies of chemistry. After completing a doctoral programme (under Börje Seller-gren) in the field of molecularly imprinted polymers at the Universities of Mainz and Dortmund, she moved to the Max Planck Institute for Colloids and Chemistry in 2005, where she now leads a research group on 'green carbon materials'.

*Note:* Photos and biographies of co-authors can be found in the appendix.

## The concept of carbon-negative products
## and a carbon-negative industry

We are still living, mentally and politically, in the 'oil age'. Overall oil production, which secures mankind's core requirements for energy and raw materials, sums up to about four billion tonnes of crude oil per year, equivalent to a cube with sides measuring four kilometres in length (official statistics of the US government, see IPM). Assuming a price of USD 100 per barrel, this translates into an economic value of USD 2.5 trillion. Crude oil, however, is running short already, and this will lead to further distribution conflicts, wars to control access to energy, economic depression, and poverty in the Third World. A reliable supply of oil is also a matter of existence for the chemical industry. Plastics, pharmaceuticals, and most objects we use in our daily lives would simply vanish without oil. The third, presumably most urgent issue associated with the oil economy is climate change and the protection of the atmosphere. As essentially all oil ends up sooner or later as $CO_2$ in the Earth system, an additional consequence of the oil economy is the generation of an excess 12.5 billion tonnes of $CO_2$ per year, with known and undisputed implications for the world's climate.

This is a typical 'dinosaur trap': the individual facts are not questioned, but governments and industrial leaders propose only marginal changes to handle the inevitable. Reducing the discussion to a debate on ways to secure cheap and available energy or to open extra energy resources is too simple by far. The problem to be solved is the simultaneous optimization of the complex interactions between the production of energy, the consumption of raw materials, and the destabilization of our atmosphere. This obviously has to occur not on a national basis but on the world scale.

One of the typical 'marginal' solutions suggested by politics is to replace minor parts of the energy and raw material stream by biomass energy products. This includes, besides direct combustion, fermentation of carbohydrates to produce ethanol fuels, the cultivation of oil seeds ('biodiesel'), or the generation of biogas via anaerobic digestion (Powlson *et al.*, 2005). The so-called first generation biofuel technologies are not unquestioned today: there are clear indications that, considering the whole supply chain, such measures may even harm more than they contribute to a solution (see Creutzig and Kammen, this volume). A detailed summary of analyses of the energy efficiencies, costs, and biological impact of such procedures was published by Gustavsson *et al.* as early as 1995 and was essentially confirmed in a new report published on behalf of the Association of the German Industry (McKinsey & Company, 2007). In the present context it is important to state that all types of biological fuel production schemes can at best only lower the further increase of $CO_2$, but cannot compensate for the already emitted $CO_2$ from fossil resources. This means that current biofuels do not help to solve the 'problem-triangle' of energy, resources and climate.

What would a really useful solution look like? It is obvious that evolutionary changes of current technology will not help us move out of this trap, but that technological transformations or technology leaps are urgently required. Systematic use of the sun for harvesting energy is certainly a transformation that could help to satisfy the energy demands of the world. However, this is not the focus of this essay. Instead, we will focus on describing how to achieve a carbon-negative energy system.

When considering climate change and the role of $CO_2$, it would be highly desirable not only to slow down further $CO_2$ emissions, but to reduce the total amount of $CO_2$ in the atmosphere. The idea is not only to provide a 'zero emission' energy system, but potentially to generate a new chemical '$CO_2$ disposal' or $CO_2$-negative industry, i.e., an industry that allows $CO_2$ to be taken out of the atmosphere and deposited securely through chemical transformation into stable substances. This thought, as simple as it is, is only rarely brought up in discussions on global sustainability (Read, 2006). It means that the search for new and efficient carbon deposits has to be reiterated also from a chemistry point of view. Optimally, material benefits for society would emerge from the disposal of carbon by creating consumer products. This type of technological transformation is discussed in the present essay.

The most important carbon converter, which binds $CO_2$ from the atmosphere, is certainly biomass. A rough estimate of terrestrial biomass growth amounts to 118 billion tonnes per year, when calculated as dry matter (Lieth *et al.,* 1975, pp. 205–6; Bobleter, 1994). As biomass contains about 0.4 mass equivalents of carbon, removal of 8.5% of the freshly produced biomass from the active geosystem would compensate for all $CO_2$ emissions from oil. Biomass, however, is just a short-term, temporary carbon sink, as microbial decomposition releases exactly the amount of $CO_2$ formerly bound in plant materials. To make biomass 'effective' as a carbon sink, the carbon in the biomass has to be fixed by 'low-tech' operations. Coal formation is obviously one of the natural conversion schemes that were active in the past on the largest scale. The sort of measure needed to protect the atmosphere is of a similar dimension: in principle, mankind has to re-create and speed up the transformation of plant material to coal, in other words, to create a new industry which converts about 10% of the world's biomass into useful carbon products and deposits.

The task to convert biomass into long-term carbon deposits seems challenging but is in our opinion in fact manageable. About 14 billion tonnes of biomass per year are produced in agricultural cycles, of which 12 billion tonnes per year are essentially thrown away as by-products. Examples of such product-by-product pairs are grains and straw, orange juice and peel, or oil seed and the rest of the plant. Even in an industrial country like Germany, the treatment of highly defined waste biomass such as from sugar-beads (4.3 million tonnes sugar per year), rape-seed production (3.5 million tonnes oil per year), or clarification sludge (3.0 million tonnes per year) could potentially lower German $CO_2$ emissions by about 10%. Most impressive are the big contributors: for every 100 million tonnes of Brazilian sugar produced per year, about 1 billion tonnes of bagasse (fibre left over after sugar extraction) are thrown away and burned. Considering that only one product of one country could significantly contribute to reductions in $CO_2$ emissions, the use of such waste products seems promising. It is important to stress that not the main but the by-products of agro-industry and foodcrop cultivation are used. This means that there is no competition between food and energy production, yet rather a synergy between the two consumption pathways.

Besides laying the 'raw material base', the 'technology base' also has to be created. Work on 'carbonization' is still a rare, but luckily growing, research topic. Geological coalification, i.e., the transformation of plant material to coal, is not the 'hot charring', as practiced by a charcoal burner, but rather a more effective 'cold' coalification, which occurs on the timescale of some hundred (peat) to hundred million years (black coal). Due to its slowness, it is usually not considered in renewable energy exploitation schemes or as an active sink in the global carbon cycle.

Different technical solutions have been tested to imitate coal formation from carbohydrates employing faster chemical processes. Classical 'hot charring', as practiced by a charcoal burner, is technologically restricted to a high-value starting product such as dry lignocellulosic materials (essentially wood). All other plant waste, especially leaves, fine fragments, and all wet plant and bacterial waste are not directly suitable for classical charring. Nowadays, a great variety of pyrolysis[1] technologies, including hydrous pyrolysis[2], are available which can transform biomass feedstock into biochar[3], gases, and/or liquids. There are also more modern biomass technologies such as biomass-to-liquid (BtL) to transform biomass into biofuels. These, however, require high input in equipment, process management or feedstock treatment, and they may even release significant amounts of greenhouse gases.

---

[1] Pyrolysis refers to the chemical decomposition of material through extreme heat.
[2] Hydrous pyrolysis refers to pyrolysis in the presence of water. Water reduces the required energy to break down components during pyrolysis.
[3] Biochar is a charcoal produced from any kind of biomass. For examples of biochar production technologies see http://www.pronatura.org/projects/green_charcoal.pdf, http://www.eprida.com, http://www.enertech.com/techno logy.

## Hydrothermal carbonization

Application of 'geological' conditions, i.e., weakly acidic pH values and exclusion of oxygen in closed deposits at high pressures and moderately high temperatures in water, leads to so-called hydrothermal carbonization (HTC) (see Fig. 1). HTC is an especially promising process as regards conditions, costs, efficiency and even ecology. Modern versions release practically no greenhouse gases and allow close to 100% binding of the carbon from the biomass in the final product. First experiments were carried out by Bergius, who described the hydrothermal transformation of cellulose into coal-like materials as early as 1913 (Bergius *et al.*, 1913). More systematic investigations were performed by Berl and Schmidt, who alternated the source of biomass and treated the different samples in the presence of water at temperatures between 150 °C and 350 °C. Their series of papers published in 1932 summarized contemporary knowledge about the emergence of biocoal synthesis (Berl *et al.*, 1932 a; Berl *et al.*, 1932 b). Later, Schuhmacher, Huntjens and van Krevelen (1960) analysed the influence of acidity on the outcome of the HTC reaction and found large differences in the decomposition schemes, as identified by the carbon to hydrogen to oxygen ratios of the final product.

$$\text{HTC:} \quad \underset{\text{carbohydrate}}{C_6H_{12}O_6} \quad \rightarrow \quad \underset{\text{'biocoal'}}{C_6H_4O_2} \quad + \quad \underset{\text{water}}{4\,H_2O}$$

$$\text{Charring:} \quad \underset{\text{carbohydrate}}{3\,C_6H_{12}O_6} + \underset{\text{oxygen}}{16\,O_2} \rightarrow \underset{\text{'biochar'}}{C_6H_2O} + \underset{\text{combustion gases}}{14\,CO_2 + 4\,CO} + \underset{\text{water}}{17\,H_2O}$$

**Fig. 1.** Chemical principle of hydrothermal carbonization (HTC) as opposed to classical charring. HTC: under temperature and catalysis, carbohydrates (here glucose) are converted into biocoal and water only. Charring: carbohydrates are partly burned in presence of oxygen ('pyrolysis'), leaving a char residue and combustion gases. The sum formula of biocoal and biochar are simplifications and depend largely on the reaction conditions. The carbon efficiency (i.e., the proportion of carbon that is converted into the end product) of HTC is close to 1, while in biochar formation carbon efficiency is only about 0.20–0.35 due to the presence of oxygen.

A renaissance of such experiments started recently with reports on the low temperature ($\leq 200\,°C$) hydrothermal synthesis[4] of carbon spheres using sugar or glucose as a starting product (Wang *et al.*, 2001; Sun *et al.*, 2004). Recently, it was found that the presence of metal ions can accelerate this type of reaction. This catalysation shortens the reaction time to some hours and directs the synthesis towards

---

[4] Hydrothermal synthesis refers to the synthesis of material from liquid solutions.

various morphologies and carbon structures (Qian *et al.*, 2006; Yu *et al.*, 2004; Cui *et al.*, 2006). It was also investigated whether the presence of ternary components in complex biomass (such as orange peel or oak leaves) alters the properties of the synthesized carbon structures (Titirici *et al.*, 2007; Titirici *et al.*, 2007 b). Unexpectedly, it was found that the presence of these components improved the properties of the end products for certain applications: benefits such as a smaller structural size of carbon dispersions and porous networks, higher hydrophilicity of the surfaces and higher capillarity emerged. These properties are especially important if biocoal is used in soil applications to increase water and nutrient storage capacity.

This acceleration of HTC for coalification makes the process a considerable, technically attractive alternative to other currently discussed carbon sequestration techniques (such as biomass burning combined with carbon capture and storage), applicable at the required scale of billion tonnes of carbon sequestered per year.

To summarize the outcome of the scientific optimization trials, catalysed HTC just requires heating of a biomass dispersion under weakly acidic conditions in a closed reaction vessel for two to twenty-four hours at a temperature of around 200 °C. This is indeed an extremely simple, cheap and easily scalable process.

HTC also has a number of other practical advantages. Once activated, HTC is a spontaneous, exothermic process. It liberates 10 to 30% of the chemical energy stored in the carbohydrates throughout dehydration (depending on conditions; this is due to the high thermodynamic stability of water). The exothermic character was already described in the first work on HTC by Bergius who warned of the violent character of the reaction! HTC also inherently requires wet starting products or wet biomass as effective dehydration only occurs in the presence of water. Since coal binds water only marginally, the final carbon can easily be filtered off the reaction solution. This way, drying schemes or more demanding isolation procedures can conceptually be avoided (even when using very wet starting products such as freshly harvested algae). Under acidic conditions and below 200 °C, most of the original carbon is recovered as solid biocoal. Carbon structures produced with HTC, either for deposit or material use, are therefore highly $CO_2$-efficient. Large-scale technical solutions for HTC have been developed but are not yet available on the market.

### The vision of decentralized $CO_2$-sequestration plants and potential $CO_2$-negative products

The simple, cheap, and scalable process of HTC in principle allows the layout of machines operating in a communal or decentralized context, and even mobile, container-type machines can be considered. For rough numbers supporting this vision, it is to be remembered that HTC is inherently exothermic and therefore an energetically 'free' process, but requires that the biomass is heated to 200 °C at the beginning. The latter can in principle be combined with the cooling of the coal and water mixture at the end. Since this type of heat management can only be efficiently implemented for a machine of a certain size, energy management plus machine investment costs define the optimal level of decentralization. In our opinion, a 'low-tech' realization will have the size of a relatively large container, which could convert 2000–10 000 tonnes of biomass per year to coal. Around 2000 tonnes of biomass are typically produced on a land area of 200 hectares (or 2 km²), which means that bioenergy generation and carbon sequestration including transport pathways could easily be a decentral or rural measure. HTC can therefore be considered as a communal, agricultural or forestry task rather than a typical industrial operation, with many machines working in parallel. To compensate the amount of $CO_2$ produced globally by burning fossil fuels each year, about two million HTC machines would be needed (much less than the number of waste water treatment plants in the world or the number of new cars sold in Germany every year).

But what can be done with all this biocoal? Biocoal generated by HTC is a product with a spectrum of possible uses. Biocoal is, for instance, a high quality energy carrier, which is easy to store and rather safe to handle and transport at the same time. Its calorific value is typically between 24 and 32 MJ/kg, which is much higher than that of low quality coal. In contrast to fresh biomass, storage is not complicated by the risk of mould, ignition or decomposition. It is also an advantage that biocoal is artificially produced: the HTC process can be directed to produce coal fuels with special properties, for instance, a very low ash content, a sulphur-free character, or a very fine particulate morphology. Thus, it can be burned for local energy or heat demand or used for industrial operations such as steel manufacturing, where high quality coal is needed and marketed. Such operations are clearly meaningful for less developed countries as they can replace expensive energy imports and can create a distinct base of wealth through trading biocoal at local levels. For the chemical industry, HTC coal (as all coal) can be transformed via gasification and the Fischer-Tropsch process into oil intermediates, thus keeping the chemical industry running like it does today. The Fischer-Tropsch process, however, is rather inefficient: only about 50 % of the primary chemical energy ends up in liquid fuel. Nevertheless, this can be economically meaningful, assuming an oil price close to USD 100 per barrel. However, this application of biocoal 'only' satisfies the need of the chemical industry for raw materials and the demand of the transportation sector for liquid fuels. All these operations are $CO_2$-neutral and can replace fossil resources, but they are definitely not $CO_2$-negative.

For the desired $CO_2$-negative products, biocoal has to be applied in long-lasting, large-scale material applications. Employing it as a construction additive to improve concrete building materials or pavements (where currently waste products of the oil industry are used) is certainly one option. Even more promising is its use as 'sorption coal' for the purification of drinking water and the improvement of soil.

'Carbonaceous soil' is presumably the largest active carbon sink of the Earth system. The highest carbon concentrations in the soils are generally found in the northern, colder latitudes rather than the tropics.[5] The only exceptions are the Amazonian dark earths, called 'terra preta', which have up to 70 times higher soil carbon contents than the surrounding soils (Glaser, 2007). Interestingly, the organic matter of these soils does not originate from natural biomass litter but from large amounts of charred materials, the residues from biomass burned many hundreds of years ago by pre-Columbian Indians (Sombroek *et al.*, 2004). The 'terra preta' soils are highly fertile: they exhibit high nutrient storage, retention capacity and base saturation (Titirici *et al.*, 2007 a; Titirici *et al.*, 2007 b) due to the physical sorption and textural properties of the charcoal. These carbon fractions have remained in the soil because they are not easily decomposed (Lehmann *et al.*, 2003; Glaser *et al.*, 2002).

Soil researchers have already proposed the 'terra preta' concept, which involves using artificial biocoal to enrich soil, creating a potential carbon sink of global dimensions and improving soil quality and plant growth at the same time. Biocoal production is more effective at sequestering carbon than the natural carbon fixation by afforestation, which is accepted as a carbon offset measure under the Kyoto Protocol (see Liverman, this volume). In contrast to fixing carbon in soil biomass, fixing it in the form of coal is a lasting solution: lignite or black coal (contrary to peat) is hardly biodegradable. The question of potential destabilization of coalified carbon is currently being assessed in more detail (Cheng *et al.*, 2006).

The combination of biocoal production for energy and 'terra preta' use may therefore be seen as a perspective for mitigation of climate change and restoration of abandoned land. Instead of clearing the rainforest for questionable palm-oil production (Pearce, 2005, p. 19), a 'carbon-reinforced rainforest' would produce even more energy, stored in wood or coal, while being $CO_2$-negative and supporting biodiversity at the same time. A non-linear benefit results from a 'biological amplification' of the original chemical efforts. It is estimated that 10 tonnes of biocoal per hectare are sufficient to remarkably improve depleted soil. Consequently, larger amounts of carbon can be bound in the growing biomass, which can then be used as a $CO_2$-neutral energy source. The scientific development of methods to adjust biocoal properties might accelerate and improve this process and thereby secure the productivity of farmland even under altered climatic conditions. The demand for such carbonaceous soil additives easily sums up to billions of tonnes per year and also represents a high economic value.

---

[5] http://biocharfund.com/index.php

## Economic and socio-economic impacts

Is this solution economically feasible? The question is especially pertinent in the case of applying HTC biocoal as a soil additive, given that the generated carbon is essentially just 'thrown away'. We have calculated that spending just 10 % of our current expenses on oil might be sufficient to compensate the global annual emissions of fossil $CO_2$ by biocoal production. This calculation assumes carbon fixation costs of USD 75 per tonne, a target that in our opinion can be met. (HTC is essentially just heating an aqueous dispersion, a process that generates part of the energy itself). These cost estimates do not take into account the added value for the geosystem or agriculture. Lehmann (2007) concluded that biochar sequestration by classical charring technology in conjunction with bioenergy generation from pyrolysis becomes economically attractive when the value of avoided $CO_2$ emissions reaches USD 37 per tonne (equal to about USD 130 per tonne biochar). This is cheaper than the presumed costs for carbon capture and storage technology (Enkvist *et al.,* 2007). The economic attractiveness might be further improved if biocoal is sold as a soil conditioner, as it is already done with peat for ornamental gardens in home improvement stores.

The cost of using biocoal as a soil additive would have to compete with the cost of using it as fuel or as raw material for the Fischer-Tropsch process. Within subsidy schemes like the German Renewable Energies Act biocoal is classified as a renewable fuel. Therefore, biocoal from waste would probably first be used in heavily subsidized power stations. Balancing or lowering subsidies to allow for the use of biocoal in soil applications is a potential political countermeasure that would also save taxpayers' money.

As discussed above, biocoal generation can be considered a communal, agricultural or forest operation. The end-products of HTC, i. e., biocoal and fertilizer (gained as a side product from the mineral part of the plants), have to be marketed where they compete with other fuels or other fertilizers. If the market is regulated properly, the small-scale technology of biocoal production seems to be extraordinarily eligible for developing countries. The combination of high amounts of low value biomasses, large areas of poor and abandoned soils, high growth potential, and high relevance of bioenergy in the tropics particularly fit the biocoal approach. Current non-sustainable markets could therefore easily be transformed into sustainable ones, especially in tropical regions. The classical biochar concept has already been adopted by organizations like Pro Natura International[6] or the Biochar Fund, which is dedicated to fighting hunger, energy poverty, deforestation and climate change.[7]

---

[6] http://www.pronatura.org/index.php?lang=en&page=index
[7] http://biocharfund.com/index.php

HTC in combination with modern biomass production schemes (such as agro-forestry and agro-industrial cultivation of algae) may lead to significantly higher productivity on agricultural soils, restoration of abandoned areas, and an expansion of bioenergy options. 'Slash and char' instead of 'slash and burn' (Steiner, 2007) not only reduces anthropogenic $CO_2$ emissions by providing biochar as a long-term carbon sink, it also improves soil fertility and yield potential. Biocoal from HTC potentially allows farmers in many eco-regions (not only in the humid tropics) to escape from the cycle of declining productivity and soil degradation, which result from shortened fallow periods. Continuous cultivation or cultivation with only very short fallow periods may be possible (Steiner, 2007). Permanent cropping with higher yields and income instead of shifting cultivation might significantly change economics and politics of agriculture.

In this way, HTC may represent a technology leap out of the 'problem triangle' composed of accelerating climate change and the growing demand for energy and raw materials. Optimally, this new technology would allow for a transition without violating social and human-rights issues, exerting a major economic impact and strongly benefitting poor countries rich in biomass and other rural areas of this planet.

## Summary

This essay presented the concept of a '$CO_2$-negative industry' based on agricultural and forest waste, which, in principle, has the potential to counterbalance $CO_2$ emissions produced by using fossil fuels. In this way, passive utilization of the atmosphere as a sink could be replaced by 'atmospheric management' that can deliberately control the $CO_2$ level. Bioenergy and bio-raw-material production might also resolve a number of energy and resource problems, even though it will not be enough to meet all of our energy needs. For a complete solution to our energy problem we will still need to transform our fossil-fuel-based industry into a renewable energy system. In the vision presented here, waste biomass is converted in a highly decentralized fashion on the community scale, potentially by hydrothermal processes, into valuable carbon products that are safe and have long natural lifetimes. We considered the case of biocoal as a soil additive, a use which holds promise for applications worldwide and potentially to brings about 'biological amplification' through increased soil fertility. There are also a large number of other options for the use of biocoal that are worth analysing, such as the purification of drinking water by sorption coals or the improvement of building materials by carbon additives. These options could reach a scale and importance similar to that of soil applications.

However, the most important message is that such technology truly has the potential to be implemented, as it does not hurt or violate current political or economic interests. The creation of an *additional* industrial scheme that compensates the imbalance caused by currently applied processes while creating *additional* value and products is usually accepted. The reason is that it is in line with the impetus of society, and it does not ask for cutbacks or modification of behaviour. Clearly, it does not change the 'name of the game' but sustains further economic growth.

## References

Bergius, F. and Specht, H. (1913). *Die Anwendung hoher Drucke bei chemischen Vorgängen und eine Nachbildung des Entstehungsprozesses der Steinkohle.* Halle a. S.

Berl, E. and Schmidt, A. (1932 a). *Justus Liebigs Annalen der Chemie.* **493**, 97–152.

Berl, E., Schmidt, A. and Koch, H. (1932 b). Über die Entstehung der Kohlen. *Angewandte Chemie*, **45**(32), 517–19.

Bobleter, O. (1994). Hydrothermal degradation of polymers derived from plants. *Progress in Polymer Science,* **19**(5), 797–841.

Cheng, C. H., Lehmann J., Thies, J. E., Burton, S. D. and Engelhard, M. H. (2006). Oxidation of black carbon by biotic and abiotic processes. *Organic Geochemistry,* **37**(11), 1477–88.

Cui, X., Antonietti, M. and Yu, S.-H. (2006). Structural effects of iron oxide nanoparticles and iron ions on the hydrothermal carbonization of starch and rice carbohydrates. *Small,* **2**(6), 756–9.

Enkvist, P. A., Nauclér T. and Rosander, J. (2007). A cost-curve for greenhouse gas reduction. *The McKinsey Quarterly,* **1**, 35–45.

Glaser, B. (2007). Prehistorically modified soils of central Amazonia: a model for sustainable agriculture in the twenty-first century. *Philosophical Transactions of the Royal Society B,* **362**(1478), 187–96.

Glaser, B., Lehmann, J. and Zech W. (2002). Ameliorating physical and chemical properties of highly weathered soils in the tropics with charcoal – a review. *Biology and Fertility of Soils,* **35**(4), 219–30.

Gustavsson, L., Börjesson, P., Johansson, B. and Svenningsson, P. (1995). Reducing $CO_2$ emissions by substituting biomass for fossil fuels. *Energy,* **20**(11), 1097–113.

*IPM – International Petroleum Monthly.* Available at http://www.eia.doe.gov/ipm/supply.html.

Lehmann, J. (2007). Bio-energy in the black. *Frontiers in Ecology and the Environment,* **5**(7), 381–7.

Lehmann, J., da Silva, J. P., Steiner, C. *et al.* (2003). Nutrient availability and leaching in an archaeological anthrosol and a ferralsol of the central Amazon Basin: fertilizer, manure and charcoal amendments. *Plant and Soil,* **249**(2), 343–57.

Lieth, H. and Whittaker, R. H., eds. (1975). *Primary Productivity of the Biosphere.* Berlin.

McKinsey & Company (2007). *Cost and Potentials of Greenhouse Gas Abatement in Germany.* Available at www.mckinsey.com/clientservice/ccsi/pdf/Costs_And_Potentials.pdf.

Pearce, F. (2005). Forests paying the price for biofuels. *New Scientist,* **188**(2526).

Powlson, D. S., Riche, A. B. and Shield, I. (2005). Biofuels and other approaches for decreasing fossil fuel emissions from agriculture. *Annals of Applied Biology,* **146**(2), 193–201.

Qian, H.-S., Yu, S.-H., Luo, L.-B. *et al.* (2006). Synthesis of uniform te@carbon-rich
    composite nanocables with photoluminescence properties and carbonaceous
    nanofibers by the hydrothermal carbonization of glucose. *Chemistry of Materials,*
    **18**(8), 2102–8.
Read, P. (2006). Reconciling emissions trading with a technology-based response to
    potential abrupt climate change. *Mitigation and Adaptation Strategies for Global
    Change,* **11**(2), 493–511.
Schuhmacher, J. P., Huntjens, F. J. and van Krevelen, D. W. (1960). Chemical structure
    and properties of coal XXVI. Studies on artificial coalification. *Fuel,* **39**, 223–34.
Sombroek, W., Ruivo, M. L., Fearnside, P. M., Glaser, B. and Lehmann, J. (2004).
    Amazonian dark earths as carbon stores and sinks. In J. Lehmann, D. C. Kern,
    B. Glaser *et al.,* eds., *Amazonian Dark Earths: Origin, Properties, Management.*
    Dordrecht, 125–39.
Steiner, C. (2007). *Slash and Char as Alternative to Slash and Burn.* Göttingen.
Sun, X. and Li, Y. (2004). Colloidal carbon spheres and their core/shell structures
    with noble-metal nanoparticles. *Angewandte Chemie* (International Edition),
    **43**(5), 597–601.
Titirici, M. M., Thomas, A. and Antonietti, M. (2007 a). Back in the black: hydrothermal
    carbonization of plant material as an efficient chemical process to treat the $CO_2$
    problem? *New Journal of Chemistry,* **31**(6), 787–9.
Titirici, M. M., Thomas, A., Yu, S.-H., Müller, J.-O. and Antonietti, M. (2007 b).
    A direct synthesis of mesoporous carbons with bicontinuous pore morphology
    from crude plant material by hydrothermal carbonization. *Chemistry of Materials,*
    **19**(17), 4205–12.
Wang, Q., Li, H., Chen, L. and Huang, X. (2001). Monodispersed hard carbon spherules
    with uniform nanopores. *Carbon,* **39**(14), 2211–14.
Yu, S. H., Cui, X. J., Li, L. L. *et al.* (2004). From starch to metal/carbon hybrid nano-
    structures: hydrothermal metal-catalyzed carbonization. *Advanced Materials,* **16**(18),
    1636–40.

# Part V

A global contract between science and society

# Chapter 28

# Promoting science, technology and innovation for sustainability in Africa

---

## Mohamed H. A. Hassan

Mohamed H. A. Hassan, a native of Sudan, obtained his PhD in plasma physics from Oxford University in 1974. He taught at the University of Khartoum, Sudan, and was Dean of the School of Mathematical Sciences. His research areas include theoretical plasma physics, and the physics of wind erosion and sand transport. Hassan is President of the African Academy of Sciences. As Executive Director of the Academy of Sciences for the Developing World (TWAS) and the InterAcademy Panel (IAP), Hassan has promoted capacity building in science and technology in developing countries through a variety of South-South and North-South collaborative programmes. In collaboration with other partners, TWAS actively advances the scientific understanding of climate change vulnerability and adaptation options in developing countries.

This essay focuses on the challenges and opportunities for promoting science-, technology-, and innovation-based sustainability in Africa. That continent's recent history is punctuated by initiatives that began with high hopes; initiatives that were characterized by lofty declarations and detailed blueprints for action; initiatives, that received warm, enthusiastic applause at their inception, but were soon forgotten, only to be resurrected in the context of subsequent initiatives that followed a similar trajectory of hope and disappointment. The Potsdam Symposium took place during Africa's most sustained period of economic growth in decades. Between 2000 and 2003, Africa's annual gross domestic product (GDP) increased by 3.7%. Between 2004 and 2006, the continent's annual GDP growth accelerated to 5.6%. This has spurred an unprecedented sense of hope on the continent. Could things be different this time? Will Africa finally chart a course to sustained development? One that is designed and implemented by the people of Africa? One that achieves an unparalleled level of sustained economic progress?

Today there is growing consensus that progress will take place only if Africa designs and implements its own developmental agenda. Partners are encouraged to join the continent's efforts in sustainable development. But Africans are now determined to take the lead and to decide for themselves what is best for Africa. And they may finally be acquiring the resources, knowledge, and power to do just that. But we also need to remember that, for all of the good news, dark shadows of despair stubbornly persist. More than 40% of sub-Saharan Africa's population – nearly 300 million people – continue to live in extreme poverty. Africa is the only continent where not a single country will meet all of the eight Millennium Development Goals (MDGs), and where most countries will not meet a single one. Poverty, disease, and degradation continue to plague the continent. As a result, progress remains tentative and perhaps unsustainable. Tensions are high. In many countries, the spectre of lawlessness and violence is constant. In short, we must not confuse aggregate economic growth with economic and social well-being. One can clearly exist without the other. Africa's situation is a case in point.

Yet, there are reasons for hope. In addition to steady annual growth in Africa's GDP, another promising trend deserves our attention: For the first time in more than a quarter century, African leaders are embracing indigenous capacity building in science, technology and innovation (STI) as strategic elements for economic growth and social well-being. If they succeed – and if such skills can become part of Africa's entrepreneurial spirit – then it may indeed be possible for the continent to chart a permanent path to sustainable development.

## The STI landscape

Africa's limited but encouraging progress in science and technology capacity building cannot be fully appreciated without examining broader developments in science and technology capacity building in the developing world. The reality is that some developing countries have invested more in STI, while others have lagged behind. This has led to another development gap. In addition to the historic gap between developed and developing countries, there is now a South-South divide. Today, a more refined categorization of countries has emerged that better reflects their relative strengths in STI.

First, there are countries with strong STI capacity. These number about 25, largely consisting of countries that belong to the OECD (the Organization of Economic Co-operation and Development). They enjoy across-the-board strengths in all areas of science and technology, and have the capacity to convert scientific and technological knowledge into products and services that boost their economies. These countries are rich in STI, and they are financially well-off.

Second, there are countries with moderate STI capacity. These countries, which number about 90, include some of the largest countries in the developing world – China, India and Brazil. But the list includes others as well: Argentina, Chile, Malaysia, Mexico and South Africa, to name just a few. It is a diverse group with wide-ranging capabilities. The majority of these countries are competent in a select number of fields. But broad pockets of weakness remain. The scientific infrastructure (including classrooms and laboratories), although improving, still lags behind the quality of instruction and equipment found in countries with strong STI capacities. The ability of these 90 countries to bring their scientific knowledge and technical know-how to the marketplace is weak, although recent indicators suggest that this transition is becoming less problematic in a few countries. In February 2007, for example, the World Intellectual Property Organization (WIPO) reported that while the United States still leads the world in patent applications, Asia is rapidly catching up. China filed nearly 4000 patent applications in 2006, more than double the year before.

But there is also a third category of countries, and these countries have weak STI capacity. A survey conducted by the Academy of Sciences in the Developing World (TWAS) has identified 80 such countries, the majority of them in Africa. These countries have very limited capacity in any field of science and technology. They have poor teaching facilities and substandard laboratories. And they have scant ability to convert their knowledge and expertise into products and services, especially products and services that can compete in the international marketplace. These countries also lack the capacity to participate in cutting-edge scientific endeavours. Many of their most promising young scientists have migrated to other countries to pursue their careers. Moreover, in most of these countries there is minimal government support for STI. More generally, there is the absence of a culture of science.

## Africa and the MDGs

Expanding the reach of STI to countries that have been left behind is one of the most critical problems of our time. But it is by no means the only one. In our interconnected world, where the Internet and airline travel have truly transformed our planet into a global community, no country remains unaffected by the problems that beset other countries. That is the message of the United Nations MDGs. The goals set targets to address the world's most pressing problems – problems that impede sustainable well-being in the developing world, and that threaten global peace and prosperity: poverty; hunger; the spread of infectious diseases; poor education; gender inequality; and lack of access to safe drinking water, sanitation, and energy.

Experts agree that the MDGs will not be met unless special attention is paid to the well-being of Africa. More than 40% of all Africans do not have access to safe drinking water. More than 70% do not have access to electricity. Twenty-five million Africans are infected with HIV. Ninety percent of the world's malaria infections occur in Africa. And more than 30 million African children go to bed hungry every night. Africa may be poor, but it is not small. Its land mass, which is more than 20% of the Earth's land mass, covers an area larger than Australia, Brazil, Europe, and the United States combined. And Africa may be weak, but it is home to some 920 million people. That's more than three times the population of the United States and twice the population of the European Union.

Africa, in short, may be poor and weak, but it cannot be ignored. In many respects, the future of our planet lies with the future of Africa. Africa is where global attention must be focused if we are to make progress in meeting the MDGs. But that still leaves the question of what tools must be employed for our efforts to succeed. The MDGs will not be met without strong capacity to generate and utilize STI, and without vigorous and sustained international partnerships to help build this capacity. As the MDGs indicate, the vast majority of these problems are related to poverty, inadequate education, poor health, and degraded environmental conditions, all of which undermine Africa's ability to meet the basic human needs of the majority of its people.

Other global issues that affect the developed and the developing worlds in equal measures are growing in significance. Climate change is at the top of this list. But there are also issues related to energy security, access to adequate supplies of drinking water, and the over-exploitation of such natural resources as fisheries and forests. Reducing the gap between rich and poor countries, and ensuring that the most critical global issues are tackled with tools that only global STI can provide, are daunting challenges. These challenges will not be met without a critical mass of well-trained scientists in all countries.

## Brain drain and brain mobility

Today, experts estimate that more than half of the scientists who have been educated and trained in universities in sub-Saharan Africa have migrated to the United States. Experience has shown that brain drain cannot be stopped unless the most talented scientists find favourable working conditions in their homelands. Once a scientist has established roots in another country, it is difficult to lure him or her back home.

Science is a global enterprise. Excellence in science has always depended on the ability of scientists to associate freely with their colleagues around the world. Such movement not only benefits international science, but also serves to deepen international understanding – a welcome by-product in today's troubled world. Yet, as we all know, the free movement of scientists, especially to the United States, has been severely restricted since the events of September 11. The scientific community fully recognizes that security interests take precedence over scientific exchange. Nevertheless, it also realizes that scientific exchange is an important instrument in the fight against ignorance, suspicion, despair, and terrorism. The US State Department, urged by the US National Academy of Sciences and others, has recently taken steps to ease the difficult process of entry into the United States for scientists travelling from abroad. But many scientists, particularly from Africa and the Islamic region, hope that more can be done. A major challenge impeding both international scientific cooperation and scientific capacity building in many countries of Africa is this: How can governments in scientifically advanced countries be persuaded to ease visa restrictions for African scientists to ensure their full participation in global science and R&D programmes?

The Internet and other forms of electronic communication have revolutionized the way in which scientific information is distributed and, increasingly, reviewed, edited, and published (see Sulston, this volume). These trends have had an enormously positive impact on global science. Never before have scientists enjoyed access to such an extensive amount of current information. Never before have scientists been able to communicate so easily and directly with colleagues in other parts of the world. And never before has international scientific collaboration been so easy to plan, organize, and implement. But African countries, particularly the continent's least developed countries, do not have sufficient resources to build and maintain up-to-date electronic communications systems. Broadband width is still too narrow in much of Africa, and expensive on-line subscription rates still prevent many African scientists from accessing the most current literature.

### African leaders show the way

These obstacles have led Africa's leaders to make increasing commitments to both research and development and regional cooperation in science and technology. For example, at the African Union (AU) Summit, held in Addis Ababa in 2006, African leaders discussed regional strategies for the promotion of science and technology and announced that 2007 would be the year of 'African scientific innovation'.

Political leaders in Africa have on several occasions expressed support for science and technology. But their meetings were followed by meagre results and disappointment. The level of commitment – and enthusiasm – expressed at the AU Summit in Addis Ababa seemed different and likely to lead to concrete results in the following years. Leaders at the AU Summit strongly recommended that each African country should spend at least 1% of its GDP on science and technology. Such a recommendation had been made several times before. Following the AU Summit, however, it actually began to be fulfilled. Several African countries, most notably those that have also embraced democracy and good governance, have increased their investments in science and technology. These countries include Ghana, Kenya, Nigeria, Rwanda, South Africa, Tanzania, and Zambia. Yet, their number is still too small.

At the AU Summit, the president of Rwanda, Paul Kagame, announced that his country has dramatically boosted expenditures on science and technology, from less than 0.5% of GDP in previous years to 1.6% starting in 2006. He also announced that his country would increase investments in science and technology to 3% of GDP over the next five years. That would make Rwanda's investment in science and technology, percentage-wise, comparable to that of South Korea, and higher than that of most developed countries. A nation that was teetering on the verge of collapse less than a decade ago, and that still lives in the shadow of genocide, has embarked on a path to science-based sustainable development.

### Working with Africa

What makes the prospects for building science and technology capacity in Africa even more encouraging is that Africa is not alone in this effort. Over the past several years, there have been increasing commitments by governments in the developed world to support STI in low-income countries, and especially in Africa.

At the G 8 Summit in Gleneagles, Scotland, in 2005, G 8 member countries unanimously pledged to provide USD 5 billion to help rebuild Africa's universities and an additional USD 3 billion to help establish centres of scientific excellence in Africa. The decision was greeted with enthusiasm in Africa and throughout much of the world. Yet, in 2007, G 8 member countries had officially authorized only

USD 160 million of funding, targeted for the creation of networks of centres of excellence proposed by the AU's New Partnership for Africa's Development (NEPAD). Equally distressing was the fact that little of this money had actually been transferred to Africa. The 'Science with Africa' initiative must continue to urge G 8 countries to fulfil the pledges that they made in Gleneagles and that were reconfirmed in subsequent meetings. Upcoming summits will provide yet another opportunity for the world's leading economic countries to live up to their word.

The World Bank, through the Science Institutes Group (SIG), headquartered at the Institute for Advanced Study in Princeton, New Jersey, has provided loans for the creation of scientific centres of excellence – so-called Millennium Science Institutes – in Brazil, Chile, Turkey, and Uganda. The institutes offer scientists from developing countries an opportunity to conduct world-class research and to pursue cooperative projects with colleagues in a broad-range of scientific fields. Several foundations have also given substantial support to science-poor countries in Africa through programmes that emphasize scientific and technological capacity building. Many of these efforts have focused on education and training for young scientists in the world's least developed countries. Rising levels of scientific excellence in developing countries – most notably, Brazil, China, India, and South Africa – have opened up new opportunities for South-South collaboration in education and research. These include the following:

- Agreements have been signed between TWAS and the governments of Brazil, China, and India, providing more than 250 scholarships a year for graduate students and postgraduate researchers in poor developing countries to attend universities in the donor countries. TWAS pays for the airline ticket. The host countries pay for all other expenses, including accommodation. This is the largest South-South fellowship programme in the world.
- Brazil's pro-Africa programme supports scientific and technological capacity building in the Portuguese-speaking countries of Angola and Mozambique. The programme includes research collaboration activities with Brazilian institutions.
- China's Development Fund for Africa, approved in 2006, will provide USD 5 billion over a five year period to assist African countries in achieving the MDGs through cooperation with China.
- The India, Brazil and South Africa (IBSA) tripartite initiative, signed by the respective ministers of science and technology, will provide funds to engage in joint problem-solving projects that focus on developing products with commercial value.

## Agenda for action

In light of these trends, what must African countries and their partners do to promote STI? First, African countries must institute educational reforms that make
science more interesting and attractive to young people. This means devising a
more hands-on approach to scientific study in the classroom, emphasizing 'learning by doing' rather than the rote memorization that has historically characterized
the teaching of science, especially biology in Africa. The initiative *La main à la
pâte*[1], launched by the French Academy of Sciences a few years ago, has become
a much-emulated strategy for educational reform in science. The results have been
encouraging, providing a blueprint for success that others can follow.

Second, African governments must support programmes to increase scientific
literacy among both children and adults. Rapid advances in science mean that science education must be a lifelong endeavour. The media can play a vital role in this
effort. For example, the London-based electronic portal SciDev.Net, which is supported by a host of aid agencies and foundations, and which receives valuable assistance from Science and Nature magazines and TWAS, has helped raise global
awareness of science and economic initiatives in the developing world.

Third, African universities must be reformed and strengthened. Each African
country must have at least one world-class research university that sets national
standards for quality education and research, and attracts the best and brightest
students. World-class universities in Africa can play a critical role in advancing
science and technology, both in Africa and internationally.

Fourth, African countries must train a new generation of problem-solving scientists, and turn science into a demand-driven exercise in which research questions
are often determined by critical social and economic needs. The 'sustainability science' initiative, launched by a group of scientists several years ago, has proven to
be a valuable first step in drawing science closer to society. But much more needs
to be done.

Fifth, African countries must build and sustain scientific centres of excellence.
This is especially important for the poorest developing countries where a culture of
scientific excellence has yet to take hold. The G 8 pledge made in 2003 to provide
USD 3 billion over 10 years to help build scientific centres of excellence remains
an unfulfilled promise.

Sixth, African and other developing countries must learn to share their 'successful experiences' in the application of science and technology to address critical
social and economic needs. The developing world's efforts in this regard have been
largely hidden from view, but thanks to the work of such organizations as TWAS

---

[1] The closest English equivalent to this French expression is 'hands-on experience'

and the UNDP's Special Unit for South-South Cooperation, information about developing-world, science-based initiatives that have successfully addressed critical issues related to poverty, public health, and the environment, are now reaching larger audiences both in the developed and developing world.

Seventh, African countries must bolster their merit-based science academies. These academies often include a nation's most prominent scientists. Yet, they have often been relegated to the status of genteel men's clubs, and have failed to play a prominent role in national discussions related to science-based policy issues. The Network of African Science Academies (NASAC), the InterAcademy Panel on International Issues (IAP), and other institutions are actively seeking to change this mindset and to strengthen the capabilities of academies, especially when it comes to interacting with policy-makers.

Eighth, African countries must follow the example of other countries in establishing and supporting science foundations that provide merit-based, competitive grants to scientists and scientific institutions. In Africa there is only one nation – South Africa – with such a foundation in place. More countries should adopt this strategy.

Ninth, for too many years Africa has lamented the loss of scientists who were trained in their own countries but who subsequently pursued their careers in the North. As China and India have shown, this brain drain can be turned into a 'brain gain' by devising effective strategies to engage a nation's scientific diaspora for the benefit of their home countries. Scientific exchange programmes, visiting professorships, and joint research projects are examples of South-North scientific cooperation that can be advantageous for both scientifically proficient and scientifically lagging countries.

Tenth, the majority of African countries do not have sufficient resources and expertise to build and maintain up-to-date electronic communication systems. The 'Science with Africa' initiative should help African scientists gain electronic access to the most current scientific literature.

## Conclusions

What does all of this rush of activity add up to? Is it just another episode of fleeting interest in countries and people that have been left behind? Or are we entering a new era marked by sustained investment in STI in Africa? I believe that we have more reason for optimism than cynicism. Indeed, I believe that we may be witnessing a transformational moment in the promotion of STI for sustainability in Africa. But for us to seize this moment, we need to develop and implement an action agenda designed to sustain – and expand – broad-based efforts for capacity building in STI in Africa.

The tripolar world of science and technology – anchored in the United States, Europe, and Japan – is being transformed into a multipolar world of science marked by the growing capabilities of Brazil, China, India, Malaysia, South Africa, and others. The critical issue is this: As the list of developing countries that gain strength in science and technology grows in the coming years, will Africa also join the fold?

The chances for success have rarely been brighter. At the same time, the consequences of neglect and indifference have rarely been more troubling. Africa, with the help of the international community, must seize the moment. If it doesn't, the promise of Africa will again remain unfulfilled with consequences that extend far beyond the continent. This course should not only boost Africa's economy and build the continent's scientific and technological capacity. It should also be designed to help reduce poverty and improve the lives of the hundred of millions of impoverished Africans.

# Chapter 29

## Information flow: the basis for sustainable participation

### John Sulston

John Sulston was born in 1942 in Great Britain. He began his studies in organic chemistry at Cambridge University, where he also obtained a PhD in the field of molecular biology. In his research, Sulston observed the cell division and differentiation in the development of tissues of the millimetre-long worm *Caenorhabditis elegans.* He was able to show that specific cells undergo programmed cell death as an integral part of the normal differentiation process. Sulston also identified the first mutation of a gene participating in the cell death process. In 2002 he was awarded the Nobel Prize in Physiology/Medicine, together with Sydney Brenner and H. Robert Horvitz, for their discoveries in relation to 'the genetic regulation of organ development and programmed cell death'. Professor Sulston was one of the founders of the Wellcome Trust Sanger Institute, where he led a team of several hundred scientists in the United Kingdom's contribution to the Human Genome Project. Since retiring as Director of the Institute in 2000 he has worked to ensure that information on genetic data remains freely accessible.

In tackling climate change we are participating in a game of prisoner's dilemma[1]. In this game each of two prisoners is invited to testify against, and thus betray, the other. If both prisoners testify, they each receive half the maximum sentence; if only one testifies, he goes free while the other receives the full sentence; however, if both remain silent, then both receive light sentences.

By sharing and acting upon our knowledge we have the opportunity to mitigate climate change. The great danger is that each of us tends to betray the group by striving for advantages over others, and if we persist on this course we and our planet will suffer dire consequences. The good news is that the climate game is a repeated version of the dilemma, in which the 'prisoners' have the opportunity to increase trust by seeing how the other responds. It is essential that we exploit this opportunity by promoting information flow in an equitable fashion. Only in this way will the necessary level of trust be attained for everyone to give up a little, so that we can collectively survive and thrive. Such levels of trust come easily to small tight-knit groups; the challenge is to develop mechanisms to achieve trust on a global scale.

The practice of science involves two sometimes conflicting types of activity. One consists of research and discovery – ranging from hypothesis-driven, problem-oriented research to the systematic amassing of data. The other type of activity is the open dissemination of information. Science has developed mechanisms to encourage both. The result is that we can all 'stand on the shoulders of giants' – or more mundanely, we all contribute to a rich mulch of knowledge out of which the new shoots of discovery grow vigorously.

It is particularly important that fundamental knowledge is placed in the public domain, so that all may share this information and use it for different purposes. Equally importantly, this approach engenders trust. However, there are many opposing forces that work against sharing knowledge and resources, and present a grave threat to effective cooperation. Because combating climate change is inherently a joint activity, it is especially important to promote sharing of information in this area.

Let us first consider various networks that are important for information sharing.

### Examples of information networks

The entire process of scientific communication, involving informal contacts, conferences, and peer reviewed publication, is essential for science. It assures accuracy, since errors or falsification usually come to light quickly, and is the basis for attribution of credit.

---

[1] A classic example in game theory. For more information see http://en.wikipedia.org/wiki/Prisoner's_dilemma

Until recently, most major **scientific journals** were accessed by subscription, which included a healthy profit margin for the publisher. This worked reasonably well for well-funded scientific communities in the wealthy countries, but excluded less well-endowed scientists and civil society from access. With the arrival of electronic versions of these publications, and the possibility of linking them for easy literature searching, researchers began looking for ways to circumvent the barriers associated with for-profit mechanisms (see Fig. 1). Consequently, a movement towards open access publishing is under way, in which the researcher pays the costs of publication, and access is free to all. This trend is not without its problems. One is that whilst it provides less well-off researchers with access to the work of others, a special fee exemption needs to be made for them to publish their own work. This may become harder to arrange as the number of scientists in developing countries grows. The existing for-profit publishers have mounted a strong rearguard action to protect their position. In addition to independent commercial publishers, their ranks include many learned societies, who have traditionally derived a substantial part of their income from publication. In cases where information has been published in the traditional way, organisations such as SciDevnet help to provide access for scientists in developing countries.

**Public databases** are central to many fields. For example, three large databases (in USA, UK and Japan) collectively provide a repository of basic biological information. They began by storing DNA sequences, and are progressively extending their role to cover a wide range of data, including other sorts of nucleotide sequences, protein sequences and structures, higher order assemblies, and software tools. Data may be associated with peer reviewed publication, or may be entered in raw form. The databases can be accessed freely by all users, and may interact with other publicly funded sources to cover specialized applications. From time to time these databases come under threat from entrepreneurial rivals, but so far they have survived. A continuing difficulty in Europe is that the EU Research Framework Programme is so far unwilling to support large-scale infrastructure in life sciences, and so EU funding for this purpose has to come through individual research projects. The resulting instability is a constant threat. A further problem in Europe is the Database Directive, which gives excessive rights to proprietors, protecting not only the form of the database but also the actual data within it no matter how it was obtained. James Boyle of Duke University and others have shown that the Directive was a misguided step, not only endangering information flow, but also failing to benefit database proprietors. The importance of public databases is illustrated in Figure 1.

Part of the information stored in these databases came from the Human Genome Project. From the outset the Project's remit was to make its data publicly available, and this was reaffirmed in the Bermuda Agreement of 1996. Against some strong

## Proprietary database

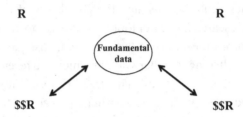

## Public database

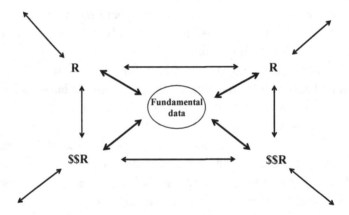

**Fig. 1.** Private and public databases. The ellipse represents the database and the arrows lines of communication. In order to preserve a viable business, the proprietor of a private database must insist that knowledge is not shared with others, otherwise it would leak out and the database would lose its financial value. This is not a satisfactory structure for holding data of fundamental importance. In addition to the obvious inequity of access for those who cannot pay, the resulting inhibition of communication is devastating to science. An important feature of the public database is that the user has access to all the data at once, to search at will and to compare with other databases. This allows the operation of novel and powerful algorithms, which would be blocked by, for example, a pay-per-view system.

opposition, the Project succeeded in its aim, and the outcome has been highly influential in keeping much biological information in the public domain. The Human Genome Project has occasionally been criticized for giving away data to profiteers, but that was unavoidable. The huge gain is that comparative analyses of data are straightforward, without barriers of any kind, as a result of which the value of information is enhanced exponentially as more is added.

**Meteorological data** is partially privatized, particularly in Europe, thus limiting information exchange. This is clearly of major significance for climate research. The same situation applies to **geographic maps.** For example, in the UK the privatisation of the Ordnance Survey (the national mapping agency) has had a very negative effect both on fair use by individuals and on the development of value-added products. These are both examples of a protectionist trend that misguidedly seizes the opportunity to collect fees while ignoring both the hidden costs of collection and the long-term consequences. All information relevant to climate change should be made openly accessible; the Global Earth Observation System of Systems (see Potsdam Memorandum, this volume) will help in this aim.

**Free software** holds its own in the commercial as well as the academic world. Its extensive use in, for example, the banking system illustrates the compatibility of open structures with profitable activity.

The **World Wide Web** needs no introduction. It is interesting to note that the Internet originated in military requirements, but was transformed into the remarkable communications system that we have today by the work of Tim Berners-Lee. Its efficacy and independence is a model as well as a vehicle for exchanges on many important contemporary issues, including global climate change.

**NGOs** play a crucial role in many areas of human endeavour. Their role in climate change is already apparent. The term civil society has come into use to describe the coherence and importance of this powerful yet loose grouping. Its significance arises from the fact that direct government power stops at national boundaries, whereas the power of transnational corporations does not. Since governments (and their collective embodiment in the UN) are very susceptible to lobbying by well-endowed vested interests, NGOs are vital in providing a democratic balance.

As the most representative multinational forum, the **United Nations** ought to be the ultimate vehicle for information sharing, and indeed it is of immense (though widely underrated) importance. It has problems of manipulation by powerful interests, and is frequently accused of excessive bureaucracy and even corruption, not unlike many governments. In the long run an important part of our global agenda should be to build on the success of the UN and to improve transparency and trust in that organisation.

The **international patent system** is of increasing importance in information flow. For many it is *the* way of sharing. However, it is a double-edged sword that can block exchanges as well as facilitating them. For example, information and materials provided in scientific publications are, or should be, available for others to use, but this freedom is being increasingly interfered with by the growing demands for intellectual property (IP) rights from those who fund science, or even from the scientists themselves. In matters of public importance, such as climate change, the acquisition of IP should be minimized.

In considering the performance of the above-mentioned types of information network, and the role of different institutions, we must remember that underlying their policies is the power of their constituents – shareholders in the case of quoted corporations, and voters in the case of democratic governments. Public institutions are not in themselves to blame for poor outcomes. Their leaders, certainly, are in a position to exercise some influence, but ultimately it comes down to the ballot box; so an enormously important task is to inform and persuade citizens everywhere of the need for strategic change. Personal changes in carbon footprint have become quite popular in developed countries, and should be supported further, through initiatives such as the Product Carbon Footprint project conceived at the Potsdam Institute. However, of themselves personal changes achieve little. They need to be complemented and framed by strong and pro-active government climate policies. Citizens play a crucial role and must be convinced of the need to vote for strategic change.

### Incentives and licensing mechanisms

In order to be realistic, economic models need to embrace the concept of dual reward rather than focussing solely on financial incentives. Depending on individual inclination and circumstances, scientists may seek one or both of the following reward mechanisms:

- Personal attribution. This is supplied by ordinary scientific publication, and, if properly set up, by attribution to databases. Personal evaluation by peers is of considerable importance as well, and can be reliable in tight-knit communities, but is capricious and susceptible to misuse on the larger scale. Therefore, scientists are entitled to expect some formal attribution to further their careers, and indeed we cannot expect efficiency unless those who are most effective are recognized as such.
- Financial reward. Scientists differ greatly in their requirements for financial reward. Most are chiefly rewarded by the chance to perform valuable scientific research, by success in their research objectives, and by the benefit to society that results from their efforts. However, some are additionally motivated by the possibility of rich returns from licensing their discoveries for profitable development.

Those who fund science may similarly be driven by a mixture of motives. They will seek a successful outcome in any case, but if the resources are derived from investment then profit is required as well. In the field of climate change, as in medicine, the awarding of prizes for successful innovation is being tested; proponents of this type of incentive point to historically successful prize systems, such as in navigation and flight.

In order to grant recognition and financial reward equitably a range of licensing arrangements has been devised. Here are some examples, ordered from most free to most restrictive:

**Free release.** This arrangement is the norm for non-commercial work. It was employed on a large scale by the Human Genome Project. It is the right model for fundamental information about the natural world.

**General public licence (GPL).** This arrangement was devised by Richard Stallman at the Free Software Foundation. By acquiring, using and modifying software under a GPL, the user agrees to make the source code available so that others can do the same. There is not – as commonly thought – any prohibition on fees, which explains why the commercial use of GPL software is increasing.

**Conditional open access.** A wide variety of licenses is being devised, with varying constraints, for example the demand for fees from high-income countries and their waiver for low-income countries. The non-profit corporation Creative Commons[2] provides analogous licenses for the mitigation of copyright.

**Exclusive rights patents and copyright.** These instruments form the backbone of our present IP system, which is essential to the global economy as currently organized. However it is widely accepted that the system is not working optimally, though powerful vested interests resist change. At times debates seem quite ideological, with any attempt at rational discussion drawing accusations of weakening the system on which our wealth depends. Ways forward are being debated at the World Intellectual Property Organisation (WIPO), as discussed in the next section.

[2] http://creativecommons.org

---

### Some problems for benefit sharing

- Excessive desire for personal attribution
- Perceived loss of incentive if IP is not retained
- Focus on short-term profit
- Exclusive patent rights as revenue source, leading to blockages
- Government perceptions and requirements, securing IP regardless
  of efficacy
- Competition for international trade, excessive IP requirements in trade
  agreements
- Inequality, leading to imbalance in negotiating power, including legal
  representation
- Unforeseen consequences of the free market
- Excessive reliance on 'corporate responsibility', which, on account of
  bottom line effect on share price, can make only a negligible contribution
- Digital rights management, which with modern electronic implementation
  is eliminating traditional 'fair use'.

---

### Elements of trust

As we have seen, our global game of prisoner's dilemma can only move forward through increasing trust between the participants. What are the important aspects of that process?

### *Benefit sharing*

Benefit sharing is a key element of trust. It is particularly important in a world where wealth is very unevenly divided, because in the absence of such proactive measures benefits will be unevenly divided as well, and will continue to drive a vicious circle of deprivation and mistrust. A recent example of such failure was Indonesia's quite understandable objection to providing its avian flu samples (the sharing of which is important to all of us) to the World Health Organisation (WHO) until safeguards were in place to ensure that Indonesia would share in vaccines that might be derived from the samples (Indonesia had observed that earlier samples had gone into the profitable activities of US and Australian corporations, whence the products were sold at prices that were unaffordable for developing countries.) Such an outcome is extremely destructive, because prediction of future epidemics, and the development of drugs and vaccines that may mitigate them, depend crucially on sharing knowledge of novel virus strains as they arise. The WHO should ensure that all countries, not just the most industrialized, benefit from sharing, thereby balancing openness with opportunity. But at the moment it is unable to do so, as it is constrained by the financial interests of its major contributors and transnational corporations.

Sharing of goods and services is carried out through the world trading system. In order for the sharing process to be equitable, we need to ensure that trade rules are equitable. A big step in the opposite direction was taken by the TRIPS (trade-related intellectual property rights) Agreement in 1995. This laid down a timetable for all nations, rich and poor, who wish to be members of the World Trade Organisation (WTO) to adopt stringent rules on IP. Since membership is important for access to markets, there is great pressure to sign up. Many developing nations adopted the rule, including, for example, India in 2005, and only the least developed nations remain outside. There is a strong sense that developed countries are pulling up the ladder: when they were in the same position as developing countries are in today they paid little attention to one another's IP rights. China has benefited greatly by ignoring the WTO throughout its recent growth period, and now joins from a position of strength.

It is important for the provision of healthcare as well as fair trade that there is flexibility within the TRIPS Agreement. Some key steps in that direction were taken in the Doha Agreement of 2001. The measures do not work very well, but in principle (through compulsory licensing) allow developing nations to avoid paying high prices for medicines.[3] Further steps have been taken by the adoption of a 'development agenda' at the WIPO, and by discussions at the WHO following the far-sighted report of the Commission on Intellectual Property Rights, Innovation and Public Health chaired by the former Swiss President Ruth Dreifuss.

The Indonesian experience over avian flu highlights the problem of 'biopiracy', in which novel genetic resources are appropriated for gain by the economically powerful. In some cases (notorious examples include the neem tree and basmati rice), centuries-old prior art[4] has been ignored by patent examiners in wealthy countries, on the grounds that it is not formally documented. Unwillingness to confront this injustice is a major obstacle to achieving a harmonized world IP system. Under the Convention on Biological Diversity, rules for benefit sharing of genetic resources are being constructed, but progress is slow because of lobbying by vested interests in the wealthy countries.

Multinational agreements are the best way to achieve benefit sharing, but only if they are honestly negotiated and fully respected. The problem is that they are not. Seeing signs of democratization of the trading rules, OECD countries (the

---

[3] One practical problem is that developed countries are liable to retaliate with trade sanctions if a developing country uses compulsory licensing, as has happened recently in the case of Thailand. Another is that for countries lacking their own manufacturing capability, the measures are prohibitively cumbersome.
[4] This term is used when previously existing knowledge bars a patent.

USA in particular with the EU as an eager supporter) have been attempting with some success to negotiate bilateral agreements with individual developing countries and to set up so-called 'free-trade areas'. The incentive in such arrangements is the granting of enhanced access to rich markets; the drawback is that usually there is an insistence on 'TRIPS+' standards of IP, in which much of the flexibility is removed. The existence of a meshwork of special agreements weakens the multilateral structure on which trade should be built. At present, we are moving towards trade that is neither free nor fair, but which rather smacks of imperialism. There are ominous echoes of the network of alliances that preceded the First World War. In order to deal with global problems, including climate change, there is a need to halt the trend towards bilateral agreements.

Unequal benefit sharing is increasing the gap between rich and poor in many ways. One very serious consequence is that opportunities for personal progress are diminished in the developing countries, with a consequent increase of legal and illegal migration. The 'brain drain' of the most talented individuals is of course particularly damaging to a country, and attempts to reverse that by investment in education, research and industry are very desirable (see Hassan, this volume). It is also vital that personal attribution is attainable in less well-endowed scientific institutes. Economists often equate incentives with financial reward, but scientists are motivated at least as much by the personal excitement of discovery and invention, and will tend to prefer working where they have both adequate resources for research and the ease of recognition that comes from being in the mainstream.

### Access to knowledge

It is fashionable in the rich countries to refer to modern society as the 'knowledge society'. Whether that is accurate or not, it is a fact that enclosure and protection of knowledge is epidemic: the scope of IP is being continuously extended, and more stringent means to prevent its unauthorized use are being introduced. We have already looked at some mitigating measures that are being taken in the area of publication. We must recognize that these measures and the transfer of technological expertise are a necessary part of developing trust. Access to knowledge is one of the most important rights for developing economies.

Easy access to knowledge is important for individuals as well as institutions, in rich as well as poor countries, and this aspect is considered by Susanne Kadner (this volume).

## *Integrity*

Another aspect of trust has to do with confidence in the accuracy and completeness of shared information. Science is self-checking, in that results are constantly queried and elaborated upon, so that errors eventually come to light. But this takes time. In the short term, accuracy of scientific information depends on the integrity of scientists, backed by peer review of publications. Apart from the inevitable errors, recent studies have revealed a steady trickle of falsified results, and occasionally a major scandal hits the headlines. It must be said that these cases represent a tiny fraction of overall scientific effort, and in view of the greatly increased number of scientists working today this fraction is probably not increasing. Nevertheless, concern about the accuracy of scientific research in a global setting is leading to the establishment of more monitoring systems to discourage misconduct. This is just as well, because in a newsworthy field such as climate change there is a great deal of commentary at very varied levels of professionalism. It is important for people to have access to sources of information that are not only open but also trustworthy.

Integrity is even more important at an institutional level than it is for individuals. Regrettably, systematic disinformation is characteristic of the lobbying and advertising industries, and, to varying degrees, of the political process. As an example of the former, the tobacco industry has for decades invested heavily in denying the link between smoking and lung cancer. While its activities have been greatly restricted in wealthy societies, it is now unashamedly peddling its wares among the poor. A prominent example of disinformation from political sources was the denial by the Mbeki government in South Africa that HIV caused AIDS. Corporate misbehaviour creates numerous impediments to tackling climate change. For example, the oil industry has funded objections to the identification of human activity as a primary cause, far beyond the point of balanced debate; the producers of bottled water lobby against the use of tap water; manufacturers of baby formula encourage mothers not to breast-feed their infants; and the food industry encourages overconsumption, resulting in increased obesity as societies become wealthier.

## *Sharing of natural resources*

A key element of benefit sharing is the equitable division of natural resources. Mostly it is left to the rather primitive mechanisms that we have just touched on. We need to do better than that if we are to deal with sharing of water, food, and the Earth itself without conflicts even more destructive than we have at present.

A striking example of failed benefit sharing is our collective inability to prevent the fishing industry destroying fish stocks, and therefore its own livelihood. Information flow is a vital step in sharing resources. But whereas information can be shared indefinitely without loss to anyone – indeed with gain as value is added to it – natural resources are consumed. We may divide the fish as equitably as we please, but if we collectively take them out of the sea faster than they can reproduce we are in the end left with nothing. Despite adequate information, understanding and communication, fishery after fishery has collapsed. Perhaps the most spectacular crash to date is that of the Grand Banks, largely under Canadian jurisdiction, in the early Nineties; it has not recovered. In Europe we are struggling with the declining North Sea fishery, once enormously rich and now a fraction of its former size. Modern technology saves effort and makes fishing safer, but it cannot solve this prisoner's dilemma for us any more than science and technology will of themselves solve the problem of climate change. Establishing sustainable fisheries is a model exercise for the EU. If we can solve this socio-economic problem, perhaps we shall find the mechanism to tackle climate change.

The nature of the problem becomes apparent when we contrast the above situations with one fishery that has probably been made sustainable. In the Seventies Iceland confronted the UK and Germany in order to establish the right to control its coastal waters (the so-called 'cod wars'), and then set up legal constraints to preserve the fish stocks. Here a small country with much to lose has achieved what countries such as the UK and Spain, let alone the EU as a whole, have so far been unable to do. Trust comes more easily to the small group; we have to find ways to foster it in the global group.

Underlying all the problems in the sharing of natural resources (including the Earth's atmosphere) is the issue of excessive human population, which is seldom discussed explicitly because it is so contentious. A detailed discussion is beyond the scope of this essay, but the issue is touched upon by Wolfgang Lucht and Walter Kohn in this volume. Here I would simply note that, in some way or another, we must start assessing the issue of population coolly, morally and humanely. Otherwise all our other efforts will be in vain.

## Conclusion

Large-scale manipulation of climate, even assuming that we have the technology and expertise to do so, can only be carried out in an atmosphere of transparency and trust. The hyper-competitive stance that has been the norm in international relations, while effective for short-term gains and understandably driven by the demands of the ballot box, will be disastrous for the problems that now face us. For the free flow and effective use of scientific information, we need to put in place settlements that are agreed by rational negotiation rather than by power struggles. In short, if we are to make progress, the globalisation of trade must be accompanied by the globalisation of justice.

# Chapter 30

# Educating and motivating global society

---

## Susanne Kadner

Susanne Kadner has a research background in biology, chemistry and oceanography. She has worked for the Parliamentary Office of Science and Technology (POST) in London, the German Advisory Council on Global Change (WBGU), and as G8-consultant for Hans Joachim Schellnhuber in his position as German Chief Government Advisor on Climate and Related Issues. Susanne now works in the Technical Support Unit of the Intergovernmental Panel on Climate Change's Working Group III.

*Note:* This chapter is a commentary on chapter 29.

'Information flow: the basis for sustainable participation' is the title of John Sulston's chapter in this book. It is probably what the creators of the Aarhus Convention had in mind when they drafted the document in 1998. This UN convention, whose formal title is 'Convention on Access to Information, Public Participation in Decision-making, and Access to Environmental Justice in Environmental Matters', is founded on the belief that citizen involvement can strengthen democracy and environmental protection (POST, 2006). Then-Secretary-General Kofi Annan went as far to describe it as 'the most ambitious venture in the area of environmental democracy so far undertaken under the auspices of the United Nations' (UNECE, 1998). On ratification of this document, more than forty states acknowledged that access to information is an essential prerequisite for public participation in environmental decision-making processes, and that sustainable development can only be achieved through the involvement of all stakeholders.

Each of us makes numerous decisions related to climate change every day. Whether as individuals, consumers or voters, our behaviour will ultimately influence the paths of greenhouse gas (GHG) emissions. As with any interconnected global problem, a fair understanding of the intricacies is required to make the right (i.e., intended) choices. This understanding is not always easily attained, especially with the complex challenge of climate change. While the UNFCCC website provides, for example, free access to the most recent data on national greenhouse gas emissions and removals, it is still a challenge for the lay person to understand the complex relationship between the emission reductions agreed under the Kyoto Protocol, and why the use of energy-saving light bulbs does not contribute to a reduction in these emissions.[1]

Obviously, the goal is not to turn everyone into a climate expert. Some people might even argue that understanding a problem does not necessarily mean acting on it (Chess and Johnson, 2007). However, the measures required to avoid dangerous climate change (Schellnhuber *et al.*, 2006) will affect everyone, whether through voluntary changes in lifestyle at a personal level, or through policies imposed by

---

[1] In the initial phases of the European Emission Trading System (ETS), which was set up according to the specifications agreed under the Kyoto Protocol, each EU member state receives a set amount of emission allowances. In total, the amount of allowances adds up to an agreed cap in greenhouse gas emissions. National Allocation Plans (NAPs) regulate the distribution of these allowances between the different energy-intensive industries of a country, such as electricity generators, oil refineries, and manufacturing plants. Each industrial installation may either use up its allowances through emitting the permitted amount of greenhouse gases or may sell the allowances on the market ('cap and trade'). When households save energy through the use of energy-efficient bulbs, less electricity is used and thus, initially, less $CO_2$ is emitted. However, the electricity provider can now sell its surplus allowances to enable other industries to emit more GHGs. In other words, energy-saving measures under the cap-and-trade system do not save any $CO_2$ emissions. They do, however, contribute to the development and spread of energy-efficient technologies and applications. These will become important during the later stages of the ETS, when industries will have to pay for their emission allowances and will pass the costs for this on to the customer, which will make – in our example – the use of energy-efficient light bulbs financially rewarding.

governments. To facilitate these changes, we need to ensure that individuals, consumers and voters understand the issues at stake. It is therefore not enough to merely provide the necessary information. We also need to ensure that people understand the implications of this information; i.e., that people understand, accept, and facilitate concrete action on all levels, from the personal to the national and global. After all, the motivation to act will most likely result from individuals feeling part of both the problem and the solution.

Who is the appropriate messenger for this information? Which people are credible, unbiased, knowledgeable, and dedicated enough to communicate the required changes? Clearly, scientists should be an essential part of this group of messengers. It may have been for this reason that the Nobel peace laureate Al Gore, in his speech to the annual meeting of the American Association for the Advancement of Science (AAAS) in 2009, told the assembled scientists, 'Keep your day job, but start getting involved in this historic debate. We need you.' (AAAS, 2009).

When faced with the choice of where to become active, there is certainly a broad range of options for the dedicated scientist. The political arena is clearly an important field where the advice of scientists is greatly needed and, thankfully, frequently sought. However, the practice of advising politicians and government officials can prove challenging as they often demand a single and simple answer rather than accepting that, due to the uncertainties intrinsic to science, a range of outcomes or solutions is possible. Concerned about their credibility, in particular amongst their peers, many scientists may thus shy away from this challenge (Cole and Watrous, 2007). Policy advice on the other hand, where a range of options is presented and discussed in order to support the legislative process, may seem a more attractive avenue of support. Many advisory bodies and national scientific societies (such as the Royal Society of London or the AAAS) have taken on the task of supporting policy development in key areas. Their tools are reports, policy briefs, and statements or letters of concern, although some doubt the impact of the latter (Meyer, 2007). However, policy advice also demands participation in formal, sometimes slow and rather institutionalized, processes that leave little room for personal engagement. The most renowned example of a scientific body providing advice to policymakers is the IPCC, which publishes its main reports every six to seven years after a lengthy process of scrutinizing discussion.

To speed up changes, scientists could (and should) also communicate to the public in a more direct manner. Here, an often used – albeit slightly problematic – communication channel has been the mass media. While newspapers, magazines, radio, television, and the Internet reach a broad audience, the results have not always been satisfying. One cause lies in the traditional model of news reporting, where a balanced approach is used to present more than one side of an argument with the aim of giving the audience the opportunity to form its own opinion. However, scientific agreement on anthropogenic climate change has now reached a level where the balanced approach no longer serves its previously well-intentioned purpose. Today, the scientific consensus on climate change should receive the relative weight it deserves, particularly in the face of dissenting claims from a number of reasonably well-known climate sceptics with sometimes dubious scientific backgrounds. The declining 'news value' of climate change also adds to the problem, as opinions that counter the general trend are favoured over 'more-of-the-same' prognoses (for example, that sea levels are rising some millimetres faster than anticipated). Due to the salience and selling points of such 'controversial articles', these reports distort the insights of science and impair the dissemination of knowledge to the public sphere. One way to counter this is for scientists to better support journalists in understanding their area of expertise, for example through workshops that explain the latest research findings in their specific discipline. In addition, scientists should be more helpful in providing clear analyses and statements, or perhaps even personal perspectives which are so important for news reporting (Ward, 2008).

Education, as another communication channel, may offer the opportunity for a more direct and possibly more rewarding experience of knowledge transfer. Teaching in schools and universities facilitates the dissemination of knowledge and understanding through the student's personal network of family and friends (Pratt and Rabkin, 2007). However, at present climate change is not represented as a specific topic in the curricula of most schools and it is only due to the personal effort of committed science teachers that it is covered at all. But for students to understand the climate crisis from many different perspectives, from physics and biology to economics and social studies, it is important to integrate it into the official curricula. Another very important aspect is science education itself, as students need a better understanding of scientific methodology and probabilities in order to judge uncertainties. By stimulating scientific discussions in class, teachers enhance the understanding that discussion of an issue does not imply that there is doubt of its existence *per se*.

Finally, a whole range of other opportunities exists for scientists to directly interact with the public. Books, blogs, public talks and public conferences, open days or contributions to documentaries and museum exhibitions are just a few possibilities for disseminating knowledge in a direct, undiluted and unbiased manner. A major challenge here is to adapt well to the different types of audiences, and to not underestimate the power of the narrative. Numerous studies have shown that certain forms of communication can make information more memorable. Suggestions range, for example, from using scenarios and analogies to evoke relevant personal experiences (Marx *et al.*, 2007), to visualizing the consequences of inappropriate actions with respect to greenhouse gas emissions (Stoll-Kleemann *et al.*, 2001).

There are clearly many challenges for scientists in appropriately communicating their knowledge. However, their main task must certainly remain the generation of knowledge in the first place. It may, in fact, be too much to demand that they should also know when, where, how, and to whom it can best be communicated. It thus strikes me that scientists would greatly profit from mediator organizations that facilitate the adequate transfer of scientific knowledge to politicians, media and the public. Of course, many NGOs already cooperate with scientists and, regarding certain issues, have largely taken over the role of informing and educating the public. Environmental NGOs, however, mostly attract those who already favour their position. One could argue that this is merely a case of preaching to the converted. Therefore, what I have in mind is something like the Union of Concerned Scientists (UCS),[2] a US non-profit science advocacy organization that is supported by numerous professional scientists and many private citizens. While this organization focuses on other environmental issues as well as climate change, its science-based activities that aim for responsible changes in government policy, corporate practices, and consumer choices seem to have the right mix to communicate scientific knowledge in an efficient and credible manner.

The most important point, however, is that a mediator organization does not have any agenda other than broad outreach and communication of sound scientific findings. The advantages of such an organization for Europe with the focus on climate change and sustainability questions are clear: scientists can concentrate on their scientific projects, while the mediator organization is responsible for communicating their findings effectively. The tasks of such an organization would include, for example, identifying the windows of opportunity to introduce one's results in policy-relevant decision-making processes, functioning as a contact point for journalists to select the right individuals for certain news stories, or even bringing scientists and artists together to help create a vision of a carbon-free future. In addition, it could help identify the target audience in public talks, and support scientists in tailoring their messages adequately without interfering in any way with the contents. Combining the knowledge of scientists with the invaluable skills of trained, respectable and committed communicators would help to increase both social pressure and political action.

As outlined by many of the authors of this book, the challenges before us are demanding, and may even appear daunting. We therefore need support from as many individuals as possible to facilitate the required societal changes. Due to their knowledge, scientists bear a particular responsibility in this context. They will need to improve their communication of the problems and solutions in order to support the public in dealing with climate change in an educated way. Whether

[2] www.uscusa.org

through closer cooperation with interested journalists or through more direct engagement with the public, the possibilities for action are ample. Of course, as Sir Crispin Tickell pointed out in an editorial in *Science* in 2002, 'Making unwelcome changes now to avoid possible consequences in an uncertain future is a difficult proposition to sell to anyone' (Tickell, 2002). But I am convinced that – with adequate support – scientists can contribute greatly to our current debate and convey precisely this message!

## References

AAAS – American Association for the Advancement of Science (2009). *At AAAS Annual Meeting, Gore Urges Scientists to Join Political Effort on Climate Change.* Available at http://www.aaas.org/news/releases/2009/0215am_gore.shtml, accessed 1 March 2009.

Chess, C. and Johnson, B. B. (2007). Information is not enough. In S. C. Moser and L. Dilling, eds., *Creating a Climate for Change: Communicating Climate Change and Facilitating Social Change.* Cambridge, 223–33.

Cole, N. and Watrous, S. (2007). Across the great divide: supporting scientists as effective messengers in the public sphere. In S. C. Moser and L. Dilling, eds., *Creating a Climate for Change: Communicating Climate Change and Facilitating Social Change.* Cambridge, 180–99.

Marx, S. M., Weber, E. U., Orlove, B. S. *et al.* (2007). Communication and mental processes: experiential and analytic processing of uncertain climate information. *Global Environmental Change,* **17**(1), 47–58.

Meyer, D. S. (2007). Building social movements. In S. C. Moser and L. Dilling, eds., *Creating a Climate for Change: Communicating Climate Change and Facilitating Social Change.* Cambridge, 451–61.

POST (2006). Århus Convention. *postnote 256.* London. Available at http://www.parliament.uk/documents/upload/postpn256.pdf, accessed 1 March 2009.

Pratt, L. G. and Rabkin, S. (2007). Listening to the audience: San Diego hones its communication strategy by soliciting residents' views. In S. C. Moser and L. Dilling, eds., *Creating a Climate for Change: Communicating Climate Change and Facilitating Social Change.* Cambridge, 105–18.

Schellnhuber, H. J., Cramer, W., Nakicenovic, N., Wigley, T. and Yohe, G., eds. (2006). *Avoiding Dangerous Climate Change.* Cambridge.

Stoll-Kleemann, S., O'Riordan, T. and Jaeger, C. C. (2001). The psychology of denial concerning climate mitigation measures: evidence from Swiss focus groups. *Global Environmental Change,* **11**(2), 107–17.

Tickell, C. (2002). Communicating climate change. *Science,* **297**(5582), 737.

UNECE – United Nations Economic Commission for Europe (1998). Aarhus Convention: Convention on Access to Information, Public Participation in Decision-Making and Access to Justice in Environmental Matters. Available at http://www.unece.org/env/pp, accessed 1 March 2009.

Ward, B. (2008). *Communicating on Climate Change: An Essential Resource for Journalists, Scientists and Educators.* Narragansett.

# Chapter 31

## Democracy and participation

––––––––

## Achim Steiner

Achim Steiner, a German national, was born in Brazil in 1961. During his studies at the Universities of Oxford and London, he specialized in development economics, regional planning, international development, and environment policy. In 2001 he was appointed Director General of the International Union for the Conservation of Nature (IUCN). In 2006 the UN General Assembly elected Steiner as Executive Director of the United Nations Environment Programme for a four-year term. He is also serving on a number of international development advisory boards.

*Note:* This chapter is a commentary on chapter 29.

John Sulston has compared the challenge posed by climate change to a repeated version of the 'prisoners' dilemma', in which the prisoners have the opportunity to increase trust by seeing how each responded in prior rounds of the game. His conclusion is that we need to increase information flow in order to build trust and hence prevent defection, in this case from collective agreements to reduce emission of greenhouse gases. I agree with Sulston's conclusion, but would go further and say that we also need to increase the flow of *knowledge* stemming from wider sharing of that information.

Sharing information on climate change and helping each other to truly understand and trust in the necessary actions necessitates close cooperation among all players of 'the game'. Such cooperation is essential if we are to avoid the future that scientists warn will be our fate if we fail to act. It is worthwhile taking a closer look at the capability of democratic institutions to foster cooperation in times of crisis. Cooperation is, paradoxically, both made more complicated and easier by democratic, more participatory forms of government because they place a high value on the individual and fundamental human rights

On the one hand, the spread of democratic governments and greater control by more individuals over decisions that affect their lives is one of the great achievements of the late twentieth century. The Nobel laureate Amartya Sen characterized *development as freedom* in his 1999 book of the same name. He also noted that in order to develop we must account for the 'worsening threats to our environment and to the sustainability of our economic and social lives'. Democracy promotes the flow of information, helping to create the informed and engaged citizenry that is needed to tackle collective challenges such as climate change. Democratic development and the strengthening of institutions that safeguard individual human rights are enshrined in the Universal Declaration of Human Rights, which last year celebrated its sixtieth birthday. They also form the core of the United Nations system, and shape our everyday work at the United Nations Environmental Programme (UNEP). It must be our collective hope that development processes will continue to bring freedom and inclusion to more people.

On the other hand, by giving more people a role in decision making, democracy can make it difficult to reach a consensus. The prisoners' dilemma grows more complicated when there are many prisoners and fewer opportunities to see firsthand the benefits of cooperation, a point Sulston makes in noting that trust comes more easily to small tight-knit groups than it does at the global scale. This does not imply a world in which a benevolent autocracy is the basis for reaching decisions – not only in relation to climate change – for the good of all, but rather means that, in Professor Sulston's words, 'an enormously important task is to inform and persuade citizens everywhere of the need for strategic change'. The challenge in a democracy is to ensure that citizens are informed and educated enough to be able to understand the issues at stake, and empowered to act accordingly even when the benefits of a choice may seem remote. This is a huge challenge in an information age that so often seems characterized by a glut of information, much of which can appear contradictory, self-serving, or just plain wrong. There is as yet no good equivalent in popular journalism to the peer review and vetting processes that, as Sulston observes, help weed out bad science. The result is that disinformation about climate change being a hoax continues to circulate in the media and public discourse, and time and energy is wasted debating whether observed climate phenomena are actually natural variations of yet undiscovered natural cycles; this time and energy could be better spent finding solutions to climate change.

We urgently need better ways of validating complex science, and communicating its inherent uncertainties to the public. People must be able to understand not only the magnitude of the problem but also the benefits of acting to curb emissions or taking steps to adapt to coming changes, even if this means making short-term sacrifices for the long-term common good. In democratic societies, the willingness to support actions for the collective good is communicated through the ballot box to those responsible for negotiating international agreements, making public policy, and enacting laws and regulations. The ballot box is a great achievement of democratic societies, but also a challenge to society if citizens base their vote on misinformation.

In addressing the problem of climate change it is important to get beyond merely communicating the issues. Communication needs to be accompanied by the development of new, and the reinforcement of existing, mechanisms that foster the inclusion and participation of wider and more informed constituencies in the policy-making process. Such mechanisms are essential to provide the 'basis for participation' that Sulston is hoping for. They are also essential in increasing the flow of knowledge and understanding stemming from the wider sharing of information that I see as a crucial extension of Sulston's argument.

Non-governmental organizations (NGOs) have a particularly valuable role to play here, and not only because they provide, as Sulston points out, a balance to powerful vested interests that attempt to influence governments. The best NGOs are very good at communicating information, encouraging participation, and rallying public support for change. A strong and vibrant NGO community is usually evidence of a democratic and open society. However, NGOs are also important in another way. What we might call 'fact building' for policy is less the result of a pure, rational quest for what is technically correct, and more about the establishment of facts within networks. This is a characteristic that NGOs share with scientists and other knowledge-based communities, groups Peter Haas has defined as 'networks of knowledge-based experts or groups with an authoritative claim to policy-relevant knowledge within the domain of their expertise'. The reach and influence of such networks and their stability vis-à-vis mainstream institutions, both at the national and international levels, help generate the political will needed to ensure that appropriate responses to climate change are adopted, initially nationally and eventually internationally.

Recognising the importance of NGOs and other non-state actors in shaping and communicating opinion, UNEP set up its Major Groups and Stakeholders Branch to enhance participation of civil society in our work. We value the perspectives that groups as diverse as trade unions, local authorities, indigenous people, youth, and the scientific and technological community bring to the table; the valuable research and advocacy functions they perform, and their role in helping foster long-term, broad-based support for UNEP's mission. These partners help us implement our work programme in a number of ways. They adapt our global efforts to national or local realities and form a valuable liaison function between UNEP and local communities. Major Groups provide the scientific, policy and legal expertise necessary for effective implementation, and act as watchdogs, helping foster accountability in governments. Our public awareness and outreach efforts rely to a large extent on partnerships with Major Groups, who are particularly effective in engaging the general public in an informative and educative manner. In the area of adaptation to climate change, for example, we are supporting efforts in a number of African countries to introduce strategies for coping with climate variability to farmers and other rural groups. In almost all cases governments have chosen to work with local NGOs in communicating this information.

Sulston is somewhat critical of the UN's ability to resist being manipulated by what he calls 'well-endowed vested interests'. It is not so much that the UN is manipulated; but when member states differ on important points, achieving agreement on a course of action is often difficult. National interests still matter a great deal, and in a consensus-based body such as the UN it can appear to the casual observer that discussion takes precedence over action. But this only highlights the importance of improving information flow so that governments clearly understand the long-term consequences of their positions and do not base these on short-term, narrow determinations. I agree with Sulston that as part of the global agenda we need to build on the success of the UN and improve trust in the capacity of the multi-lateral system to facilitate equitable and fair outcomes, although I would not defer this to the long term as he does.

Hence, with respect to climate change, we must aim not only to increase the flow of scientifically correct information, but also to foster understanding and application of the received information, thus enhancing responsible action by a larger constituency. Such development requires trust and close cooperation among a large number of different players, which is best achieved through focused actions with clearly defined goals. In this context, it may be worth referring to Sulston's call for a regime in which the private acquisition of intellectual property should be minimized, and funding on climate change conducted in a manner that promotes open-access publication of research findings. Providing an example from the area of public health, Sulston points out that the most successful new initiatives are being undertaken by public-private partnerships supported by charitable and government funds, and he sees scope for this model in the area of climate change mitigation. This idea is certainly worth exploring.

One reason, perhaps, for optimism regarding efforts to develop anti-malarial drugs or vaccines for diseases that disproportionately affect people in developing countries is their narrow focus. Such efforts do not aim at developing universal health care for all people but have a singularity of purpose that fosters trust by keeping the number of participants or players small and the result focused. In such a setting, confidence-building measures are more likely to be successful and to lead to mutually beneficial outcomes.

Extended to climate change, the implication is that it may well be wise to concentrate initially on a few important and achievable collective goals, such as improving the efficiency and lowering the cost of solar cells as a low-carbon energy technology. Scientific breakthroughs in this area could help build a consensus for collective action as lower-cost renewable energy technologies help reduce emissions in developed countries while allowing expanded access to energy in developing countries. Success in this one area of collective endeavour, for example, would build confidence and momentum for other cooperative challenges.

In the negotiations on a successor to the Kyoto Protocol, discussions on technology issues have matured, and there is a growing recognition that it is necessary to strike a balance between public and private interest. There are a number of means to achieve this. Two examples are increased government support for research and development of low-carbon and adaptation technologies, and support by developed countries for so-called enabling measures in developing countries, which help create the necessary markets. Hopefully, by tackling only one or a few problems at a time, we will be more successful in communicating the urgency of climate change and stimulating effective action to deal with it.

# Part V

The Potsdam Memorandum

# Chapter 32

# Potsdam Memorandum

---

**Main conclusions from the symposium 'Global Sustainability:
A Nobel Cause', Potsdam, Germany, 8–10 October 2007**

*We are standing at a moment in history when a Great Transformation is needed to
respond to the immense threat to our planet. This transformation must begin immediately and is strongly supported by all present at the Potsdam Nobel Laureate
Symposium.*

### The need for a Great Transformation

The world-wide socioeconomic acceleration after World War II has pushed our
planet into an unprecedented situation: humanity is acting now as a quasi-geological force on a planetary scale that will qualitatively and irreversibly alter the natural Earth System mode of operation – should business as usual be pursued.

As outlined by the Intergovernmental Panel on Climate Change, anthropogenic
global warming through greenhouse gas emissions is the foremost of an entire set
of emerging development, security and environmental crises which require an integrated response. Yet climate protection ambitions appear to be on a collision course
with the predominant growth paradigm that disconnects human welfare from the
capacity of the planet to sustain growth. Humanity is faced with the major challenge of making a drastic reduction in greenhouse gas emissions, which will require shifts in lifestyles in rich countries, while meeting urgent development and
growth needs in the poorer countries, the home of the vast majority of humanity
underlining the right to development. Ensuring that some nine billion people can
live a decent life requires, above all, access to affordable, sustainable and reliable
energy services, which are currently based almost exclusively on fossil fuel resources and unsustainable use of traditional fuels. The issue of 'carbon justice' and
the urgency of the matter at hand require unprecedented cooperation and rapidity
in response.

Is there a 'third way' between environmental destabilization and persisting
underdevelopment? Yes, there is, but this way has to bring about, rapidly and

ubiquitously, a thorough re-invention of our industrial metabolism – the Great Transformation. This is an awesome challenge, yet we have one comparative advantage over all previous generations: an incredibly advanced system of knowledge production that can be harnessed, in principle, to co-generate that transformation together with courageous political leaders, enlightened business executives and civil society at large.

### Crucial sustainability challenges and responses

The whole gathering placed the challenge of climate change and energy security firmly in the context of sustainable development, supported the rights of developing countries to social and economic development, and took careful account of interactions between climate policy and the challenges of development in the short, medium and long-term. In so doing it expressed its strong support for the Millennium Development Goals and the concepts of broad-based and multi-dimensional development that they embody.

A range of actions in the areas of climate stabilization, energy security and sustainable development are considered necessary, in particular, these could include:

1. In order to achieve **climate stabilization,** a post-2012 regime should comprise the following key elements:
   - A global target such as the 2 °C-limit relative to preindustrial levels or the (largely equivalent) halving of worldwide greenhouse gas emissions by 2050. A series of consistent short and medium-term emissions targets are also essential to drive investment and technology and to reduce the need for greater action later.
   - A leadership role of industrialized countries both in regards to drastic emissions reductions and development of low/no-carbon technologies in order to give poor developing countries room for urgently needed economic growth within the boundaries of a global carbon contract.
   - Carbon justice. Striving for a long-term convergence to equal-per-capita emissions rights accomplished through a medium-term multi-stage approach accounting for differentiated national capacities. An important goal would be the reduction of the total amount of greenhouse gas emissions, which is the product of per capita consumption times population, where both factors are crucial.
   - The generation of a carbon price, for instance, through an international cap-and-trade system (of systems) based on auctioning permits.
   - The establishment of a powerful worldwide process supporting climate-friendly innovation, international cooperation of R&D institutions, combined

with increased RD&D funding, integrating basic research as well, to facilitate technology transfer and cooperation.

- Major contributions to a multinational funding system for enhancing adaptive capacities.
- Scaled-up efforts to both reduce emissions from deforestation and accelerate ecologically appropriate reforestation by creating new incentives for communities and countries to preserve and increase their forests.
- Ensure reductions of non-$CO_2$ greenhouse gases.

2. Energy demand is projected to grow dramatically. Efficiency and a range of readily-available low carbon technologies are the key to offset the growth for energy services. In order to attain **energy security,** consistent with environmental integrity, an international strategy should have the following foci:
   - Systemic efficiency revolution and productivity increase including fuel switching, combined heat power and an energy saving lifestyle which is necessary but not sufficient.
   - Portfolio approach consisting of a systematic exploration of the economic and technological potential of all of the relevant mitigation options.
   - Design of investment strategies based on the portfolio approach; e.g., intelligent systems, grid infrastructure, storage technologies, demand-side measures, and deployment of renewables such as solar that has huge potential already now. Upfront investments, in addition to carbon finance, are needed to support emerging technologies and increase their market share; e.g., feed-in laws.
   - Rapid implementation of demonstration projects for advanced solar energy and carbon capture and storage to foster ingenuity and drive down costs.
   - Stabilizing long-term expectations of investors at capital markets and establishing microcredit institutions in developing countries aimed at financing low-carbon technologies.

## A global contract between science and society

There is overwhelming evidence that we need to tap all sources of ingenuity and cooperation to meet the environment and development challenges of the twenty-first century and beyond. This implies, in particular, that the scientific community engages in a strategic alliance with the leaders, institutions and movements representing the worldwide civil society. In turn, governments, industries and private donors should commit to additional investments in the knowledge enterprise that is searching for sustainable solutions.

This new contract between science and society would embrace many elements, yet three of them are critically important:

1. A multi-national innovation program on the basic needs of human beings (energy, air, water, food, health etc.) that surpasses, in many respects, the national crash programs of the past (Manhattan, Sputnik, Apollo, Green Revolution etc.).
2. Removal of the persisting cognitive divides and barriers through a global communication system (ranging from international discourse fora to a truly world-wide web of digital information flow). Part of this would be the emerging 'Global Earth Observation System of Systems (GEOSS)' that could especially provide early warning about imminent natural or social sustainability crises.
3. A global initiative on the advancement of sustainability science, education and training. The best young minds, especially those of women, need to be motivated to engage in interdisciplinary problem-solving, based on ever enhanced disciplinary excellence. The ambition is to win over the next generation for laying the cognitive foundations for the well-being of the generations further down the line.

# Chapter 33

# The Potsdam Memorandum: a remarkable outcome of a most important conference

---

## Klaus Töpfer

Klaus Töpfer was born in 1938 in Waldenburg, which then belonged to Germany but is now Polish. He studied economics in Mainz and Frankfurt and earned his doctorate at the University of Münster in 1968. After serving as a government official, professor, and adviser on development politics, he became Minister for the Environment and Health in the state of Rhineland-Palatinate in 1985. In 1987 Töpfer became Federal Minister for the Environment, Nature Conservation and Nuclear Safety under Chancellor Helmut Kohl. From 1994 to 1998 he served as Federal Minister for Regional Planning, Civil Engineering and Urban Development. In 1998 he was appointed Under-Secretary-General of the United Nations, General Director of the United Nations office in Nairobi, and Executive Director of the United Nations Environment Programme. In 2009 he was appointed Founding President of the Institute of Advanced Sustainability Studies in Potsdam.

*Note:* This chapter is a commentary on chapter 32.

It was certainly a historic event that took place in Potsdam, Germany, in October 2007: The Potsdam Nobel Laureate Symposium entitled, 'Global Sustainability: A Nobel Cause'. The conference was convened at a remarkable venue, a baroque palace built by the Prussian King Frederick the Great, reflecting the atmosphere of a monarchist epoch. This era also gave rise to the First Industrial Revolution, a revolution based on the technical innovations of the steam engine and railway systems, inducing the first major use of fossil fuels. It was also a social revolution as reflected in the Stein-Hardenberg reforms of the Prussian administrative system. All of this culminated in the collapse of the monarchy in Germany and the difficult start of democracy.

The Potsdam Memorandum, which was adopted at the end of this remarkable symposium, starts out by stating: 'We are standing at a moment in history when a Great Transformation is needed to respond to the immense threat to our planet'. Shortly after this symposium the dramatic crisis of the financial institutions hit the world like a tsunami, provoking drastic consequences for economies worldwide. The quotation above reflects the double challenge we are facing. More than ever before, the relationship between economic development and stability, and the integrity of the ecosystems in our world are becoming evident. This global economic crisis is a declaration of bankruptcy of the 'short-term world', an economic paradigm focused solely on quarterly results, with a reward system directly echoing this short-termism. It is also a declaration of bankruptcy by a society that subsidizes its 'wealth' by externalizing the main part of the costs linked to production and consumption, imposing them on coming generations, on human beings living far away, and on nature's capital. These costs involve the exploitation of the environment, as well as financial debts and burdens.

A further visionary conclusion of the Potsdam Memorandum was to emphasize the relationship between the right to development, mentioned in the Rio Principles as early as 1992, and the stabilization of ecosystems, especially the fight against climate change. The Potsdam Memorandum stresses that 'Humanity is faced with the major challenge of making a drastic reduction in greenhouse gas emissions, which will require transforming lifestyles in rich countries, while meeting urgent development and growth needs in the poorer countries, the home of the vast majority of humanity, underlining the right to development'. Grasping this challenge should be the foremost priority of global society and should lead to political actions at all levels. It requires moving beyond short-termism, and appreciating our responsibility for the medium- and long-term consequences of actions and reactions in our world. In his epochal book The Principle of Responsibility, the German-Jewish philosopher Hans Jonas formulated a new categorical imperative, an ethical approach to decision making in our technological society: 'Act in a way that the consequences of your actions are compatible with the permanence of real human life on Earth'. This is a categorical imperative for a world committed to sustainable development. It is the alternative to a 'throw-away society', which was and still is a reflection of our short-sighted political and economic systems.

More than ever before, we require a new paradigm for economic and political action. The Potsdam Memorandum rightly calls for a '*third way* between environmental destabilization and persisting underdevelopment'. At the moment we are confronted with a myriad of signals indicating that the responses to the financial and economic crisis are again based on short-term reactions. The measures taken mainly aim at preserving existing structures; they clearly do not start the journey to a 're-invention of our industrial metabolism', nor do they lead 'the way to the Great Transformation', as called for in the Potsdam Memorandum. Analysis of the economic stimulus packages decided upon by nearly all governments around the world to overcome the economic crisis shows that those hundreds of billions of dollars and euros are mainly being spent stabilizing demand for the old structures of roads and cars, and backing the purchasing power of the consumer. Only a few countries have taken the path towards a 'Great Transformation', towards a world with higher energy-efficiency and a massive decarbonization of energy supply. South Korea stands out for having committed around 80% of its economic stimulus funds to measures in line with a 'green economy'. In China the corresponding share is around 30%, in Germany it is as low as 13%, in the United States around 11%. The message of the 'Great Transformation' requires that the financial crisis must be taken as an opportunity to kill two birds with one stone. This means responding to the short-term financial and economic crisis in a way that supports long-term sustainability of the global economy and society. The new 'industrial metabolism' must be the focus of global attention if we are to overcome the crisis of our age.

It is ethically wrong that the poorest of the poor again have to bear the main burden of crises that were caused by those living and acting in the so-called developed part of the world. The facts that Muhammad Yunus mentioned in his speech on the occasion of the 2006 Nobel Peace Prize ceremony in Oslo must be addressed: 'The world's income distribution gives a very telling story. Ninety four percent of the world's income goes to 40% of the population, while 60% of people live on only 6% of world income. Half of the world's population lives on two dollars a day. Over one billion people live on less than a dollar a day. This is no formula for peace.' Development is becoming synonymous with peace in this globalized world.

Tackling the double challenge of honouring a right to development and successfully combating climate change urgently requires a 'global contract between science and society'. This message was a most important conclusion to the symposium. It reflects the huge opportunities arising from science and technology in our world. The acceleration of scientific discovery, which is unprecedented in history, has given us deep insights into the patterns of nature and life. These insights form an important basis for successfully realizing the 'Great Transformation'. The necessary scientific understanding must be further deepened by investing further billions of dollars and euros in research and technological development. It is most apt that the Nobel laureates in Potsdam called for a new 'Apollo Program', to leverage innovations and technologies that allow for the fulfilment of basic human needs without exceeding the Earth's capacity for renewal.

In his 'Berlin speech' of 2009, the German President Horst Köhler called for the next industrial revolution to be an 'ecological industrial revolution'. The turnaround that he called for comprises a revolution of efficiency in energy and resource use. It must also put an end to the externalization of social and environmental costs, and address the categorical imperative of responsibility, including responsibility towards future generations. Beyond an unprecedented boost to investment in science and technology, the Potsdam Memorandum also calls for a 'removal of the persisting cognitive divides and barriers through a global communication system'. A new general understanding of the interrelationship between science, society and politics must be established. The founding idea of the Intergovernmental Panel on Climate Change (IPCC) was to involve governments in a process led by climate scientists. This intergovernmental practice must be broadened to counteract the growing gap between the insights of science, their acceptance by society, and their implementation by politicians. Again, the crux of the matter is to accept responsibility. When Hans Jonas formulated his new categorical imperative for the technological society he did not in any way deny the need for technical progress. Today, at this historic time, an increasing number of 'science outlet centres' is needed to advance mutual understanding between science, society and politics.

Science and technology form without any doubt the basis for the 'Great Transformation'. However, a change in consumption patterns in the developed world is also urgently needed. The Potsdam Memorandum called for 'transforming lifestyles in rich countries', taking into account that the lifestyle of the global rich is highly subsidized – voluntarily and involuntarily – by people in other parts of the world and by future generations.

The Potsdam Memorandum, concise as it is, represents indeed an historical document. It focuses on the dramatically destabilized economic and ecological world of today. It not only describes the problems and formulates the challenges; this memorandum also suggests the solutions. The utmost must be done to apply these recommendations to day-to-day decisions in this crisis-stricken world. The 'reinvention of our industrial metabolism', the 'Great Transformation', the 'global contract between science and society', the categorical imperative for the technological society – these are not abstract, academic considerations. They must become the cornerstones of our common endeavour to pass on a sustainable world to our children and grandchildren.

# Appendix

# Glossary

*Absorption spectrum*. The fraction of electromagnetic wavelength that is absorbed by a given material from a range of frequencies. The absorption spectrum of → *photovoltaic* cells indicates the wavelength fraction of incident radiation that is converted into electrical energy.

*Adaptation*. Adjustments in natural or human systems in response to actual or expected climatic changes. They are intended to decrease negative effects or exploit potentially beneficial opportunities.

*Additionality*. Projects approved under the → *Clean Development Mechanism* need to show that their planned → *greenhouse gas* reductions would not be implemented without the extra incentive provided by → *carbon credits*; i.e., that they are additional to existing or planned → *mitigation* efforts.

*Afforestation*.\* Direct human-induced conversion of originally forested land that has not been forested for more than 50 years through planting, seeding, and/or human-induced promotion of natural seed sources.

*Agent-based modelling*. Numerical simulations of actions and interactions of autonomous individuals with the aim of better understanding the functioning of a complex system as a whole.

*Agro-ecological zone*. A land unit that is defined based on its soils, land form, climate, and ecosystems. Based on these characteristics, each unit has specific potential or constraint for land use.

*Annex I countries*. The group of countries listed in Annex I of the → *United Nations Framework Convention on Climate Change*, including all of the OECD countries and economies in transition; all other countries are referred to as Non-Annex I countries. Annex I countries have committed themselves to the target of reducing their → *greenhouse gas emissions* individually or jointly to 1990 levels by the year 2000. In support of this commitment, most Annex I countries agreed to legally binding emission reduction targets through the → *Kyoto Protocol*.

*Anthropogenic*. Caused by or produced by human beings.

*Anthropogenic emissions*. Emissions of → *greenhouse gases*, greenhouse-gas precursors, and airborne particles (aerosols) that are associated with human activities such as burning of fossil fuels, → *deforestation*, land-use changes, livestock, and fertilization.

*Asymptotic stable state*. Term used in the mathematical theory of dynamic systems. It describes a state (e.g., in the simplest case the value of one variable) that a system approaches after some time (asymptotic) and that it returns to if perturbed (stable).

*Bali roadmap*. A series of steps designed to help reach an effective post- → *Kyoto* treaty at the UN climate conference in Copenhagen in December 2009. These steps were agreed upon at the UN climate conference in Bali, Indonesia, in December 2007.

*Baseline*. A reference point from which one can determine if observed patterns are changing with time. In the context of → *mitigation* efforts, the baseline is the amount of → *greenhouse gas* emissions at a certain point in time against which efforts of countries to decrease emissions are measured.

***Bidirectional grid***. Grid with connections that allow electricity to move in both directions. It constitutes the basic infrastructure required for distributed electricity generation from a diverse mix of → *renewable energy* sources.

***Biofuel***. Any liquid, gaseous, or solid fuel produced from plant or animal → *biomass* (e.g., ethanol and biodiesel). First-generation biofuels are mostly produced from food crops such as soy and sugar cane using conventional technologies. Second-generation biofuels are derived from ligno-cellulosic material (i.e., the non-food material of crops, such as wood, stems, or leaves) through chemical or biological processes. Third-generation biofuels are made from algae.

***Biogeochemical cycles***. The cycling of chemical elements or molecules through the biotic (living organisms) and abiotic (water, land and air) compartments of an ecosystem. In effect, these cycles represent closed loops although they may take from a few days to millions of years to complete.

***Biomass***. Organic matter consisting of or derived from living organisms, including products, by-products, and waste derived from these organisms (e.g., wood or straw). Biomass counts as a → *renewable energy* resource.

***Biopiracy***. The practice by some corporations, especially within the pharmaceutical industry, of appropriating the traditional knowledge and genetic resources of others (e.g., developing nations and indigenous peoples) without sharing the benefits.

***Bretton Woods System***. Agreement among the world's major industrialized countries on how to manage the world's commercial and financial relations, organized by the International Monetary Fund (IMF) and the World Bank. Established in the mid-twentieth century, it was the first example of a single monetary order among independent nations.

***Business as usual (BAU)***. Scenario which assumes different demographic, social, economic, technological and environmental developments but in which no additional → *mitigation* initiatives, such as implementation of the → *United Nations Framework Convention on Climate Change* or the emissions targets of the → *Kyoto Protocol,* are included.

***Cap and trade***. Market-based approach for reducing emissions of pollutants. A central authority (such as a government or international body) defines a maximum permissible amount of pollutant that can be emitted. Based on this cap, companies or other groups receive emission permits/allowances (see → *carbon permits*). Should companies need to increase their allowances, they can buy these from those who pollute less. In theory, trading can occur between companies, at national and international levels, and enable emission reductions to take place where they are cheapest.

***Carbon budget***. Concept that refers to the total cumulative amount of → *carbon dioxide* emissions that are admissible over a given period of time to attain a specific → *mitigation* target (e.g., limiting global warming to below 2 °C above preindustrial levels). This concept is applicable since, due to the long lifetime of carbon dioxide in the atmosphere, the temperature increase is largely independent of the temporal pathways of emissions (see → *stock-pollutant*). Once a global carbon budget is determined it can be distributed among nations (e.g., based on criteria of equity that aim to achieve equal per capita → *emission rights*).

***Carbon capture and sequestration/storage (CCS)***. → *Mitigation* technique that captures → *carbon dioxide* emissions from major sources such as coal-fired power plants and stores it underground in geological formations or in the oceans. This process is currently being tested worldwide and is not yet available on a large scale.

***Carbon credit***. Credit for a reduction of → *greenhouse gas* emissions. Each carbon credit is equal to the reduction of one metric tonne of → *carbon dioxide* or → *carbon dioxide equivalent*. These credits can be obtained by investing in projects belonging to the → *Clean Development Mechanism* or other certified carbon reduction schemes.

***Carbon dioxide (CO₂)***. Molecule composed of one carbon and two oxygen atoms, and one of the main → *greenhouse gases* in the Earth's atmosphere. $CO_2$ concentration is measured in parts per million (ppm); i.e., the current concentration of 389 ppm (June 2009) indicates that there are 389 molecules of $CO_2$ in our atmosphere per 1 million molecules. During the last 800 000 years, $CO_2$ concentration has never exceeded 300 ppm. The recent increase is known to be largely → *anthropogenic*.

***Carbon dioxide equivalent***. The concentration of → *carbon dioxide* that has the same global warming potential as a given mixture of carbon dioxide and other → *greenhouse gases*.

***Carbon footprint***. The total amount of → *greenhouse gas* emissions produced directly and indirectly by an individual, organization, event, or product. It is mostly expressed in tonnes or kilograms of → *carbon dioxide* or → *carbon dioxide equivalents*.

***Carbon intensity***. The amount of carbon emitted per unit of energy produced, often measured in grams of → *carbon dioxide* emitted per megajoule of energy, or per Gross Domestic Product (GDP).

***Carbon leakage***. The increase in → *carbon dioxide* emissions outside a country or region, occurring as a direct consequence of a climate policy that caps emissions within this country or region. Companies operating under the umbrella of an → *emissions trading scheme* may, for example, relocate their production sites abroad to avoid costs incurred for emitting → *greenhouse gases*.

***Carbon offset***. Compensation for carbon emissions that are impossible or too costly to be avoided at a certain location or from a certain emitter. Such compensation can be achieved through the purchase of → *carbon credits* that certify emission reduction or → *carbon sequestration* elsewhere (see also → *Clean Development Mechanism*).

***Carbon permit***. Emission entitlements allocated by a government to individual companies, allowing them to emit a specific amount of carbon. If actual emissions are greater than the permitted amount, the company has to offset its emissions by purchasing surplus carbon permits from other companies or by obtaining → *carbon credits*.

***Carbon permit auctioning***. A method of distributing → *carbon permits* among carbon emitters, allowing the market to set the price for carbon.

***Carbon sequestration***. Biological or → *anthropogenic* removal of gaseous → *carbon dioxide* from the atmosphere and its long-term storage in terrestrial or marine reservoirs. Biological carbon sequestration can be enhanced through protection of ecosystems and improvements in agricultural techniques. Anthropogenic sequestration can be achieved through technologies such as → *carbon capture and storage*. Carbon sequestration, especially in forests, is an important component of → *mitigation* policies (see → *Reducing Emissions from Deforestation and Degradation*).

***Carbon sink***. A medium – such as oceans, soils, and ecosystems – that removes carbon from the atmosphere and stores it for a prolonged period of time.

***Carbon tax***. An environmental tax on → *carbon dioxide* emissions with the goal to reducing these. This method is often suggested as an alternative to → *cap and trade*.

**Carbon trading**. See → *cap and trade*.

**Cellular automata**. Class of models that consist of a regular grid of cells. The models are based on rules that define the characteristics of the cells (e.g., taking the state 'on' or 'off') and the interaction among single cells (e.g., switching the state of a neighbouring cell). These models are particularly well fit to study the dynamic behaviour of complex systems consisting of many interacting, spatially well-ordered parts.

**Cellulosic ethanol**. See → *biofuels*.

**Certified Emission Reduction (CER)**. → *Carbon credits* given out for emission reductions that are achieved through the → *Clean Development Mechanism*. Certified Emission Reductions count towards the emission reduction goals agreed upon under the → *Kyoto Protocol*.

**Clean Development Mechanism (CDM)**. An agreement allowing → *Annex I countries* to achieve part of their → *greenhouse gas* limitation and reduction commitments in developing (i.e., Non-Annex I) countries. By investing in carbon reduction projects that support → *sustainable development*, Annex I countries can obtain → *carbon credits* that count towards their reduction commitments, while avoiding more expensive emission reductions in their own countries. An important aspect of CDM is → *additionality*.

**Climate justice**. Climate justice requires a reduction in the inequality between rich countries that have emitted most of the → *greenhouse gases* and poor countries that have contributed very little to climate change but suffer most from it. It also acknowledges the priority of poor countries to develop economically and to overcome poverty. The suggested means of achieving climate justice include transfer of technology and finances, greater efforts at → *mitigation* in industrialized countries, and equal carbon → *emission rights* for all people on Earth.

**Complete sectoral coverage**. Inclusion of all relevant sectors, such as transport, industry, and households, in regulations that limit → *greenhouse gas* emissions.

**Compressed air system**. → *Energy storage system* in which, during periods of low demand, surplus energy is used to compress air; during periods of high demand decompressing air supplies energy.

**Concentrating solar power (CSP)**. Concentrating solar rays through mirrors or lenses into smaller beams. These concentrated light beams are then used as a heat source for a conventional power plant or focused on → *photovoltaic* surfaces to produce electricity.

**Decarbonization**. Changing the economy, industry or any other part of society in such a way that less and ultimately no → *carbon dioxide* is emitted. Decarbonization mainly refers to a reduction in the burning of fossil fuels through increased → *energy efficiency* and expansion of renewable energies.

**Demand-side action**. → *Mitigation* measures that address the consumers, aiming at, for example, lowered energy demand through, for example, increased → *energy efficiency* and energy-saving behaviour of individuals (see also → *supply-side action*).

**Deforestation**. The natural or → *anthropogenic* process that converts forested to non-forested land (see → *afforestation* and → *reforestation*).

**Digital rights management**. Access control technologies that limit the usage of digital contents and devices. May be used by hardware providers, publishers, copyright holders and individuals to inhibit the unforeseen or undesired use of digital content.

***Direct solar insolation***. Solar energy received at a surface perpendicular to the sun's rays, without the energy received from scattering or reflection of solar radiation by particles in the atmosphere. It is measured in watts per square metre or in kilowatt hours per square metre per day.

***Discount rate***. The discount rate describes how future assets (bonds, capital stocks, investments, etc.) are devalued just because their pay-off lies in the future. A high discount rate implies a high devaluation of future consumption; a discount rate of zero reflects that present and future consumption are equally valued.

***Dispatchable electricity generation***. Power generation facilities that can be turned on or off at any time depending on demand (powered, for example, by fossil fuels, → *hydroenergy,* or → *biomass*).

***Emission reduction pathways***. Patterns of decrease in → *greenhouse gas* emissions over time. They can be modelled under various scenarios of economic growth, population growth, and changes in → *energy efficiency* and energy systems.

***Emission rights***. See → *carbon permit.*

***Emissions trading scheme***. See → *cap and trade* and → *European Union Emissions Trading System.*

***Energy efficiency***. Using less energy to provide the same level of energy service. It is formally expressed as the ratio of useful output of a system, conversion process or activity to its energy input.

***Energy flux***. Transfer rate of energy through a medium. This rate varies for different energy sources; for example, solar and wind energy fluxes are determined by the time of day and the prevailing meteorological conditions.

***Energy storage system***. A system that stores energy for later use. At relatively small scales, batteries can be used as storage systems. At larger scales, energy can be stored by, for example, pumping water uphill or by using → *compressed air systems.*

***European Union Emissions Trading System (EU ETS)***. Established in 2005, this system is so far the largest multi-sector and multi-country emission trading scheme in the world. Its aim is to reduce emissions from large installations in the energy and industrial sectors (see also → *cap and trade*).

***Exothermic***. A type of chemical reaction that releases energy, usually in the form of heat.

***Externality***. A market externality is the impact (positive or negative) of a market transaction on a third party that is not directly involved in the transaction. In terms of climate change, this means that the price paid for energy does not reflect the costs associated with damages resulting from global warming caused by energy production.

***Forest degradation***. Reduction in the quality of forested habitat without direct reduction in forest areas.

***Fullerene***. Molecule that is entirely composed of carbon, and arranged in spheres, ellipsoids or tubes. They are important in many technological applications, from electronics to nanotechnology.

***Game theory***. Description of behaviour in which one's own success depends on the behaviour of others, with each player trying to maximize his or her personal success (in contrast to → *social planner*). Game theoretic principles apply to economics, sociology, computer science, politics, and biology.

***General Circulation Model (GCM)***.\* A numerical representation of the climate system based on the physical, chemical and biological properties of its components, their interactions and feedback processes, and accounting for all or some of its known properties. GCMs are applied as a research tool to study and simulate the climate, and for operational purposes, including monthly, seasonal and inter-annual climate forecasts.

*Geoengineering*. In the context of climate change, the concept of geoengineering usually refers to proposals that envisage a large-scale engineering of the environment to combat or counteract the effects of rising → *greenhouse gas* concentrations. Examples include the injection of sulphate particles into the atmosphere to increase reflection of incident solar radiation as a means of cooling the climate; or the iron fertilization of phytoplankton to strengthen → *carbon sequestration* in the oceans.

*Geothermal energy*. A → *renewable energy* source derived from the Earth's internal heat that originates from the formation of the planet, radioactive decay of minerals, and solar energy absorbed at the Earth's surface.

*Global Earth Observation System of Systems (GEOSS)*. Initiative that aims to improve the relevance of Earth observations for addressing global environmental challenges. The envisaged result is a global public infrastructure generating comprehensive, near-real-time environmental data, information and analyses for a wide range of users.

*Global environmental space*. Concept in → *sustainability* studies that looks at each resource – such as water, air, forest and agricultural land – separately and tries to assess which level of human activity can be supported by the ecosystem without causing irreversible damage. In the context of climate change, the remaining global environmental space refers to the maximum admissible amount of → *greenhouse gas* emissions if unmanageable climate change is to be avoided.

*Greenhouse gases (GHGs)*. Molecules in our atmosphere – mainly water vapour, → *carbon dioxide,* methane, nitrous oxide, and ozone – that absorb and re-radiate infrared (heat) radiation. The → *anthropogenic* increase of these gases causes a rise in global mean temperature with associated changes in the functioning of the climate and entire Earth system.

*High-voltage direct current (HVDC)*. Technology to transmit electricity over long distances. The advantage of using HVDC power lines compared to the usual alternating current (AC) transmission is a smaller loss of electricity and thus cheaper transmission over large distances.

*'Hockey-stick' pattern*. Recent sharp increase in the concentration of → *greenhouse gases,* especially → *carbon dioxide,* in the atmosphere following a long period of relative stagnation during the past thousands of years. The resulting graph resembles the form of a hockey stick.

*Hybrid vehicle*. Vehicle that uses two different power sources, most commonly a combustion engine and an electric motor. Fuel usage, and thus → *carbon dioxide* emissions, of such vehicles are lower than those of standard vehicles.

*Hydroenergy*. A → *renewable energy* source derived from flowing water, such as rivers or oceans (see → *ocean energy*).

*Hydrological regime*. Patterns and amount of water flowing through a system, from input (e.g., through precipitation or groundwater inflow) to outflow (e.g., through evaporation, photosynthesis or rivers). Changing conditions (floods or droughts) can severely impact the hydrological regime of ecosystems and the survival of species, including humans.

***Increasing returns (to scale)***. Economic term referring to circumstances in which each unit of variable input added to a system leads to a disproportionate increase in output. Increasing returns can be a result of learning effects (see → *learning curves*).

***Industrial metabolism***. Constitutes the totality of human-controlled processes that convert raw materials through energy and labour into finished products and wastes. By studying the flow of materials through society, the understanding of the sources and causes of → *greenhouse gas* emissions can be improved.

***Intellectual property rights***. The right of ownership of intellectual or artistic inventions, such as trademarks, copyright, and patents. It is protected by the World Intellectual Property Organization.

***Intergovernmental Panel on Climate Change (IPCC)***. An international body of scientists, established in 1988 by the United Nations and the World Meteorological Organization. Its task is to publish reports that summarize the most recent scientific findings on all aspects of climate change. These reports serve as a major resource for political decisions on climate change policies.

***Kyoto Protocol***. An international treaty, established as a protocol to the → *United Nations Framework Convention on Climate Change,* with the goal of 'stabilization of → *greenhouse gas* concentrations in the atmosphere at a level that would prevent dangerous → *anthropogenic* interference with the climate system'. It came into full force in 2005, and has been ratified by 183 nations. Currently, negotiations focus on a post-Kyoto treaty, intended to take effect after the first commitment period of the Kyoto Protocol, which is due to expire at the end of 2012.

***Leapfrog.****\** The ability of developing countries to bypass intermediate technologies and jump straight to advanced clean technologies. Leapfrogging can enable developing countries to move to a low emissions development trajectory.

***Learning curve***. The learning curve describes how the average production costs in a specific industrial sector decrease as a function of total installed capacity. For example, average production costs of a solar panel have decreased as the number of panels installed worldwide has risen rapidly over recent decades.

***Lumped load pattern***. Increased → *energy flux* into an electricity grid at specific times of the day. For example, an expansion of → *photovoltaics* would increase the energy load at noon-time. To smooth this energy 'lump', a → *smart grid* is needed, which activates energy-demanding procedures at times of high load.

***Marginal cost***. Additional cost of producing one more unit of a good. If the quantity increases greatly, marginal costs may include the cost of building a new factory or power plant.

***Marginal damage***. Additional damage that is caused by a unit increase in → *greenhouse gas* emissions (often measured in dollars per additional ton of → *carbon dioxide equivalent* emitted). Assessing the marginal damage of emissions forms part of an economic cost-benefit analysis of climate change.

***Microcredit***. A small loan with particular terms and conditions, most often extended to people in poverty, helping them to start a business. These people often lack steady employment or a verifiable credit history and therefore cannot meet even the most minimal qualifications to gain access to traditional credits.

***Mitigation***. Implementation of policies to reduce → *greenhouse gas* emissions and enhance → *carbon sinks* with the aim to reducing the extent of climate change.

***Near-surface temperature***. To estimate global near-surface temperature, air temperature data, measured over land at the standard height of 1.5 metres above ground and recorded at various weather stations across the globe, are merged with water temperature measurements in the upper metres of the oceans.

*Nonlinear behaviour.* Attribute of a system, in which the change of one component induces a more (or less) than proportionate change of another component. It can make the system dynamics difficult to predict because simple changes in one part of the system can have complex effects throughout.

*No-regrets policy.** Such a policy would generate net social benefits whether or not there is climate change associated with → *anthropogenic* emissions of → *greenhouse gases*. In the context of emissions reductions, no-regrets policies refer to options whose benefits equal or exceed their costs to society, excluding the benefits of avoided climate change (e.g., reduced energy costs and reduced emissions of local/regional pollutants).

*Nuclear fission.* Reaction in which the nucleus of an atom splits into smaller parts, often free neutrons and lighter nuclei, giving off large amounts of energy. This energy is captured in nuclear power plants or released in nuclear bombs.

*Nuclear fusion.* Reaction in which several nuclei merge to form one nucleus. This reaction releases or absorbs energy depending on the weight of the nuclei. It occurs naturally in stars and is currently explored as a means to create electricity through 'controlled fusion'.

*Ocean acidification.** Increased concentrations of → *carbon dioxide* in sea water causing a measurable increase in acidity (i.e., a reduction in ocean pH). This may lead to reduced calcium sequestration rates of calcifying organisms such as corals, molluscs, algae and crustaceans.

*Ocean energy.* A → *renewable energy* source acquired from ocean waves, tides, currents, and temperature and salinity gradients.

*Opportunity cost.* The difference between the potential gains of one choice versus those of an alternative that was rejected. Opportunity costs may arise for emitters of → *greenhouse gases* in a → *cap and trade* system because → *carbon permits* cannot be sold to obtain money for other investments.

*Peak oil.* The time at which the rate of the world's (or a nation's) oil production is at its maximum and after which oil supply decreases, reflecting diminishing exploitable reserves.

*Peak shaving.* Providing additional electricity from generators at times of high demand to avoid shortage of supply and dampen cost increases.

*Photon.* The basic unit of electromagnetic radiation such as light.

*Photothermal/photovoltaic energy.* A → *renewable energy* source derived from the sun's energy. This energy is either converted into heat (photothermal energy) or directly into electricity (photovoltaic energy).

*Primary energy.* The energy that is contained in raw fuels (such as coal, gas, oil, nuclear or renewables) before it is converted into secondary energy (electricity or heat).

*Redox system.* A chemical system in which both reduction and oxidation (i.e., transfer of electrons) occur. Batteries used as → *energy storage systems* often operate on the principles of redox reactions.

*Reducing Emissions from Deforestation and Degradation (REDD).* Recognizing that forests are an important → *carbon sink*, parties at the thirteenth meeting of the

→ *United Nations Framework Convention on Climate Change* in 2007 accepted that forest protection is an important contribution to → *mitigation.*

***Reforestation.**** Direct human-induced conversion of non-forested to forested land through planting, seeding and/or the human-induced promotion of natural seed sources on land that was previously forested.

***Remote sensing technologies.*** Equipment and software that permit the measurement of characteristics of the environment from a distance, such as satellite use to measure vegetation cover.

***Renewable energy.*** Energy derived from resources that are not depleted through their use, such as for solar, wind, water, → *ocean* and → *geothermal* energy, or resources that can be re-grown, such as → *biomass.*

***Renormalization group.*** A mathematical method of viewing a physical system at various scales, like a microscope with different magnifying lenses, thus allowing scientists to investigate different components of a system and their interactions.

***Research and Development (R&D).*** The process of revising old ideas, techniques and products and developing new ones to improve future performance and returns. International collaboration and investments in R&D are specifically encouraged by the → *United Nations Framework Convention on Climate Change* to help developing nations to → *leapfrog* to a low-carbon society and to better adapt to the changing climate.

***Salination ingress.*** Salt water flowing into areas that previously contained only freshwater.

***Savannization.*** Process in which forested areas turn into savannah, a grassland ecosystem with interspersed trees and shrubs, due to fire, → *deforestation,* or a drying climate.

***Secondary energy.*** See → *primary energy.*

***Semiconductor.*** Substance that partially acts as a conductor of electricity, and partially as an insulator. Silicon is one of the most widely used materials to produce semiconductors, for example in the construction of → *photovoltaic* cells.

***Smart grid.*** Electricity network that uses digital technology (e.g., → *smart meters*) to efficiently match intermittent and decentralized renewable power production with varying energy demands (e. g., by turning on washing machines during times of peak energy load or switching off refrigerators during times of electricity undersupply).

***Smart meter.*** Advanced electrical meter that enables a two-way communication between the supply and demand side. The aim is to match energy generation with consumption patterns by informing consumers about the differences in market prices over the course of the day (see → *smart grid*).

***Spillover effects.**** The effects of domestic or sector-specific → *mitigation* measures on other countries or sectors. Spillover effects can be positive or negative and include effects on trade, → *carbon leakage,* transfer of innovations, and diffusion of environmentally sound technology. They represent a type of market → *externality.*

***Stock-pollutant.*** Waste material that is absorbed very slowly by the environment and thus accumulates in air, water, soils, and ecosystems over time (such as heavy metals, non-biodegradable plastic, and → *carbon dioxide*).

***Supercapacitor.*** Electric double-layer capacitors, exhibiting an unusually high energy density compared to common capacitors (or condensors). Their electricity storage potential lies between that of capacitors and batteries, while their quick charging rates make them superior to batteries.

**Supply-side action.** → *Mitigation* measures that address the production side, aiming, for
example, at a switch from → *carbon intensive* forms of energy generation to zero-
carbon technologies and → *renewable energies* (see also → *demand-side action*).

**Sustainable development.**\* Concept that defines a process of change in which the exploitation of resources, the direction of investments, the orientation of technological
development and institutional change are all in harmony and enhance both current
and future potential to meet human needs and aspirations. Sustainable development
integrates political, social, economic and environmental dimensions.

**Sustainability.** See → *sustainable development*.

**Tipping point.** The point in time or level of external forcing at which a system undergoes
an abrupt and/or irreversible change in response to a relatively small perturbation.
Global warming may have the potential to trigger this type of change in a number
of regional-scale features of the Earth system (so called tipping elements). Examples
include the melting of the Arctic sea-ice, of the Greenland ice-sheet, or of perma-
frost soils.

**United Nations Framework Convention on Climate Change (UNFCCC).** An international treaty to develop strategies for climate change → *mitigation* and → *adaptation*
which was ratified in 1994 by 192 countries. An extension of this treaty is the
→ *Kyoto Protocol*.

**United Nations Millennium Development Goals (UN MDGs).** An eight-point road map
on how to end poverty in the world, with measurable targets and a clear deadline in
2015.

---

\*Definition adapted from glossary of IPCC 2007 Fourth Assessment Report working group III; available at http://
www.ipcc.ch/pdf/assessment-report/ar4/wg3/ar4-wg3-annex1.pdf.

# Co-authors' biographies

**Markus Antonietti** was born in Mainz, Germany, in 1960. He began his studies in chemistry at the University of Mainz, where he also obtained his PhD in 1985. He was appointed Professor of Physical Chemistry at the University of Marburg in 1991, and became Director of the Colloid Department at the Max Planck Institute of Colloids and Interfaces in Potsdam in 1993. His research focuses on the synthesis and properties of functional polymers, and the techniques for characterizing them. In recent years, he has launched a major programme on 'sustainable chemistry'.

**Tariq Banuri** began his career in the Civil Service of Pakistan, went on to receive a PhD in economics from Harvard University, joined the United Nations as a Research Fellow at the World Institute for Development Economics Research, and was the founding Executive Director of the Sustainable Development Policy Institute in Pakistan. He worked as Senior Fellow and Director of the Future Sustainability Program at the Stockholm Environment Institute, and recently joined the United Nations as the Director of the Division for Sustainable Development. He has served on national as well as international forums for policy, advocacy, and research, including as a Coordinating Lead Author on the Nobel prize-winning Intergovernmental Panel on Climate Change.

**Nico Bauer** completed his PhD thesis at the Potsdam Institute for Climate Impact Research (PIK) under the supervision of Professor Ottmar Edenhofer in 2005. He returned to PIK in May 2007 after working at Fondazione ENI Enrico Mattei in Italy, and the Paul Scherrer Institute in Switzerland. He is currently involved in the development of modelling tools for PIK's Research Domain III, and is Co-Chair of the research groups on energy system and macroeconomic modelling. He is also involved in several other scientific projects, contributes to policy relevant reports, and supervises PhD students.

**Markus Haller,** born in 1975, studied energy and process engineering at the Technical University of Berlin. In 2007, he joined the Potsdam Institute for Climate Impact Research as a PhD student. He works on the representation of the power sector of developing countries in integrated assessment models, with a focus on the integration of renewable energy sources into rapidly growing electricity systems.

**Hermann Held** is Co-Chair of the research domain 'Sustainable Solutions' at the Potsdam Institute for Climate Impact Research (PIK). To advance mitigation strategies, he develops and applies methodologies that aim at an optimal mix of options under conditions of risk and uncertainty. He completed his PhD in physics with a fellowship from the Max Planck Society, followed by an Alexander von Humboldt fellowship at the University of California at Berkeley. In 1999 he joined PIK in order to merge his system science and environmental management interests. He lectures on climate science, economics and statistics.

**Matthias Kalkuhl** studied applied system science at the University of Osnabrück, and mathematics at the University of Granada. Since 2008 he has worked as doctoral student at the Potsdam Institute for Climate Impact Research. His research interests are dynamic game theory, policy instruments modeling, and the economics of exhaustible resources.

**Gerhard Knies** has worked in elementary particle physics research at DESY, CERN, SLAC, and the University of California. In 2003 he founded the Trans-Mediterranean Renewable Energy Cooperation and later initiated the DESERTEC Industrial Initiative, which aims at using clean power from deserts to provide energy, water, and climate security worldwide.

**Elmar Kriegler** works in the PIK research domain 'Sustainable Solutions'. His research focuses on assessing the technological and climatic risks of climate change mitigation policies. He also works on the coupling of climate, energy and economy models under conditions of uncertainty. Kriegler studied physics at the University of Freiburg, obtained his PhD in physics at the University of Potsdam, and was a Marie Curie fellow at Carnegie Mellon University in Pittsburgh.

**Anders Levermann** trained as a physicist in Marburg, Berlin and Kiel, Germany. He completed his PhD in theoretical physics at the Weizmann Institute of Sciences, Israel. Since 2003 he has worked at the Potsdam Institute for Climate Impact Research, and in 2007 he also became Professor of Dynamics of the Climate System at Potsdam University. His research focuses on tipping elements of the climate system. He is head of the flagship activity, Tumble, which investigates the risk of tipping for Greenland, Antarctica, the Atlantic overturning, and monsoon circulations.

**Johan Lilliestam** holds a MSc degree in environmental science and physics from Göteborg University, and a MA in environmental management from Freie Universität Berlin. He is currently working on his doctoral thesis at the Potsdam Institute for Climate Impact Research in the research domain 'Transdisciplinary Concepts and Methods'. His research focuses on electricity costs and security of supply in a SuperGrid Europe.

**Hermann Lotze-Campen** studied agricultural science and agricultural economics at Kiel (Germany), Reading (UK) and Minnesota (USA) Universities. He holds a PhD in agricultural economics from Humboldt University, Berlin. He previously worked for a European aerospace company and as a policy consultant. At the Potsdam Institute for Climate Impact Research Lotze-Campen leads a research group that works on the interactions between climate change, agriculture and food production, land and water use, and adaptation options through biomass energy production and technological change.

**Michael Lüken,** born in 1979, studied physics in Jena, Austin (USA) and Heidelberg. Since 2006, he has been a PhD student at the Potsdam Institute for Climate Impact Research. He is currently working on model-based assessment of mitigation strategies within a long-term and multi-regional perspective. His PhD thesis will focus on the implications of climate policy for welfare distribution among different regions of the world.

**Jennifer Morgan** joined the World Resources Institute as its Director for Climate and Energy in September 2009. Previously, she was the Global Climate Change Director of the environmental NGO 'E3G'. In 2007, she worked with the German Chancellor's chief advisor Hans Joachim Schellnhuber, and in 2008 advised former Prime Minister Tony Blair. Prior to joining E3G in 2006, Jennifer Morgan led the Global Climate Change Programme of WWF. She holds a BA in political science and Germanic studies from Indiana University, and an MA in international affairs from the American University, Washington, DC.

**Dieter Murach,** born in 1953, studied forestry in Göttingen, receiving his PhD in 1983. Since 2001 he has been Professor of the Ecology and Economics of Forestry at the Eberswalde University of Applied Sciences. He is also Dean of Silviculture at the State Forestry Office. His research focuses on dendromass production, the ecology of forests, and biocoal applications.

**Robert Pietzcker** joined the PIK research domain 'Sustainable Solutions' as a doctoral student after working briefly as a consultant with McKinsey & Company. In his doctoral thesis he will analyze the representation of capital inertia in hybrid energy-economic models, focusing on the example of decarbonizing the transport sector. Previously, Pietzcker studied physics at the University of Freiburg and at McGill University in Montreal before graduating from the University of Jena.

**Karsten Sach,** born in 1959, is Deputy Director-General at the German Federal Ministry for Environment, Nature Conservation and Nuclear Safety, with responsibility for international cooperation. Prior to assuming that office in 2004, he led the department for International Cooperation, Global Conventions and Climate Change. Since 1999 he has been the chief German negotiator at the Conferences of the Parties of the United Nations Framework Convention on Climate Change. Since September 2008 he has been Chairman of the Management Board of the European Environment Agency and since January 2009 additionally Chairman of the International Renewable Energy Agency (IRENA). Sach studied law and obtained his PhD at the Albert-Ludwigs-Universität in Freiburg in 1993.

**Katrin Vohland** is an expert in biodiversity research, mainly studying the inter-relationships between biodiversity, climate change, and ecosystem services. She has worked in the Amazon on speciation of arthropods and on sustainable pathways to protect forests, and in Africa on the contribution of rainwater harvesting to landscape functions. At PIK she investigated the impact of climate change on protected areas. Vohland recently moved to the Museum for Natural History in Berlin to strengthen biodiversity research in Germany and connect it more closely to international programmes and research requirements.

**Ernst Ulrich von Weizsäcker,** born in 1939, was Professor of Biology from 1972–75 at the University of Duisburg-Essen, and President of the Wuppertal Institute for Climate, Environment and Energy from 1991–2000. He served as a Member of Parliament and Chair of the German parliament's Environment Committee from 1998–2005. From 2006–08 he was Dean of the Bren School of Environmental Science and Management at UC Santa Barbara, California. Since 2007 Weizsäcker has been Co-Chair of the International Panel on Sustainable Resource Use. His main publications include 'Factor Four' (with A. and H. Lovins, 1997) and 'Factor Five' (with C. Hargroves and M. Smith, 2009).